# Contemporary
# Textile
# Engineering

# Contemporary Textile Engineering

*Edited by*

## F. HAPPEY

*Emeritus Professor, School of Textile Studies, University of Bradford, Bradford, West Yorkshire, England.*

1982

## ACADEMIC PRESS
*A Subsidiary of Harcourt Brace Jovanovich, Publishers*

London    New York
Paris    San Diego    San Francisco    Sao Paulo
Sydney    Tokyo    Toronto

ACADEMIC PRESS INC. (LONDON) LTD.
24/28 Oval Road
London NW1

*United States Edition published by*
ACADEMIC PRESS INC.
111 Fifth Avenue
New York, New York 10003

British Library Cataloguing in Publication Data

Contemporary textile engineering.
  1. Textile machinery
  I. Happey, F.
  677'.0285   TS1525

  ISBN 0-12-323750-5   LCCN 82-71018

Printed in the United Kingdom by W. & G. Baird Ltd.
at the Greystone Press, Caulside Drive, Antrim.

# Contributors

M. CHAIKIN, *Faculty of Applied Science, The University of New South Wales, P.O. Box 1, Kensington, Sydney, N.S.W. 2033, Australia.*

J. D. COLLINS, *Faculty of Applied Science, The University of New South Wales, P.O. Box 1, Kensington, Sydney, N.S.W. 2033, Australia.*

K. DOUGLAS, *USTER News Bulletin, Zellweger Uster Ltd., CH-8610 Uster, Switzerland.*

E. DYSON, *Postgraduate School of Textiles, University of Bradford, Bradford, West Yorkshire BD7 1DP, England.*

F. HAPPEY, *Whirlow, Lon Refail, Llanfair P.G., Anglesey LL61 5YN, Wales.*

J. INCE, *International Wool Secretariat Laboratories, Ilkley, West Yorkshire, England.*

J. A. IREDALE, *School of Textile Studies, University of Bradford, Bradford, West Yorkshire BD7 1DP, England.*

P. R. LORD, *Department of Textile Materials and Management, North Carolina State University, Box 5006, Raleigh, North Carolina 27650, U.S.A.*

W. PARKIN, *School of Textile Studies, University of Bradford, Bradford, West Yorkshire BD7 1DP, England.*

D. RICHARDS, *Wool Industries Research Association, Leeds, England.*

B. SCHOFIELD, *School of Textile and Knitwear Technology, Leicester Polytechnic, P.O. Box 143, Leicester LE1 9BH, England.*

W. S. SONDHELM, *10 Bowlacre Road, Hyde, Cheshire SK14 5ES.*

H. VITOLS, *Department of Mechanical Engineering, University of Technology, Loughborough, Leicestershire LE11 3TU, England.*

G. R. WRAY, *Department of Mechanical Engineering, University of Technology, Loughborough, Leicestershire LE11 3TU, England.*

# Acknowledgements

My thanks are due primarily to all my colleagues and friends who have contributed so generously in writing chapters for this book. Their collaboration has been of immense value and is much appreciated. Grateful recognition is extended to the Science Library of the University College of North Wales, Bangor, for their cooperation, particularly in the organization of inter-library loans. Acknowledgement is due to Mrs D. Bell for providing the index, and to the Staff of Academic Press for their help in the production of the book.

# Preface

Having worked in the pre 1940 and post 1945 periods of textile development I was pleased to be asked to edit this book on Contemporary Textile Engineering because of the enormous changes and developments which have taken place in the last half century. Progress has been so extensive, and even startling, that it has been difficult to choose the topics to be dealt with in one small volume. Obviously there must be omissions but it is hoped that all the major processes and developments have been covered. In selecting contributors it was my feeling that their basic training was likely to be very different. Some could be trained mechanical engineers oriented towards the study of textiles, or scientists who became interested in textile engineering and technology, and some might be textile technologists with post-graduate training in engineering. These three broad categories provided a wide range of approach and have contributed significantly to the interest and the scope of knowledge shown in the various chapters. The book is a reflection in miniature of the many types of skill which have made the modern textile industry what it is today.

I feel that the importance of the advance of textile engineering and technology as an international subject showing world-wide liaison and collaboration is a matter to ponder. The subject has no national bounds. It is a fundamental need in the establishment of stable societies in all countries as one of the three fundamental requirements of food, clothing and shelter. Textiles, as clothing, provide one of these, but food and shelter are also matters in which various parts of the textile industry are interested and to which they make a considerable contribution. In this book I have attempted to show how the subject of textile engineering has become world-wide in scope, and it can be seen that in the developing countries textiles has become a leading technology. One finds that there is a general attitude that this is an easy option on which to start industrial development. However, having seen many of these advances at first hand I feel that, although this may appear to be so now, much concentrated effort has been expended over the years to develop the industry to its present advanced technological state in order to provide for some of the requirements of the third world.

In planning the book I have been guided by two main principles, to keep the work short in comparison with the field to be covered and to include the newest advances which have been made in textile engineering. At the same time an effort has been made to trace some of the fundamentally simple beginnings through to their modern shortened and accelerated production sequences. Contributions have been obtained from experts working in various fields of textile engineering from research in processing natural and synthetic fibres to the micro-processing control through spinning to their use in weaving and knitting. The chapters have been programmed to follow on from an historical introduction on general textile development to the present day production of wool tops, the various aspects of yarn manufacture in different forms, including the influence of man-made fibres, and a chapter on waste recovery. Later chapters include open-end spinning, data logging and quality control, and advances in fabric manufacture. These cover loom design, non-woven materials and knitting technology. A final section deals with micro-processors and micro-electronic devices with particular emphasis on "silicon chip technology" in the knitting industry. Other uses of micro-processors are dealt with in the various chapters on yarn production, recycling of waste, automatic weaving and heavy fabric manufacture.

A further aspect of textile engineering not included in this book is its recent application to the provision of structures for use as dwellings or other accommodation. The earlier types of tented structures were simple cotton erections, but with the development of non-absorbent man-made fibres it has become possible to provide light frame dwellings with more than one compartment or "room". These developments are well known but the design of Inflated Structures as accommodation is less familiar. This work was "restricted" at first (1970) and was carried out in the main at the Military Vehicles and Engineering Research Establishment at Christchurch, Hants. I was associated with this and with the further extension of the work under a Home Office Committee of the British Standards Institute for the civilian field. It is to be hoped that when the "Code of Practice" for the production, use and safety of these Air Supported Structures has been formulated and published that this new aspect of Textile

Engineering will come to fruition with the production of many more such structures and their application in a variety of fields.

It is felt that the book has broken new ground in that an effort has been made to include international authors of authority to deal with original developments arising from engineering research in the textile field and their application in industrial practice.

*August, 1982*                                              F. HAPPEY

# Dedication

---

To my wife without whose help this work

would not have been completed.

# Contents

## Chapter 10—High Speed Automatic Weaving                289
*Walter S. Sondhelm*

## Chapter 11—Heavy Fabric Manufacture—Carpets and
## Needlefelts                                             341
*J. Ince*

# Chapter 1

# Introduction: Premises of Contemporary Developments in Textile Engineering

F. HAPPEY

## I. Historical Introduction

There are three epochs in the history of the modern textile industry. The first was before the 1914–18 world war, the second between the two wars and the third followed the 1939–45 war. The rate of automation and electronic control has advanced very rapidly during the last few years and entirely new materials and fibres have been introduced together with new methods to deal with them. This will be considered in subsequent chapters of this book, and to suggest that such a work could be exhaustive would be unrealistic. However it will give an outline of some of the main advances over the last 20 years and will also show how the industry has expanded internationally. As a prelude, it is of interest to glance at the previous development of textile technology.

In the first period of automation of the industry, wool and cotton were the main raw materials and water power was used. The Hargreaves Spinning Jenny provided an example as early as 1767. The Arkwright Mills, still in operation today but not under water power, stood on the banks of the Derbyshire Derwent, and Samuel Crompton combined the principles of water power and spinning jenny to produce the first "mule". The Luddite riots in 1812 in the West Riding of Yorkshire indicated the resentment at the replacement of hand loom weaving by units driven by water power. It was, however, when power changed to steam raised by coal that textile industries expanded in Yorkshire and Lancashire, the former with ample supplies of soft water for wool scouring and finishing and the latter with climatic

1

conditions ideal for the processing of cotton. In the middle third of the 19th century these processes spread to France and Germany, and equipment and expertise were introduced into the United States of America. In the latter part of that century there followed a general spread of the textile industry through Central Europe and into Asia, but this was small and the equipment was imported from the textile machine makers of Europe and the States. To illustrate this spread of textile technology, I was in Helsinki in 1972 and discussed with Prof Ahti Reijonen of Helsinki University of Technology the transfer of his textile department to the city of Tampera where the textile firm of Finlayson-Forssa had been since 1820. I quote the handbook which gives a history of their organization as follows: "The Finnish cotton industry was established in 1820 with the foundation of a textile factory at the Tammerkoski Falls in Tampera by a Scotsman, James Finlayson. There the first cotton yarn was spun as long ago as 1827".

Before leaving these early days it is of interest to draw attention to the various development calendars published. Excellent accounts of the chronological development of the textile industry is to be found in the Year Books for wool, cotton etc.[1] The 1960 issue cites a history of the wool trade from 500 BC, when Phoenician traders visited Britain, to 1953, when the London Futures Market opened. To select two of the earlier important inventions one might consider (a) the Jacquard engine and (b) the Noble comb. The former was introduced into Britain in 1820 and has been developed constantly since then. The latest developments will be mentioned in later chapters. It is now designed to operate with electronic controls for the selection of the lifting of the single warp threads; the cutting of the Jacquard lifting cards to select yarns by colour can be done by photo-electric cell control activated by the individual colours used.

The Noble comb, on the other hand, arguably one of the more ingenious but least understood textile inventions of the mid 19th century, has suffered much less modification and development over the years. It is used for the separation of top and noil in the medium length wool worsted industry, but before describing it it is convenient to consider the historical development of automatic combing to replace hand combing. This occurred between 1800 and 1850 in the case of wool and flax (hackling). The first steps towards mechanization were taken about 1800, when a machine, known locally as "Big Ben", was devised by Cartwright and was covered by patents in the 1790s. This did not prove a usable machine but pointed the way to later developments.[2] In 1846 Heilman invented a comb for cotton which worked on an intermittent action, and which was refined over the years and retained in principle for cotton combing. Later it was developed to the rectilinear comb used for cotton, short wool for dry worsted spinning (French system) and wool and man-made fibre mixtures. However this system was not

FIG. 1. Holden comb. (Courtesy Bradford Industrial Museum.)

applicable to oil combed wools where heated combs had been used since the hand combing era of worsted manufacture. The Holden comb (Fig. 1) was a development of the Heilman system and both these were more suitable for short staple fibres and have survived in the rectilinear comb. (See Chapter 2 on modern top making.) In 1853 Noble patented the comb which became standard for the processing of oil combed tops in the wool worsted industry of Yorkshire. Its function is described in Chapter 2 with illustrations but a typical plant lay-out of some years ago is shown in Fig. 2. Another combing system surviving from this period is the Lister comb (Fig. 3), which has a continuous action using heated combs and can be supplied with a constant flow of long staple sliver in sheet form. The fringe of fibres is held in a clamp and combed to remove short fibres. The comb moves away with these and the long fibres are beaten on to a rotating pinned cylinder and drawn off as a continuous top. Improvements in the design of the surviving combing systems have been made but the fundamental principles have not changed.

Until the advent of man-made fibres these methods have controlled the standard of quality of the yarns made in long wools and hairs and to some

F. HAPPEY

Fig. 2. A typical Noble combing plant in production. (Courtesy Bradford Industrial Museum. Photographer C. H. Wood (Bradford) Ltd.)

FIG. 3. Lister comb. (Courtesy Bradford Industrial Museum.)

extent in linen and regenerated fibres, woollens and worsteds and better quality cotton yarns. Details of the changes introduced in recent years are discussed in Chapter 3 on long staple yarn manufacture. Rectilinear combs were generally used for the shorter wools, but in some cases were used for very long wools, and formed the basis of the French combed tops of the continental trade. The rectilinear system has an intermittent process and has had much development work lavished on it in the rehabilitation of industry in Europe after the 1939–45 war. Some attempt was made in Bradford to study the mechanisms of the Noble comb and improve them,[3] but unfortunately the work was curtailed by lack of finance. It is to be hoped that the continuous production sequence and high standard of tops produced will be re-developed in the not too distant future.

## A. Wool sorting

At least up to the 1950s differences in manufacture affected markedly the definable qualities of different worsted materials. At this time raw wool qualities had been defined for over 100 years by a number varying from about 32 for coarse mountain wools to 100 for the finest merinos. This represented the best yarn count that could be obtained at the spinning frame for any batch of wool. Thus the Bradford system of wool sorting matched up staples of wool having the same average spinning properties. Fibre diameter was the main feature of the "handle" but many other factors were quoted as having an effect on spinning, e.g. fibre length, crimp, etc. This process of wool sorting is still retained in the specialized sections of the high class worsted trade, but the advent of man-made and regenerated fibres has caused wide changes in yarn definition.

## B. Wool grease disposal

It is well known that natural wool may contain a large amount of animal grease, as much as 50% by weight in some merinos and about 10% in the coarsest British wools. The effluents from scouring plants were originally a waste product in the West Riding, but ultimately effluent discharged into the sewage in the Bradford area was recovered to produce the valuable by-product lanolin. As shown in Chapter 2, the problem of recovery of wool grease has not yet been fully solved.

## II. Yarn Production

From the mid 19th century the carding, combing, drawing and spinning of worsted yarn followed two different paths. The Bradford system produced the main bulk of yarn processing in the Yorkshire West Riding and the French process was used mainly in Europe. The former used an oil combed top, had nine drawing operations and the fibre movement was controlled by twist. An outline of the typical processes of attenuation of the top to the roving stage from which the yarn proceeds to the spinning operation by cap, ring or flyer methods is given in a table in Chapter 3. The controlling forces are by twist changes in the drawing process and this progresses from about 1 turn/ft in the undrawn top to 48 turns/ft in the roving. It was considered in the Yorkshire trade that the lean yarns produced from the oil combed sliver were the most satisfactory for their use. Some modifications were made, particularly in the change from olive oil lubricant due to the increased cost of this commodity. Much research was carried out on mineral, vegetable and animal oils as lubricants in twisted sliver drafting and standard lubricants have been evolved.

## A. Noble combed yarns

There are two factors in Noble combing which are important in the production of tops for the manufacture of worsted cloth. First the fibres are pulled into alignment at the combing circles (see Fig. 24, Chapter 2) and secondly the crimped fibres are stretched in the steam heated pin section of the comb. Here separation occurs and the heated fibres are temporarily set and retain their aligned arrangement. The top is collected off the drawing-off rollers which produce a false twist in the top-sliver, packing it tighter as it enters the can, retaining the straightness of the fibres which have been semipermanently set in alignment. This parallelism of the fibres is retained through the subsequent drawing and spinning stages to the finished yarn. Thus the crimp has been removed from the wool and an aligned and slightly twisted roving is available to draft and spin, originally by flyer, into a lean yarn. It is of interest to consider the difference in use of almost similar Noble combed tops in the cloth manufacture of the Bradford and Huddersfield regions of the West Riding. In the former the general method was to spin yarn to its quality limit, putting in twist to a high value and producing a yarn of maximum leanness and tightness of packing. In the Huddersfield trade higher quality tops were used to spin lower counts of yarn with less twist. Thus the Bradford system produced more closely packed fibres in tighter yarns giving a firmer fabric after finishing. The Huddersfield cloths had a softer handle due to the crimp in the fibres re-exerting itself and increasing the fabric voluminosity.

## B. Other basic textile materials and processes

The author has concentrated attention on summarizing the early development of textile engineering in the worsted industry but in the woollen

TABLE I. World production of textile fibres in million kg.[a]

| Year | Cotton | Wool | Regenerated cellulosic | Synthetic | Total man-made |
|------|--------|------|------------------------|-----------|----------------|
| 1940 | 6971 (75) | 1134 (12) | 1127 (12) | 5 (—) | 1132 (12) |
| 1950 | 6647 (71) | 1057 (11) | 1612 (17) | 69 (1) | 1681 (18) |
| 1955 | 9492 (71) | 1265 (10) | 2278 (17) | 267 (2) | 2545 (19) |
| 1960 | 10 144 (68) | 1467 (10) | 2608 (17) | 702 (5) | 3310 (22) |
| 1965 | 11 605 (63) | 1493 (8) | 3338 (18) | 2052 (11) | 5390 (29) |
| 1970 | 11 370 (54) | 1593 (8) | 3433 (16) | 4692 (22) | 8125 (38) |
| 1974 | 13 624 (52) | 1481 (6) | 3515 (13) | 7471 (29) | 10 986 (42) |

[a] Figures in brackets indicate % of grand total of fibres. Minority fibres are excluded. Extracted from Man-made Fibres for Developing Countries.[4]

process automation and power transmission from carding to the finished fabric followed a parallel path, first driven by water power and later by steam. Inspection of the history of development of various sections of the textile industry shows parallel expansion of automation in processing cotton, linen, silk, jute and later the more exotic hairs such as alpaca, mohair, etc. It is of interest that during the later 19th century and even up to the development of man-made fibres 75% of fibres used in the textile industry were cotton, 12% wool and 12% cellulosic and regenerated fibres. By 1974 cotton production was reduced to 52% and wool to 6% whereas man-made and regenerated cellulosics provided 42%,[4] see Table I.

## C. The woollen system of spinning

Although the term "worsted" still legally refers to yarns from combed tops of wool spun on the worsted system the process designated "woollen" defines a method of yarn manufacture from short fibres. These may be short wool fibres (noils, etc.), recovered wool waste, shoddies, mungos from the waste and rag trade, short cotton fibres and man-made staple. The great difference between the worsted aligned fibre drafting system and the woollen process is that the short fibres are arranged randomly in a condensed slubbing. The woollen carding system consists of two parts and the final stage produces a uniform distribution of randomly arranged fibres as a web which completely covers the card surface. The web is removed in equally spaced and equally dense slivers which are condensed by rubbing into single slubbings. The yarns are spun from these with little drafting, in earlier times on woollen mules and later on ring frames. As Chapter 5 gives a comprehensive scientific study of engineering technology in this field the subject will not be pursued here. A detailed account of the earlier phases of the work is given in the yearly handbooks on wool and cotton.[1]

## III. Motive Power and its Distribution

Prior to the inventions of Arkwright, Cartwright and Crompton in the late 18th century, all yarns were hand spun and textile production was a craft process run by man or woman power; subsequently power was provided by the water wheel. However, with the discovery and application of steam power greater ingenuity was brought to bear on the modification of textile processes, and by the 1840s long shaft steam driven power units had been introduced to drive lines of machines (see Fig. 2). This type of motive power, with its obvious unavoidable waste of energy, continued until after the turn

FIG. 4. Power house in Bradford Technical College (later University) 1913–50. Supplied DC current with a backing accumulator standby to DC motors for line shaft drives. The boilers also provided steam for the scouring, dyeing, and wet processing sheds of the textile department.

of the century when the advent of electricity led to the gradual replacement
of steam by electric power. An example was shown in the Bradford Textile
School where the new building of 1912 housed a power plant in which a
classical horizontal engine driven by steam power ran a direct current
dynamo to provide current to run large motors to drive line shafts in the
different sheds of the department (Fig. 4). As power needs increased with
the installation of more equipment extra steam engines were added, and it
was only after it became impossible to obtain DC motors of the type required
owing to the general trend towards the use of AC that the system began to be
abandoned. All units in the competitive field of textile production now use
electric power for ease of application and for energy saving, as when
machines are not in use no power is needed to drive the wheels and shafting
of earlier days. The lack of atmospheric pollution is also of major importance in these days of environmental protection.


## IV. First Phase of the Man-made Fibre Industry

This part of the textile industry owes its origin, in the mid 19th century, to the
chemical studies of Schönbein (1845) on the preparation of nitrocellulose.[5]
Extensive work on the reactions to produce nitrocellulose esters up to full
nitration of 14.1%, $(C_6H_7O)(NO_3)_3$, as a tri-nitroglucose unit of the cellulose chain has been carried out since this time.[6] From an engineering point of
view, cellulose nitrate was developed as a smokeless explosive. It was found
to be soluble in ether/alcohol at nitrations of ~12% and in acetone at
~13.1–14.1%. These viscous solutions of nitrocellulose (collodion) proved
capable of extrusion and formed a basis for producing nitrocellulose
filaments (Chardonnet silk).[5] These were, of course, extremely inflammable
and therefore were treated with caustic soda to remove the nitrate groups
and to reconvert the material to cellulose. Later, this wet spinning method
was replaced by a dry (evaporative) process using the same dope formation,
as ether/alcohol or acetone could be evaporated readily from the extruded
filaments, the degree of nitration being determined by the solvent used. The
process was obviously dangerous but a new textile material as well as a
plastic emanated from the work. The plastic was to play an important part in
the later development of cellulose acetate yarns. Additionally in the explo-
sives field, gun-cotton is used as a high explosive—nitrocellulose + nitroglycerine + a stabilizer as a propellent and nitrocellulose + nitroglycerine + a
hydrogen bonding compound, e.g. carbamite, as a solventless cordite for
rocket propulsion. With the realization that cellulose could be put into
solution the use of nitrocellulose as a fibre forming material was discontinued, but it was retained as celluloid in the plastics field.

## A. Cuprammonium silk

In 1890 Despeisses patented the spinning of cellulose filaments from cuprammonium solution.[7] On extrusion the cellulose was regenerated into filaments in sulphuric acid and then pulled into very fine filaments (~1 den) by plastic extension. These were pseudo-circular, of the same order of strength as pure silk and for many years augmented this as the yarn used for "fine ladies' wear" until replaced by nylon. The 1960 Wool Year Book gives a detailed history of the rayon industry from its inception to that date.[7] It is clear that the first stages of chemical engineering were developed from the work on fibre regeneration and later from the petro-chemical industry. Even the recent high tenacity carbon fibres had been anticipated by Swann when he produced carbon fibres from cellulose nitrate in his work on electric lamp filaments. The early literature is not clear on how this was done but I assume that the fibres were denitrated before carbonization occurred.

## B. Viscose

The most important of the regenerated textile fibres originated with the work of Cross and Bevan who produced alkali soluble xanthate in 1892. In 1893, in collaboration with Little in Boston, they made a cellulose film. Two years later Cross and Bevan patented the making of cellulose acetate filaments, again for the manufacture of carbon filaments. In America in 1895

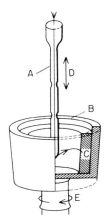

FIG. 5. Topham Box. The original centrifugal wet viscose yarn collection system. A: delivery tube of wet yarn; B: Topham spinning pot; C: yarn passing through A and collected by centrifugal force to build up cake C; D: delivery tube traverse; E: rotation spindle of pot B.

Little also succeeded in making cellulose acetate filaments. Development proceeded rapidly after this and may be summarized as follows:

1897. Fremery and Urban made textile threads from cuprammonium cellulose.

1899. Cuprammonium cellulose made in Germany. Glanzstoff Fabriken AG, Elberfeld formed.

1900. Public exhibition of first viscose yarns in Paris by Cross and Bevan. C. F. Topham invented Topham Box which collected the wet viscose yarns and held them in cylindrical form by centrifugal force. The cylinders were perforated so that the cakes of fibre were partially hydro-extracted. The cakes, of about 1 lb weight, contained water soluble sulphur relicts from the coagulating process so that they were washed and de-sulphurized, bleached and finally washed and dried (Fig. 5).

The above is a short summary of the early development of the man-made fibre industry. Further development continued, particularly in the 1920s and 1930s to form the second phase of the industry.

## V. The Second Stage in the Rayon Industry

In 1914, with the advent of war, development of viscose and cuprammonium was retarded and attention was centred on cellulose nitrate. This was used as collodion for plastic coating of aeroplane wings, for safety reasons it was replaced later by cellulose acetate dope manufactured by the British Cellulose and Chemical Manufacturing (Parent) Co. Ltd at Spondon, Derby. At the end of the war this firm turned to the development of cellulose acetate yarn for textile purposes.

### A. Cellulose acetate

The manufacture of yarn from this material demanded an entirely new technique, not only of spinning but of the acetate manufacture prior to this. The method of manufacture of acetate with an average of 2½ acetate groups per glucose ring was at this time a triumph of chemical engineering. It is detailed in many publications.[8] Secondary cellulose acetate is readily soluble in acetone/water (~10%) and is coagulated by extrusion through jets and dried rapidly by a counter current of warm air. It is then drawn through a small hole in the base of the spinning tube, collected at a predetermined speed on a take-up roller and passed to a bobbin on a cap spinning unit. It is essential to recover the acetone in this process as the acetate content of the dope is 20–25%. Hence the incorporation of a complicated chemical engineering plant to recover at least 95% acetone used. Without this the process would not be economically viable.

## VI. Fibre Structure

In the early 1920s great advances were initiated in the study of fibre structure.[6] The importance of this work cannot be over-stressed but is beyond the scope of this book. Key publications included those of Sponsler and Dore between 1923 and 1926[9] whose X-ray studies showed that macroscopic fibres were composed of molecular chains of cellulose which were glucose residues linked together in long chains. Much scientific in-fighting took place at this stage. The "how" of engineering ideas and processes had worked technically in the past and the new scientific work was beginning to provide the "why". From the knowledge acquired from the molecular studies the man-made fibres were evolved in the following years. Possibly the best review of progress up to 1932 in the field of natural and regenerated fibres was given at the Faraday Society Conference in Manchester on The Colloidal Aspects of Textile Materials in that year.[10]

At that time I was a newly qualified PhD in textile science and was about to start work at the British Celanese Corporation as a physicist. My active interest in engineering was stimulated first in the workshops of the textile department of Leeds University (1930–32), and secondly in the Celanese factory at Spondon, Derby (1932–35). In the former I saw the premises of break-spinning and in the latter I devised the first false-twist spinning unit for the production of voluminous yarns. The patent on break-spinning, BP 411 862, by A. F. Barker and C. C. Binks was applied for in December 1932 and completed in 1934, see Fig. 6.[11] After this no progress appeared to be made and in the 1940s I found that this patent and equipment had been acquired by Dr Dreyfus and the Celanese Corporation. I left the matter in abeyance until after I had left Courtaulds in 1950 to become head of the School of Textile Industries at the Bradford Technical College. In the meantime Courtaulds had taken over the Celanese Corporation and I enquired if any of the old break-spinning equipment was still available, but was told that no trace of it was to be found. Now, 50 years after the original Leeds work, break-spinning has become a great engineering innovation, some 20 years ago in Czechoslovakia and now worldwide. A detailed study of the recent developments is given by Dr Dyson in Chapter 7.

## VII. Cellulose Acetate Staple, False Twist and Bulked Yarns

During my time at British Celanese, on account of my previous experience in the woollen and worsted departments of Leeds University Textile Department, I was involved in a development programme on acetate staple manufacture. This was being carried out on a cotton carding, drawing and ring

Side elevation of dual "tuft" spinner

(a)

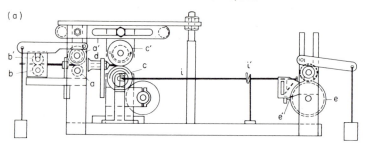

Plan diagram of dual "tuft" spinner

(b)

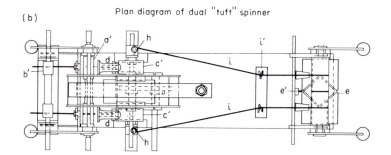

(c)

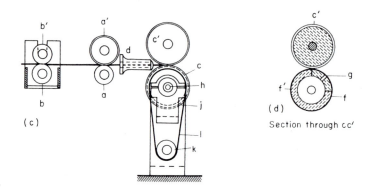

(d)

Section through cc'

(e)

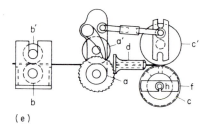

spinning system. The resulting yarns were doubled and manufactured either by weaving or knitting. Naturally the fabrics had a lean look, and whereas fabrics made from continuous acetate were elegant and acceptable, those from staple fibre were not. In contradistinction to this, viscose staple was being manufactured and found to be well received in the cotton industry. This was understandable as cotton and viscose were entiamorphic forms of cellulose (native and hydrate) and mercerized cotton was a mixture of the two. Cellulose acetate, however, had an average of $2\frac{1}{2}$ out of 3 of its OH groups substituted by acetyl radicles. This reduced its ability to absorb water and caused the yarn to electrify readily, retain the static electricity and therefore undergo great fibre repulsion during spinning. Continuous yarn was drawn through epicyclic rollers sufficient to cause fibre breakage and the resulting sliver was spun on a ring system. These methods failed to produce a voluminous yarn and at this stage I produced several reports on various systems designed to achieve this end, but without success. Only the patent literature emanating between 1933–37 from the British Celanese Corporation on the subject can give any idea of the work done on carding, drawing and spinning techniques in an effort to produce a technically acceptable cellulose acetate staple fibre yarn. I was also asked to produce a report on the influence of wool fibre crimp on the voluminosity of woollen and worsted fabrics. In this I drew attention to the basic "reversible extensibility" of wool fibres, pointing out the essential difference between celanese staple fibre and wool.

I then turned to the manufacture of acetate bulked yarns and this gave rise to the first experiments on continuous texturizing by false twist. This process consisted of setting the molecular structure by solvent or steam on one side of the twisting unit, taking out the twist of the macro-structure but retaining the physical effect of the first twist in the second stage of the retwisting. Figures 7 and 8 are drawings of the original false twisting units and the parts are numbered and described in the figures. Figure 7 is an illustration of the patents basis of BP 424 880[12] and Fig. 8 that of the simplified spindle of BP 442 073.[13] Experimental frames were built and quantities of bulked acetate yarn were made, and extensive wearing trials were carried out. After 1935 when I left British Celanese I took no further part in this work although it still continued. As a final comment on this research project it is pertinent to

FIG. 6. Working drawing of break (tuft) spinning unit of A. F. Barker and W. C. Binks. BP 411 862. The sliver passes through feed rollers a,a' and through a direction tube to the nip rollers c,c'. It passes into the orifice g, forms tufts which pass out through the nozzle h after twisting by rotation of c. The yarn i is drawn from h and wound on tandem bobbins through guides i' (Fig. 6b). Better tufting was obtained by using a ratchet wheel and pawl (Fig. 6c), worked from the nip roller of the tuft spinner.

say that the process of introducing crimp into the cellulose acetate yarn was a success, but it was unfortunate that the plastic nature of this material destroyed any chance of retaining the crimp in the yarn. The wearing trials of the fabrics made from the bulked yarn were dismal failures as the yarns stretched in wear and dimensional stability of the fabrics was not possible. The application of false twist systems to the new man-made fibres became one of the greatest triumphs of the third period of textile development referred to at the beginning of this chapter. The reason for this was very simple as these new fibres were in the main melt spun products, e.g. polyamides and polyesters, and were heat set in the false twisting process and could retain their crimp. This subject is dealt with in detail in Chapter 9 by Professor Lord.

## A. Fortisan yarn

Another important aspect of my work at British Celanese was the study of production of high tenacity yarns on the Weissenberg patents which included the stretching of acetate yarn in weak aqueous solutions of dioxan. This process was discontinued after it was discovered that dioxan was an extremely toxic substance with a probable harmful effect on the work force involved. Fortunately it was found that the acetate yarn could be stretched in superheated steam and this new process was substituted. As shown by their X-ray fibre diagrams, these acetate yarns were highly oriented. Their tenacity could be increased from 1.3 g/den (approx) to 3 to 4 g/den, but the extensibility was reduced to a mere 5% from ~25%. The acetate yarn was saponified and Fortisan (hydrate cellulose) yarns with up to 11 g/den were produced.[14] This never had extensibility greater than ~5% but had a valuable use in strong braids for inanimate parachute loads. Further X-ray work done during this period at Celanese included studies on dry spun yarns of viscose and cellulose acetate. I was informed that this work was used in the Celanese/Courtauld litigation on the Dreyfus patents on the dry spinning of acetate. The outcome of this case was of tremendous importance to industry as it released the method of dry spinning (evaporative) for unrestricted exploitation.

## VIII. The Third Period of Textile Engineering Development

After five years spent in medical physics when my textile interests were necessarily in abeyance I was invited in 1939 to take up a post as chemist-in-charge (physicist) in the Royal Ordnance Factories to build up a research unit to study the processes leading to the production of nitrocellulose and the cordites made from it. This is not a field to expand upon in a textile engineering book, but the experience gained was of value in clarifying my ideas later when I joined Courtaulds Ltd, Coventry. Added to my experience of wool, cotton and cellulose acetate I had acquired a knowledge of cellulose nitrate in the detailed study of nitration carried out in collaboration with Woolwich. This work[15] gave a clear indication of the differences in the reactions of the acetylation and nitration of cellulose and formed the rationale for the process of non-solvent triacetate manufacture after I joined Courtaulds in 1944.

During the period 1945–50 the advances made in textile engineering were phenomenal both in the development of new materials and in the introduction of new techniques for the manufacture of known and new fibres into finished fabrics. Probably the first important innovation was the introduction of the melt spinning process for man-made fibres. A third extrusion method had been added to the wet (~1890) and the dry (~1920) methods. Melt spinning was comparatively simple and there is an abundance of literature on the various early processes.[16,17] However a short summary will be given on the yarn manufacture of such polymers as nylons and polyesters. In spinning these from the polymer chips the material was filtered through sand in a molten condition at about 300°C in a nitrogen atmosphere to minimize polymer degradation. The molten polymer was pumped through jet holes by a rotary pump, cooled by a cross air flow, passed into a re-heating chamber and drawn away at the thread guide on to the take-up bobbin. Originally it was then taken through a filament stretching process (cold drawing), passed on to an output feed roller, through a thread guide and taken up on a large bobbin as drawn nylon. This gives only a broad idea of the original processes but detailed references to the equipment etc. are given.[18] The spinning speeds at this time, about 1947, were of the order of 600 m/min. This contrasted with the days of viscose spinning using the Topham Box (about 1906) when only 40 m/min were produced. In the later chapters it will be shown that speeds of more than 1000 m/min are possible at the present day. A comparative table of fibre production is given in Table I.

## A. Man-made fibres and yarn strength

In early times it was recognized that fibres owed their relevant textile properties to their strength and elongation at break. With the advent of man-made fibres yarn strengths in general had increased and it was possible to relate their strength to the orientation of the crystallites in the yarn. The best way of looking at this orientation was with the X-ray spectroscope. This method is so well known now that only brief reference will be made to it.[6] Highly oriented regenerated cellulose (Fortisan) can have a tenacity of ~10 g/den, whereas normal viscose has a tenacity of $\sim$2 g/den. However the viscose can be extended $\sim$20% before break but Fortisan extension to break is usually less than 4%. On drawing, man-made fibres generally increase their orientation considerably but retain a possible extension of about 20%. Thus the usable fibres are included in the class of comparatively strong fibres which have sufficient elasticity as well as extension at break to be capable of textile processing.

## IX. Chemical Engineering in Fibre Manufacture

After 1945 chemical engineering made rapid strides in relation to man-made fibre technology and many new polymers were synthesized. Most of these were rejected because of the high cost of the modification needed to meet the chemical engineering requirements necessary to produce yarns acceptable for weaving and knitting as well as dyeing and finishing. In melt spun polymers two properties which determine to a great extent their spinning possibilities are the melting point (MP) and glass transition temperature (GTT). The former is obvious, but the latter less so. It is the temperature at which the molecular attachments in the amorphous component of the fibre are released. This allows the fibre to shrink if free to do so, but in most cases the crystalline parts of the fibre are suspended in the amorphous component and the fibre can be drawn out to varying degrees. In particular, care must be taken in the manufacture of fibres which have a low GTT as wide variations may occur in fibre properties if this temperature is exceeded, especially with acrylics and the vinyl acetates and chlorides where the GTT may be very low, e.g. $\sim$50°C. A third important property of fibrous molecules is associated with molecular symmetry along the chain length. This is considered in detail elsewhere[19] and will not be discussed further here. In general the man-made fibres have a higher tenacity than natural and regenerated fibres (Fortisan being an exception), so that decreased tenacity is not a major drawback to their use. This is emphasized because it plays a considerable role in determining the mechanical and chemical engineering design of the processes to which the fibres may be subjected. This is particularly the case in the production of voluminous yarns by shrinkage methods as an alternative to the false twist process of bulking (see Chapter 9).

It is not proposed to expand further on the introduction of new fibres and filaments in recent years as this is summarized in Table I on world production of fibres from 1945–75. The International Conference on man-made fibres for developing countries in 1976 also included a statistical survey of the manufacturing centres of these products. As a matter of interest, in the breadth of processes involved Moncrieff[8] lists over 400 different types of fibres and filaments as being available in world markets as early as 1957. It may be argued that many of these named materials are the same, but it would be an imprudent manufacturer who risked the mixing of similar products from different sources in the production of a woven or knitted cloth without adequate knowledge of their properties.

It is obviously impossible in this introduction to trace the detailed history of all the chemical and mechanical engineering advances in the rayon industry, but two processes which show the expansion of automation in yarn manufacture from spinning dope to finished yarn appear of special importance, and deserve mention. As the Topham Box was a key invention in the extrusion of viscose rayon, so the "Advancing Reel" system of the I. R. Corporation provided another major advance (Fig. 9a). It was developed to form the cascade system of ten advancing reels providing two acid washes, two stages of yarn de-sulphurization, one sodium hypochlorite wash, one bleach wash, one emulsion surface treatment of the yarn, one drip and wash

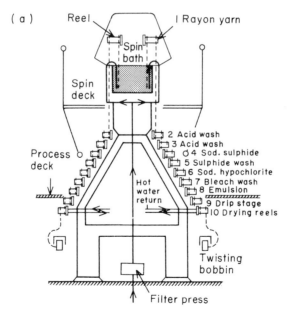

FIG. 9a. Cascade system of wet spinning of viscose. Diagrammatic representation of IRC continuous viscose spinning process. Viscose is fed to the spin-bath, coagulated and passed down via ten advancing reels of the sprocket and spider type, through baths of finishing treatments as indicated. At the ninth stage the yarn is washed and at the tenth passed round drying reels and collected on twisting bobbins on the bottom of the three level spinning unit.

FIG. 9b. Shortened Nelson viscose spinner. Extrusion baths are situated at the rear and coagulated yarns pass on to an enclosed chamber where they travel along two staggered rollers. They are finally washed and dried and carried on to bobbins shown at the front of the unit. Yarns can be seen clearly in transit between top and bottom rollers in the machine.

stage, one hot roller stage and finally a stage of bobbin twisting. This machine produced an excellent yarn immediately ready for use in approximately 3 lb cap spun packages. The Nelson system is a shortened continuous spinning machine in which the de-sulphurizing and bleaching processes are omitted (Fig. 9b). A single long pair of skewed rollers are used to transfer the yarn from the processing side of the machine to the drying unit which is still part of the traverse system. From here the yarn is guided over an oiling system and collected on large ring spinning bobbins. An example of this system on a seven spindle basis can be seen in the man-made fibre laboratory of Bradford University (Fig. 10), but the last time the author saw such a unit in industrial production was in Finland in 1972 at the viscose factory Sateri Osakeyhtio at Valkeakoski. It is of interest to note that by these continuous methods of producing stabilized, dry yarns the speeds of spinning had been increased from about 40 m/min in the production of wet cakes by pot spinning to about double that rate in the cascade and Nelson processes.

FIG. 10. Man-made Fibres Research Laboratory, University of Bradford Textile Department. Redevelopment of the old power house shown in Fig. 4 provided sufficient height for three levels of flooring to accommodate to full dimensional scale spinning and ancillary equipment for wet (A), dry (B) and melt (C) spinning, with controls. Also shown (D) is a pilot plant viscose dope manufacturing unit which can also be used for preparation of other wet-spun fibres. In the background is a nine spindle Nelson spinning unit (E) of the single advancing reel type, with a drying section and a ring spinning take up unit for the dried yarns.

## A. Cellulose tri-acetate manufacture

Fundamental studies of the acetylation of cellulose had shown that the reaction passed through the amorphous phase of the cellulose and then into the highly crystalline component to a stage of complete acetylation.[20] The development of this at the pilot plant stage led to the dry spinning production of cellulose tri-acetate (JPS) from methylene chloride. Much of the earlier technology of cellulose acetate yarn manufacture could be omitted in this method. The more important omissions included saponification of the acetate from 63% to 54% acetyl content, its precipitation at that stage, its washing and also the recovery and concentration of the acetic acid. The fully acetylated cellulose was removed in fibrous form, dissolved in methylene chloride and dry spun. The solvent proved to be non-toxic, but of course it had to be recovered.

## B. Unwoven tissues

It had become clear that in partially acetylated fibrous cellulose the tri-acetate component of the fibres was soluble in methylene chloride but that the unchanged cellulose was not.[20] It was found possible to produce carded webs of the fibres, and by competent layering techniques and subsequent treatment with methylene chloride the cellulose fibres could be held together by an adherent film of cellulose tri-acetate after removal of the methylene chloride. This provided a simple and expedient method of unwoven tissue production and was covered by BP 638 591 and US Patent 2 535 919.[21,22] A further two-component fibrous system using cyano-ethyl cellulose was devised and patented under BP 640 411.[23] It is important to stress that unwoven tissue manufacture has expanded into a major industry but the fundamentals of the engineering processes as outlined above in the original patents remain virtually the same.

## C. Alginate fibres

From about 1940 seaweeds became of interest as a source of water soluble fibres. Speakman and Chamberlain[24a,b] carried out development work on the manufacture of alginate fibres from seaweed at the University of Leeds but the work was taken over later by Courtaulds Ltd. The spinning solutions were made by dissolving the "alginic" acid from the seaweed with alkali. This actually was the guluronic acid form of the sugar and the less soluble mannuronic acid was retained in the seaweed.[25] To spin these fibres little modification of a normal viscose spinning routine was required. The alginate, however, is attacked readily by bacteria so the solution of ~9–10% sodium alginate was sterilized and extruded into normal calcium chloride solution and precipitated. After washing and drying the precipitated calcium alginate could be stored satisfactorily. It was possible to leach out the calcium with an alkali scour and the alginate again became soluble. However if calcium alginate fibres were treated with beryllium acetate, part of the calcium was replaced by beryllium and the fibres became highly resistant to soap and soda scours. Eventually soluble alginate fibres were available for use in the production of certain textile materials such as gauzes etc. and in the hosiery industry.

## D. Regenerated protein fibres and their polypeptide synthetic analogues

A summary of the work done on the chemical and mechanical engineering required to produce protein filaments was given by Wormell.[26] Protein sources included milk casein, soya bean casein, zein and on a smaller scale edestin from the castor bean, while wheat gliadin and proteins from cotton and flax seeds were considered. Animal protein filaments have also been produced from egg albumen, feather keratin, silk, gelatin and fish protein. It seems possible that in future fibres may be made from synthetic polypeptides. Considerable development work was needed to devise the means to produce the protein fibres but in most cases a wet spinning programme was required.[27] Eventually two major industrial production units were established. The first used peanut protein (Ardil) which was made and marketed by ICI Ltd, and the second produced a fibre, Lanatil, from milk casein based on the Ferretti Italian patents. Wormell developed a fibre, Fibrolane, from casein which was produced and marketed by Courtaulds Ltd in the late 1940s.[26] This however was not a stable fibre and although it had a limited success in mixture yarn manufacture, little if any is manufactured now for textile purposes. The work done, however, led us to look at the regeneration of wool. Also, an edible fibrous material was developed from the protein fibres which was flavoured and coloured and sold as "Kesp" to eke out the supplies of more conventional edible protein.

The possibility of the regeneration of wool from solution produced the next extrusion programme and patents were filed in 1949.[27] In this work the cystine link in the wool structure was reduced, the wool curd dissolved in cuprammonium solution and then extruded as filaments which showed a $\beta$ wool structure. In contradistinction to previous casein filaments, these were oriented readily in the coagulating and spinning baths. This was quite a step forward but the $\alpha$ structure of the original wool molecule had not been reproduced so that no $\alpha \rightleftharpoons \beta$ elasticity was present in the regenerated fibres. This was due to the replacement of the cystine link by a formaldehyde linkage such as that used earlier in the stabilization of regenerated casein and

FIG. 11. Skeins of regenerated wool. (a) Skein of regenerated $\beta$ keratin protein fibres from reduced waste wool. (b) Skein of regenerated $\alpha$ keratin fibres from oxidized waste wool.

other proteins. However the regeneration of wool in the $\alpha$ form soon followed.[28] Wool was oxidized by 2% peracetic acid and could then be dissolved in weak alkali and precipitated as a curd which showed an $\alpha$ protein structure similar to that of the original wool (Fig. 11). This led to the correlation of work done on the polymerization of the synthetic polypeptides, fibre formation from these and manufacture of such fibres into fabrics on a pilot plant scale. This work appears to have been abandoned, but it provided the basis for the study of synthetic protein models. These formed the premise of molecular biology which has since led to an extensive expansion of bio-engineering.

## X. Textile Engineering from 1950 Onwards

In 1950, after appointment as head of the department of textile industries in Bradford, the author's commitments were expanded widely to include many fields of interest. It was possible to plan a wider view of textile development. The fundamental research into textiles and fibre science were still of prime importance but the engineering and development sides of the work also became a major concern. The main division of these interests were : (1) studies of established processes to ascertain the basic fundamentals of the operations, thus leading to improvements; (2) introduction of new fibres and filaments; (3) introduction of new mixture materials leading to new techniques; (4) the application of new fibres to industrial use.

Not only has enormous expansion of textile production occurred during the last 25 years but the distribution of the industry has spread internationally.[4] Up to 1945 the technique of worsted, woollen and shoddy manufacture had changed little from the time of the invention of the Noble comb in 1853, but in 1955 there appeared two new inventions: the Auto-leveller by Raper, and the High Draft Spinning by Ambler. These made a great impact. Raper's automatic auto-leveller was capable of reducing the number of operations in the Bradford open drawing system from nine to four, and the number of operatives from ten to six. The floor space required was reduced from 4800 ft$^2$ to 3600 to accommodate a single line of drawing. These processes are described in Chapter 3 on long staple yarn manufacture but they were detailed in 1955 at the First International Wool Textile Research Conference as great advances in worsted spinning techniques.[29]

## A. Yarns from mixed fibres

As shown in Table I, in 1965 the total production of regenerated and man-made fibres had increased to 29% of world total fibre production, while wool had decreased slightly to about 8% and cotton retained its lead at 63%. However it was in wool/man-made fibre mixtures that the greatest outlet of mixture yarns was found. The earlier mixtures were wool/viscose staple fibre. The viscose was supplied either as cut staple carded into top form or as cut top made from continuous filament tow cut or broken into top form. The mixture tows were run through a standard process of worsted yarn manufacture, and these yarns were used for cheaper and lower grade mixture fabrics. Little difficulty was encountered in the manufacture of cotton/viscose yarns where the viscose fibres used were about the same length as the cotton fibres. In finishing, these fibres were readily compatible as they differed very little in dyeing properties as chemically they were basically the same. The mixtures of wool and nylon or polyesters presented a different problem. They were not difficult to draw and spin from separate tops to produce mixture yarns and to manufacture into finished materials but problems arose when uniform distribution of fibres did not occur as barring of fabrics resulted and dyeing difficulties arose.

The man-made fibres gave the fabrics increased strength and, especially with the polyesters, stiffened the material. The use of such yarns has become so common that their properties tend to determine the character of the resulting fabrics. Wool is weaker but retains its voluminosity and elasticity and increases the warmth of clothing as well as retaining its ability to absorb moisture, whereas nylon and the polyesters are almost hydrophobic. It should be noted that wool production has only increased slightly from 1940 to 1974, from 1134 to 1481 million pounds, but with the mixture fabrics a greater quantity of wool-containing materials has become available throughout the world. These mixed materials have had a marked influence on the "old wool" trade, or recovery of textile materials from rags as shoddy. The synthetic materials in the fabrics resulted in the old-fashioned process of rag-tearing running into difficulty because of the toughness and strength of the synthetic fibres which were strong enough to damage the rag-tearing machinery. Consequently the waste trade of the "Heavy Woollen" district of Yorkshire declined. An alternative scheme of rag cutting rather than tearing was devised by the author to overcome these problems and this is described in Chapter 6 together with some related industrial developments.

## XI. Fabric Manufacture

### A. Woollen and worsted

In considering the early development of fabric manufacture some points should be noted. First the processes of weaving were very soon changed from the narrow cloths of the hand loom to the wider fabrics of up to 60" produced on the power looms. Where yarns were too weak to stand up to the three fundamental processes of shedding, picking and beating up, the warp and in some cases also the weft were stiffened by sizing. It is, however, important to realize that much textile engineering development has taken place over long periods of time, especially design improvements in established engineering practices. This is clearly shown in the development of weaving machinery which is dealt with by Mr Sondhelm in Chapter 10 on high speed automatic weaving, by Dr Ince in Chapter 11 on heavy fabric manufacture and by the editor in Chapter 6 on fibre recycling. Generally, and in the woollen and worsted trades especially, looms were designed to weave a very wide range of fabrics. This allowed the manufacturer to vary his type of material as fashion dictated particularly in the more expensive sections of the industry. This led to the division of the wool industry into the so-called vertical and horizontal divisions of the trade. The former usually applied to the woollen industry which processed the raw wool or waste and shoddy to the finished fabric ready for sale. The horizontal system generally referred to the worsted trade because of the greater latitude allowed to the manufacturer in selecting special types and qualities of yarns for the production of specialized fabrics.

## B. Cotton

Consideration of Table I shows that cotton forms a major part of the international textile industry. In addition, early in the production of rayon, cotton formed the basic raw material in the manufacture of cuprammonium rayon and viscose. Later, however, the latter was made from wood pulp. Excellent summaries of the development and spread of the cotton industry are given in a number of reports so that they will not be dealt with here.[30] The industry is international, and initially the USA led in both the growth of the raw material and its manufacture. Later this was followed by Egypt, India, and other Far Eastern and African countries.

## C. Hosiery

Hosiery manufacture developed along different lines from the making of fabric by weaving, in the sense that the former usually incorporates the making up of the material into garments in the processing. These may be made from individually manufactured pieces which have a specific shape and are assembled to form the garment. Alternatively garments may be knitted to the shape required for the finished article, an obvious example is the knitting of hose and half-hose.

There are two basic methods of knitted fabric formation: (a) weft knitting in which a continuous yarn is used, travelling perpendicular to the direction of fabric build-up; and (b) warp knitting in which a complete warp of threads is knitted by interlacing of the parallel threads to form a continuous fabric.

Detailed information on the development of the different knitting mechanisms from the weft knitting of half-hose to the 200″ Raschel warp knitting machine for the production of long wearing ravel-resistant carpets together with the relevant industrial and patent references is given by Lancashire.[31]

One of the more interesting and economically important effects of the introduction of new fibres into the textile industry was the replacement of silk and cotton yarns in the ladies' hosiery trade by cuprammonium filament yarns (e.g. Bemberg) about 1920. These in their turn were superseded in about 1950 by nylon, first as continuous filament and later as texturized yarn. It is not proposed to expand this discussion further but additional references to industrial expansion are given in subsequent chapters. Some aspects of development work in this field are given in Chapter 12 on advances in stitch-bonding, warp and weft knitting systems and automated knitwear manufacture. This shows what advances can be made with university and industrial collaboration. Certainly much can be gained when this is encouraged.

## XII. Mixed Fibres, Tows, Tops and Yarns. The Greenfield Top (1936)

With the introduction of man-made staple fibres changes were needed to facilitate their processing with other fibres, and some examples of these technical advances are summarized in this section. The staple fibre was normally cut from continuous filament after spinning, and passed through the de-sulphurizing, washing and bleaching viscose routine and then baled ready for despatch to the spinner. In 1937 this system was replaced by production of the Greenfield Top. For this the viscose tow was spun from banks of jets with upwards of 1000 jet holes to form tows of approx 200 000 denier. After washing and drying the tows were collected from adjacent spinning heads to form a tow containing many thousands of viscose filaments. This was passed to a helical cutter and drawn on to a gill box which drew the fibres into a sliver, alignment being preserved by draft and pin control. This reached the yarn manufacturing sections of the industry as a top ready for drawing. To introduce their new Greenfield Top to the worsted and cotton industries Courtaulds Ltd set up a demonstration plant at Arrow Mill, Rochdale, to promote the tow to top process.

### A. The Pacific converter (1950)

A later advance on the Greenfield system was introduced by the Pacific Mills Corporation of Lawrence, Mass, and further developed by the Warner Swasey Co. of Cleveland, Ohio, who were also responsible for the pin drafting system for the reduction of tops to slivers for various top compositions of fibre length and materials. When melt spun man-made fibres were used there was a tendency for the ends to fuse and stick so that debonding rollers had to be introduced.[32]

### B. The turbo-staplizer

In this American system of fibre formation from continuous filaments cutting is replaced by breakage of the tow.[32] The basic principle of this machine is to stretch and break the continuous filaments into a staple of varying fibre lengths. In the case of man-made fibres the staple top is generally stretched and requires a final shrinkage process to produce "length stable" fibres. Having mentioned briefly these early methods of tow to top conversion, the reader is referred to Chapter 4 on sliver production. There a major expansion of the engineering and applied mathematical analysis of more recent methods is given. Also included is a discussion of the physical properties of yarns and fabrics caused by variations in the engineering techniques described.

## XIII.  Some Aspects of High Speed Spinning and Twisting

In the production of yarn from staple fibre a uniform rate of presentation of untwisted fibres into the twisting zone is necessary. Prior to automation this was controlled by hand. This was a slow process but uniformity of the slubbing or roving entering the spinning system was achieved. The rate of yarn production was, and still is, controlled by the draft, the turns per unit length required in the yarn and the rate of revolution of the spindle. After automatic methods were introduced, further factors affecting the stability of the spinning process were the frictional and centrifugal forces in the yarn balloon as it revolved during the introduction of twist. The three main systems of twist insertion are flyer, cap and ring spinning. (The mule is omitted here because it has become almost obsolete.) The flyer system is the most cumbersome but the strain on the yarn is minimal and mainly consists of a friction component due to the yarn being wound round the flyer or passed through a hollow passage along the length of the flyer arm. The inertia of the system confines the spindle speed to ~3000 rpm but the method is still used for heavy long staple yarns such as mohair and other coarse hairs, belting yarns and thick and medium yarns for hosiery and hand knitting. In the case of the cap and ring spinners spindle speeds of up to 8000 rpm are usual for normal staple yarns. The dry spinning of continuous filament yarns may be associated with spindle speeds of 12 000 rpm. Due to the small amount of twist involved the rate of yarn production increases proportionately. In melt spinning another factor is involved in that yarn, after consolidation, may be cold drawn and speeds of collection may be of the order of 1200 m/min. Three methods of twist insertion are important in contemporary development. Two were mentioned earlier in this chapter and their premises have been outlined in Chapter 7 on break spinning and in Chapter 9 on the use of false twist in filament texturing. The third method of "two-for-one" is mentioned in Chapter 3 on long staple yarn manufacture. With regard to break spinning, the author attended a conference in Prague in 1969 on the subject which was being developed at that time in Czechoslovakia. The machines then wound yarn into flat cheeses weighing up to 1.5 kg with a spinning rotor running at 30 000 rpm. This was about four times the rate of production of a normal classical spinning frame of that period. To complete these introductory references to twisting it is of interest to note that a "known possible method" of inserting additional twist into fibres was made an engineering reality in the introduction of the "two-for-one" twisting process. In addition to other reasons the method is of great value because it is capable of inserting twist economically and efficiently into extruded regenerated yarns and also into man-made filament yarns which are generally extruded at high speed with a minimum of twist. The method may be used for the production of single or multiple strand yarns for the tyre industry which was one of the earlier examples of its use, replacing the earlier up-twisting process.

## XIV. Engineering Research on Extrusion. From Mass Production to Micro-Techniques

Development work on filament extrusion has been carried out since the end of the 19th century, including first the wet spinning of cellulose, then dry spinning of acetyl-cellulose and later the melt extrusion of synthetics such as nylon and terylene. These three spinning techniques gave great impetus to polymer chemistry and many new polymers have become available for domestic and industrial use. The complex chemical engineering processes involved in the controlled manufacture of these materials on an industrial scale has made the assessment of their fibrous properties costly and in many cases prohibitive. Thus possible sources of useful raw material may have been discarded. However, even using expensive individual monomers, polymerization can be studied readily on a laboratory scale under controlled conditions, whereas in a large scale production unit or pilot plant this is economically not viable. Therefore in the textile department of Bradford University a microextrusion laboratory was set up in 1967[33] to study, in collaboration with RAF, Farnborough and the polymer section of the National Physical Laboratory, Teddington, certain types of fibres (Fig. 10). These included thermally stable fibres of which the American Du Pont product Nomex was the best known at that time. To give an idea of the scope of this work references are given dealing with the manufacture of heat stable fibres,[34,35] carbon fibre precursors,[36] programmes on new polymers and their fibre forming properties[37,38] and especially heat resistant fibres. Studies were also made on fibres reconstituted from waste materials but this work will be dealt with in Chapter 6 on the recycling of fibrous materials.

During the progress of this micro-extrusion work some interesting and useful engineering advances were made in the method of yarn collection.[33] Small samples of filament were wound from godets and collected on a second godet of a conventional take-up machine. The godet was designed to be wide enough to allow collection in a series of narrow bands across the width of the roller (Fig. 12). Each alternate band represented one filament section since time was required to effect experimental changes in the conditions of

FIG. 12. Banded fibre collection unit in micro-extension laboratory. A is collection drum with bands of fibres. B shows bands of fibres cut away for examination and assessment of fibre properties.

extrusion.[39] Thus a series of separate samples was obtained each representing a known set of conditions of manufacture. A more sophisticated method in take up in continuous spinning was developed later,[40] see Fig. 13. Filament was wound from a two-roll godet mechanism on to the upper of two collecting package units and after collection of sufficient yarn on this, the lower unit was rotated through 180° to the upper position.

The yarn was broken as it passed to the second package and was then

FIG. 13. Two roll godet mechanism for yarn collection. A series of yarn samples can be collected without breaking the continuity of the spinning operation. A indicates the interchange collection units.

wound on to this. The first package sleeve could then be removed and in this way a series of packages could be obtained of different yarns representing different materials or different extrusion conditions according to the problem being studied. These samples may appear small but a spinning time of three minutes at a take up of 300 m/min produced nearly 1 km of yarn which was more than adequate for preliminary investigation of new fibre properties. Much of the work carried out by this means was based on industrial collaboration and is therefore confidential and has been covered by patent, but work on other laboratory samples of new polymers was published.

## XV. Conclusion

This chapter has attempted to provide a broad view of textile engineering developments over the last two centuries in order to obtain the necessary background against which to demonstrate some of the more recent technological advances. The developments during the latter part of this period have been closely related to the author's own experience in order to give additional interest and to emphasize the wide field covered at the present time in the production of machinery for the newer technologies. In many cases it will be seen that the developments illustrated in the remaining chapters had small experimental beginnings that led eventually to major industrial developments. Taken together it is hoped that these chapters will provide a summary of the more important facets of textile engineering as it stands today. The international character of recent development has been emphasized and the influence of the "old technology" on the "state of the art" in modern industrial textile practice demonstrated.

## Acknowledgements

The author wishes to thank the City of Bradford Industrial Museum for photographs for Figs 1–3, HM Patents Office for permission to publish drawings from earlier patents, and former colleagues of the School of Textile Technology and the photographic department of the University of Bradford for Figs 4 and 10–13. Thanks are due also to the staff of the Textile Department who have supplied information from the departmental records which has been of the greatest assistance, and to the Science Library of the University College of North Wales, Bangor, for organizing loans through the Inter-Library Services.

## References

1. *Wool Year Book* (1960). Textile Mercury. Manchester.
2. James, J. (1857). *History of Worsted Manufacture in England*.
3. Dyson E. and Happey, F. (1960). Wool Research Conf. Harrogate. *J. T. I. Trans.*, Pt. I p. 1016.

4. *Man-made Fibres for Developing Countries* (1976). Sasmira, Bombay, India.
5. Schönbein, H. (1845). See *Introduction to the Chemistry of Cullulose* (Marsh, J. T. and Wood, F. C.), p. 289. Chapman and Hall, London (1942).
6. Happey, F. (1978). *Applied Fibre Science*, 1, c. 1, Academic Press, London and New York.
7. Despeisses, (1960). *Wool Year Book*, p. 461. Textile Mercury, Manchester.
8. Moncrieff, R. W. (1957). *Man-made Fibres*. National Trade Press. London.
9. Sponsler, O. L. and Dore, W. H. (1926). *Coll. Symp. Monograph*, 4, 174.
10. "Report Faraday Soc. Meeting". *Trans. Far. Soc.*, 34. Manchester (1932).
11. Barker, A. F. and Binks, W. C. (1934). BP 411 862.
12. Finlayson, D. and Happey, F. (1935). BP 424 880.
13. Finlayson, D. and Happey, F. (1936). BP 442 073.
14. Happey, F. (1978). *Applied Fibre Science*, 1, c. 1, 4. Academic Press, London and New York.
15. Happey, F. (1953). *X-ray Diffraction in Poly-crystalline Materials*, c. 26, pp. 533–41. Inst. of Physics, London.
16. Corothers, W. H. (1940). *High Polymers*, I. Interscience, New York.
17. Whitfield, J. R. and Dickson, J. T. (1941). BP 578 079.
18. Hill, R. (Ed) (1953). *Fibres from Synthetic Polymers*. Elsevier, Amsterdam.
19. Happey, F. (1976). *Man-made Fibres for Developing Countries*, pp. 178–79. Sasmira, Bombay.
20. Happey, F. (1950). *J. T. I.*, 41, T371.
21. Happey, F., Grimes, J. H. and Courtaulds Ltd (1950). BP 638 591.
22. Happey, F., Grimes, J. H. and Courtaulds Ltd (1951) US Patent 2 535 919.
23. Happey, F. and McGregor, J. M. with Courtaulds Ltd (1951). BP 640 411.
24a. Speakman, J. B. and Chamberlain, N. F. (1944). *J. Soc. Dy. & Col.*, 60, 264.
24b. Speakman, J. B. BP 545 872 and 572 798.
25. Atkins, E. D. T. *et al.* (1973). *Biopolymers*, 12, 1875.
26. Wormell, R. L. (1954). *New Fibres from Proteins*. Butterworth, London.
27. Happey, F. and Wormell, R. L. with Courtaulds Ltd (1952). BP 673 676.
28. Happey, F. and Wormell, R. L. with Courtaulds Ltd (1953). BP 692 876.
29. *First International Wool Textile Research Conference Report*, E, Pt. I (1955). Australia.
30. *Working Party Report on Cotton* (1946). HM Stationery Office. London.
31. Lancashire, J. B. (1959). *Revue of Textile Progress*, p. 267. Butterworth, London.
32. Moncrieff, R. W. (1957). *Man-made Fibres*, p. 503. National Trade Press, London.
33. Green, D. B. (1980). *Applied Fibre Science*, 3, 197. Academic Press, London.
34. Dyson, E., Happey, F., Montgomery, D. E., Sewell, J. H. and Tregonning, K. (1973). BP 1 333 084.
35. Dyson, E., Happey, F., Judge, A. and Montgomery, D. E. (1975). BP 1 392 935.
36. Blakey, P. R., Montgomery, D. E. and Tregonning, K. (1970). *J. Text. Inst.*, 61, 234.
37. Happey, F. and Green, D. B. (1977). *Eur. Polymer J.*, 13, 689.
38. Tse, W. W. (1973). *MSc Thesis*. University of Bradford.
39. Richardson, A. (1978). *PhD Thesis*. University of Bradford.
40. Atkinson, C. (1978). *PhD Thesis*. University of Bradford.

# Chapter 2

# Worsted Topmaking

## M. CHAIKIN and J. D. COLLINS

## I. Introduction

### A. Processing of wool fibres

Worsted processing utilizes longer finer wool fibres to produce strong, fine, smooth, uniform yarns for weaving or knitting into high quality fabrics. The overall processing sequence divides naturally into two sections, viz. worsted topmaking and worsted drawing and spinning. The basic functions of topmaking are to remove the various natural or acquired impurities in the raw wool and to mechanically manipulate and mix the fibres to produce a top sliver, a uniform, coherent sliver of parallel fibres free from entanglement and short fibres. Additional functions in the sequence may be to dye the fibres or to carry out a process such as shrinkproofing. The top sliver provides the raw material for drawing and spinning where the mass per unit length is progressively reduced to produce a worsted yarn.

As a first step in the topmaking sequence raw wool from bales is opened, mixed and dusted to remove as much as possible of the loosely bound contamination, mainly larger dirt particles and a limited amount of vegetable matter. Scouring is then carried out to remove the protective coating of wool grease, the dried perspiration known as suint and various forms of dirt bound to the fibres by the grease. The wool is dried and conditioned prior to mechanical processing.

The dry scoured wool is carded to remove as much as possible of the entangled burrs and other vegetable matter as well as residual dirt, to disentangle, mix and randomize the fibres and to form the fibres into a coherent, uniform sliver with fibres arranged as far as possible parallel to the sliver axis.

The fibres are then subjected to a number of preparatory gilling or pin-drafting operations (typically three) where a number of slivers are "doubled" and drafted to produce an output sliver. The objectives here are to straighten fibres and to improve fibre parallelization, to provide further fibre mixing and to reduce mass per unit length fluctuations in the sliver.

Slivers are then fed to a combing operation further to straighten and parallelize the fibres, to remove short fibres and to remove any impurities such as vegetable particles, dust and neps still remaining in the sliver. Combing affects the randomization of fibres of different fibre lengths and reduces sliver coherence. Further finisher gillings (typically two) are carried out to mix and randomize the fibres to produce the top sliver.

An important aspect of the mechanical processing is the minimization of problems such as fibre breakage, fibre loss and static electricity effects. It is essential to maintain correct processing conditions in relation to lubrication, fibre moisture content and room humidity. The older spinning system using the Noble comb uses additions of up to 3% of oil whilst the now more usual system involving the rectilinear comb uses less than 1% oil whilst maintaining high room humidity.

A sliver washing process called backwashing may be applied to the sliver before or after combing. The objective of backwashing before Noble combing was to remove dirt remaining after carding. Backwashing after rectilinear combing was aimed at reducing excess residual fatty-matter due to fibre lubricants or to faulty scouring, and a result of the process was the regularization of moisture content and the parallelization of fibres giving an attractive "ironed" finish. From a technical point of view backwashing is no longer required and its only virtue is the enhancement of the appearance of the top for sale. The majority of comber/spinners do not use the process.

At the top stage the sliver may be dyed or subjected to a process such as shrinkproofing. This usually necessitates recombing and finisher gilling to restore the original condition of the top.

### B. Processing of fibres other than wool

Staple synthetic fibres produced by conversion processes involving stretch-breaking or cutting of fibres to the appropriate lengths make considerable use of the worsted system either as blends with wool or alone. These fibres may be introduced at the carding stage from bales or at any of the gilling processes as slivers or directly as top sliver.

## II. Scouring and Scour Effluent Treatment

Prior to mechanical processing it is necessary to scour the wool to remove grease, suint and dirt. In choosing a scouring system there are three general requirements to be considered, viz. the economical removal of impurities, the minimization of fibre entanglement and the question of effluent disposal. Both aqueous and solvent systems have been used but despite considerable effort only two solvent plants were operating in 1980, one in Japan and one in the Soviet Union.[1]

A solvent system requires basically three steps, the removal of grease by a non-polar solvent, the removal of excess solvent and an aqueous treatment to remove the water-soluble suint and dirt. Solvent processing does not affect the initial fibre arrangement thus leading to minimum entanglement, higher carding production rates, less fibre loss, less fibre breakage and a superior top. A high recovery of grease (up to 90%) is possible, the potassium-rich suint salts can be recovered for use as a fertilizer and the effluent problem is minimal. On the debit side, however, there are a number of claimed disadvantages. Fibre handle tends to be harsh and brittle due to complete grease removal, there is concern about potential fire risks, the grease is very inferior to that produced by aqueous scouring, there is incomplete removal of dirt and the wool has a dingy colour. Probably the major concern relates to the economics of the process. A very high capital outlay for plant is required and the extra plant complexity seems likely to increase maintenance costs and limit the plant life when compared with aqueous scouring plants. In addition the economics depend very heavily on the world price of lanolin which is depressed at the present time. This decline in demand for lanolin and the consequent loss in revenue from grease sales is a matter of some concern to the scouring industry.[1]

Aqueous emulsion scouring is the oldest form of wool scouring and is still the conventional method. The oldest form, using soap and alkali, usually soda ash, is still in use today despite the advent of synthetic detergents. Greasy wool passes through a series of washing bowls followed by a rinse bowl and two techniques have been used in operating the system. The desuinting method, little used today, involves using only water in the first bowl to dissolve the suint salts and much of the heavier dirt particles. Subsequent bowls use additions of soap and soda to provide the necessary detergency action to remove the wool grease and remaining dirt.

The conventional aqueous system involves the counterflow method of liquor movement where fresh water is added to the final bowl and the overflow from each bowl is passed back to the previous bowl and is finally discharged from the first bowl. Detergent and alkali are used in the initial bowls whilst the last bowl is reserved for rinsing with water. With the development of synthetic detergents the trend has been firstly to use nonionic detergents with the addition of alkali and then to nonionic detergents alone. This latter method is the most common in use at the present time.

Wool is generally fed to the first bowl using a hopper feeder to produce a layer of fibre tufts. It has been recognized for some years that wool flow variability is a major problem with almost all wool scouring plants, and methods of control have been largely unavailable. Recent research into weightbelt feeding[2] has indicated that a considerable reduction in variability can be achieved resulting in a more evenly dried product, fewer stoppages, energy savings and increased production rates.

Mechanical action and transport of the wool through the scouring system has been accomplished in a variety of ways but the older system using harrow rakes (illustrated for a single bowl in Fig. 1) is still the major method in use today. Wool fed into the bowl is transported through the bowl by the action of the harrow rake which descends into the wool, rakes it towards the delivery end, is withdrawn and returned to repeat the process. Wool on exit from the bowl passes through a pair of squeeze rollers to remove excess liquor. It has been shown that the removal of grease from the wool increases with the relative speed of the solution through the wool, and the squeeze rollers, in providing a high relative liquid to wool velocity, are known to play an important part in the detergent process.

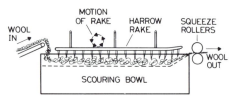

FIG. 1. Conventional scouring bowl using a harrow rake.

Considerable fibre entanglement can be generated during scouring and the Petrie-WIRA scour[3a] offers a modified harrow system which uses smaller bowls with denser wool packing and reduced harrow speed and stroke. It is claimed that this reduces fibre entanglement. The Fleissner scour offers an alternative system using several rotating perforated drums with blades attached. Wool on entering a bowl is pushed under the liquor surface using a dunking roller and floats in the solution to the first drum. This pushes the wool downward and forward forcing liquor through the wool and into the inside of the drum. The liquor in the drum is returned to the bowl in such a manner that it promotes the movement of the wool towards the delivery end of the bowl.

During the 1960s considerable efforts were placed on developing scouring processes which minimized entanglement and which would thus produce large savings by decreasing fibre loss and increase top mean fibre length. Scours based on jetting of fibres resting on a perforated belt conveyor[4] and on jetting of fibres held compressed between two perforated belt conveyors[5] were constructed. A further scour development operated on a more revolutionary concept where liquor movement through the fibres held between two perforated belt conveyors was produced by the belt movement past specially shaped wedge-nips.[6] These methods have not been adopted for a variety of reasons, the most important probably being due to the fact that with present fibre processing systems the reduction in fibre entanglement has not led to the large cost benefits envisaged with reduced fibre losses as well as a more valuable top. Interest in such developments seems to have lapsed in the 1970s as a result of the pressure towards developing efficient scour effluent treatment systems.

Wool scouring effluent can cause massive effluent problems as indicated by the comparison that a 1 tonne/hr continuously operating scour processing an average Australian Merino wool is equivalent in pollution load to the raw domestic sewage of a town of about 75 000 inhabitants.[7] Such highly polluting effluent can seriously reduce the capacity and effectiveness of municipal purification plants and authorities are placing high and escalating charges for effluent disposal and in many cases placing very severe limits on acceptable pollution loads for discharge. A recent survey[8] of world-wide charges and protective legislation shows in particular that Japan has the most stringent legislation and Japanese scourers have been forced to install expensive evaporation incineration plants to conform to these regulations.

Very substantial world-wide efforts have been made in recent times to develop effective liquor treatment systems and an IWS survey[8] has made a detailed examination and comparison of the large number of methods either in current use or at the pilot plant or semi-industrial stage. It is not possible to give a comprehensive treatment here and the following description is limited to some of the more promising systems including developments in scouring techniques, treatments external to the scour and both combined.

The scouring technique used is an important component in any treatment system since this can have a dramatic effect on the pollution load of the effluent to be treated. A major advance here has been the introduction of the WRONZ Comprehensive Scouring System,[9, 10] essentially a very efficient liquor handling system which may be applied to a conventional scour. Figure 2 illustrates schematically a comparison between the WRONZ Rationalized Scour based on this system and a conventional scour incorporating a settling tank for solids and centrifugal grease recovery. In the latter system the overflow from bowl 1 passes first to a settling tank and then to a centrifuge for grease extraction. The WRONZ system incorporates a flock catcher, a heavy solids settling tank, heat exchangers, centrifugal grease recovery and recycling of most of the liquor. Solids and grease removal are continuous,

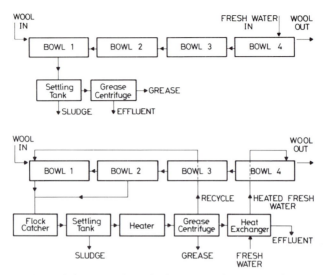

FIG. 2. Comparison of the operation of a basic scouring system (top diagram) with that of the WRONZ rationalized scouring system (bottom diagram).

addition of chemicals and counterflow are carefully controlled and heat recovery is carried out prior to the controlled discharge of the effluent to the drain. Compared with conventional systems, much less water is used, energy consumption is minimized, additional grease is recovered and there is a dramatic decrease in the pollution load.

A later development has been the introduction of mini-bowls[10,11] further enhancing the above advantages of the WRONZ system and in addition using less space, attaining equilibrium faster and reducing capital cost and maintenance. A single bowl of the normal six bowl system is illustrated in Fig. 3. The system, by recycling concentrated liquors, gives a highly efficient removal of both dirt and grease thus yielding a lower volume, less contaminated effluent and hence reducing the capacity and cost of a secondary effluent treatment.

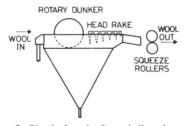

FIG. 3. Single bowl of a minibowl scour.

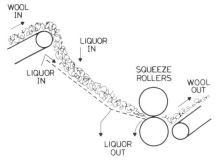

FIG. 4. Lo-Flo wash plate.

Lo-Flo scouring[12] developed by the CSIRO in Australia is based on the concentration–destabilization principle that if the natural suint and salt contaminants are allowed to build up in the early stages of scouring the wool grease/dirt emulsion can be cracked by heating and centrifuging. In order to allow the wool to be scoured in extremely contaminated liquor containing high levels of suint and grease a special washing bowl called a wash plate as illustrated in Fig. 4 has been developed. The essential features of a Lo-Flo installation are illustrated in Fig. 5 where three wash plates are placed before a number of conventional bowls. Liquor from wash plate 1 is passed sequentially through a decanter centrifuge (for dirt removal) and then through a disc centrifuge (for grease removal) and returned. High recoveries of grease are achieved but dirt removal has presented some problems, additional dirt removal being achieved in the later conventional bowls.

An attempt to improve the removal of dirt has led to the development of Mini-Flo scouring,[8] a combination of the Lo-Flo and Mini-bowl concepts. Figure 6 illustrates the projected Mini-Flo scour incorporating two wash plates, three mini-bowls and two conventional bowls. Flow-down from wash plate 1 is evaporated to produce an additional sludge for disposal.

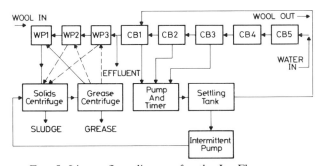

FIG. 5. Liquor flow diagram for the Lo-Flo scour.

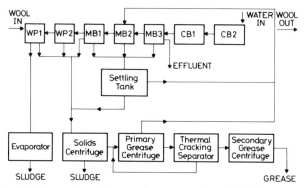

FIG. 6. Liquor flow diagram for the projected Mini-Flo scour.

Many treatments external to the scour have been developed involving a variety of principles including acid cracking, chemical flocculation, biological breakdown, centrifuge-evaporation, ultrafiltration/evaporation, evaporation/incineration, solvent extraction and solvent destabilization.[8] Biological breakdown by anaerobic/aerobic treatment in lagoons allowing natural breakdown and final spray irrigation is a relatively cheap and effective solution but it is only applicable in rural areas with abundant land. Evaporation and evaporation/incineration systems provide a completely effective treatment but these are extremely expensive both in capital and energy costs.

The University of New South Wales, Australia has developed two highly promising systems, the UNSW Hot-Acid Flocculation treatment[7,13] and the UNISAS biological treatment.[7,14] Process flow diagrams for these processes are depicted in Figs 7 and 8 respectively. In the hot-acid flocculation process,

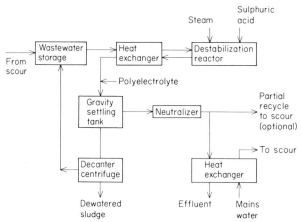

FIG. 7. Operation of the UNSW hot-acid flocculation treatment system.

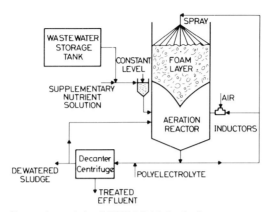

FIG. 8. Operation of the UNISAS biological treatment system.

waste liquor from the nozzle discharges of the primary grease separator is mixed with centrate from the decanter centrifuge and passed to the reactor where the solution is destabilized by boiling in the presence of acid. The resultant solution flows to a gravity settling tank, the settled material being pumped to a decanter centrifuge which extracts a dewatered sludge. The preferred treatment is to recycle 50% of the neutralized effluent back to the scour.

The UNISAS process is a technologically updated variation of the conventional activated sludge process offering a speeded up and hence much cheaper alternative to the conventional process. Wastewater is passed to the aeration reactor where micro-organisms break down the effluent. Material composed primarily of micro-organisms is continuously passed to a decanter centrifuge concentrating it for recycling or disposal as a dewatered sludge. This permits the process to work with a high micro-organism concentration in the aeration reactor and this together with efficient oxygen transfer produces rapid biological breakdown of the effluent. The effluent from the process is suitable for river discharge or for recycling within the mill.

In choosing an effluent treatment system many factors must be considered. Table I gives a comparison of approximate relative capital costs and reductions in pollution load for a number of combinations of in-scour and external treatments. Comparisons of operating costs have not been given since these will depend on such factors as whether an additional treatment is required prior to discharge, the cost of effluent and sludge disposal, the value of grease recovered, the additional operating expenses and the cost of capital. In addition in a multi-scour plant a reduction in capital costs for an external system may be achieved by building a larger plant to handle the total effluent. An in-scour treatment such as Mini-Flo would necessitate the

TABLE I. Comparison of scour effluent treatment systems.[8]

| Scour | Effluent treatment | Relative capital cost | %Pollutant reduction |
|---|---|---|---|
| Basic scour | Sewer | 1.0 | 0 |
| Basic plus settling and grease removal | Sewer | 1.2 | 30 |
| WRONZ rationalized | Sewer | 1.3 | 50 |
| WRONZ Mini-bowl | Sewer | 1.5 | 60 |
| Mini-Flo prototype | Sewer | 1.8 | 65 |
| Projected Mini-Flo | Sewer | 2.0 | 90 |
| WRONZ rationalized | UNISAS | 1.6 | 99 |
| WRONZ rationalized | Hot-acid | 1.7 | 90 |
| WRONZ rationalized | Evaporation/ incineration | 2.5 | 100 |

replacement of the existing scouring plant, and a breakdown in the effluent treatment could mean that the entire plant would have to be stopped.

## III. Drying

After scouring it is necessary to dry fibres from regains of around 40 to 60% to the normal regain of around 15%. In principle drying is carried out by circulating heated air through fibres held in position by a moving supporting surface which is perforated to allow the free flow of drying air through the wool. In practice dryers contain a number of drying sections and air flow through these sections is by countercurrent flow to ensure maximum moisture removal. Cold dry air is drawn through the hot wool as it leaves the dryer cooling the wool and gaining heat. The air then is moved progressively through each section until it is exhausted at the entry end of the dryer. At each stage it is heated and passed through the wool gaining moisture so that in the first section the wet wool entering the dryer meets the hottest and most saturated air.

Transport of the wool through the dryer is achieved either by supporting it on a perforated conveyor or by a series of suction drums where wool is held on the drum by the radially inward pressure of the moving air. This latter system is the main system used since it combines minimum floor area with high air speeds which do not disturb the drying material.

Since a perforated support acts as an additional resistance to air flow it has always seemed logical to provide a support with maximum hole area so that minimum resistance to air flow is achieved. Thus, for example, conventional drum dryers use up to 48% open area. Recently a new dryer, the Unidryer,[15] has been developed which completely reverses this apparently logical approach by employing open areas of only about 10% in the supporting plate and this has produced a significant improvement in drying efficiency. This apparent anomaly can be explained by the observation that the reduced open area counteracts the usual tendency of air to channel through less dense parts of the wool bed resulting in greatly improved uniformity of air flow through a non-uniform bed of wool fibres.

The Unidryer uses higher bed loadings than can be achieved in drum dryers and the better utilization of drying air results in a doubling of drying rate per unit of bed area, the reduction in volume of circulating air by a factor of three and a 15% reduction in power requirement compared with drum dryers. The basic Unidryer is illustrated schematically in Fig. 9. It is essentially a two section machine incorporating a novel system which maintains air pressures in the entry and exit sections of the dryer close to atmospheric pressure so that sealing problems leading to heat loss are eliminated.

The claimed advantages of the new dryer compared with conventional drum dryers are savings in capital cost and floor area (40% less) due to its smaller size, easier adaptability to direct gas firing with more than 30% savings in heating costs, easier adaptability to full automatic control, 15% less power consumption, applicability to a wide range of materials, a reliable and inexpensive conveying system and the ability to minimize drying costs by using the optimum material thickness.

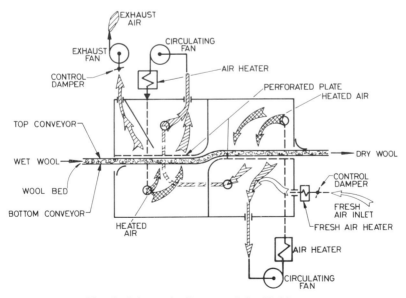

Fig. 9. Schematic diagram of the Unidryer.

Automatic control of drying to reduce heat losses in air exhausted from a dryer and to reduce final wool regain fluctuations has been limited due to the lack of reliable measuring equipment for air humidity and fibre regain and by difficulties resulting from the slow reaction of conventional dryers to changes in operating conditions. The NOCH Unitrol Humidity Control System[15] is a recent development measuring exhaust humidity by an automated wet and dry bulb technique and making appropriate adjustments to the flow of fresh air entering the dryer. The system, in addition to the conservation of fuel, counteracts long term changes in dryer moisture load. The inventors of this system also report progress towards a reliable wool regain measuring device to enable control of short-term fluctuations in drying load.

## IV. Worsted Carbonizing?

Carbonizing is the only process which completely eliminates vegetable matter contamination from wool. In principle burry wool after scouring is passed through sulphuric acid. Drying concentrates the acid on the burr and a high temperature baking process results in the embrittlement of the burr by the action of the acid. Crushing reduces the vegetable matter to an easily removable powder and the wool is neutralized with soda ash.

Unfortunately in practice the process leads to chemical and mechanical damage to the wool as well as fibre entanglement giving increased fibre breakage particularly in carding; this significantly reduces mean fibre length and increases fibre loss as noil. For these reasons carbonizing is generally restricted to woollen yarn manufacture and is used for those wools where vegetable matter cannot be effectively handled in the carding process.

Significant advantages should occur for worsted processing if the above problems could be eliminated. The most obvious gain would be that some burry wools presently carbonized could be upgraded and processed through the worsted system. Another possible gain would lie in the modification and improvement of the carding process, since the function of burr removal would be eliminated, or even the replacement of the carding process by a process with a far more gentle action on the fibres. It is worth noting that a major stumbling block in previous attempts to modify substantially or eliminate carding has been the presence and removal of vegetable matter.

Significant progress has recently been made towards solving these problems[15, 16] by improvements to acidizing, crushing and neutralizing. A rapid acidizer (illustrated schematically in Fig. 10) has reached the industrial prototype stage. In the acidizer, wool is carried quickly at an accurately controlled rate through a series of weirs and squeeze rolls by two porous

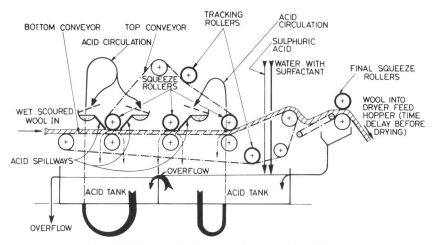

Fɪɢ. 10. Schematic diagram of the rapid acidizer.

conveyors which prevent fibre movement and hence entanglement. A 7% acid solution (stronger than normal) containing surfactant is used and in the limited time available (45 seconds) burrs are fully wetted on the surface by acid whilst fibres do not have time to soak up their full quota of acid. A "rest" period of several minutes is provided before drying and excess acid on the fibre surface is absorbed by the fibres. The process is based on research which indicated that surface acid is largely responsible for localized damage.

Experimental work on crushing has indicated[16] that fibre damage occurs

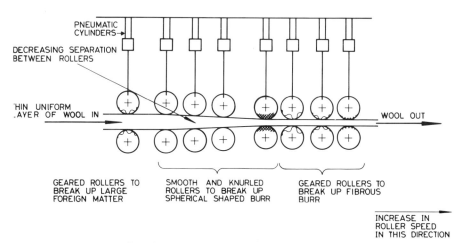

Fɪɢ. 11. Pilot plant crusher for crushing without damage.

C

when fibres are trapped between burrs and the fluted rollers and that damage is caused by the rubbing action brought about by the top and bottom rollers moving at different rotary speeds. This action is essential for crushing with thick layers but this speed differential is unnecessary if a sufficiently thin layer of wool is used. Figure 11 illustrates a pilot plant crusher, designed on the basis of these findings, which has proved effective in crushing without damage.

Conventional neutralizing with soda ash is too slow to produce a flat-bed plant similar to the acidizer. This problem has also been overcome by using a weak (0.2%) ammonia solution in the presence of surfactant reducing the time from 3 minutes to 30 seconds. A rapid neutralizer has been successfully tested at the pilot plant stage.

## V. Carding

After scouring and drying, carding is carried out to remove as much as possible of the entangled burrs and other vegetable matter as well as residual dirt trapped between the fibres, to disentangle, mix and randomize the fibres and to form the fibres into a coherent uniform sliver with fibres arranged as far as possible parallel to the sliver axis. The basic actions are produced as a result of the relative motion of interacting rollers densely covered with wire teeth. This wire clothing may be of three general types as illustrated in Fig. 12. Flexible wire clothing consists of flexible wire embedded in a flexible backing, semi-rigid clothing consists of stiff wire or metal inserts in a less flexible backing whilst metallic clothing is completely rigid consisting of a long steel ribbon with a saw-tooth type pattern as indicated.

The basic working unit for disentangling and straightening is illustrated in Fig. 13. Fibres are carried forward on the teeth of the cylinder past the stripper to the worker. The relative surface speed of the cylinder is faster than the worker producing a point to point action. Some of the fibres on the cylinder are caught in the teeth of the worker and the relative motion of the rollers separates these fibres from those on the cylinder, partially straightening the fibres in the process. Fibres transferred to the worker are lifted from

FLEXIBLE

SEMI-RIGID

METALLIC

FIG. 12. Types of card wire clothing.

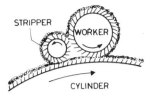

Fig. 13. Action of the worker and stripper combination.

the worker teeth by the faster moving teeth of the stripper. These fibres are in turn transferred back to the cylinder by the faster moving teeth on the cylinder.

The carding machine contains a number of cylinders with a number of worker/stripper units on each cylinder. Thus fibres are subjected to a number of carding actions which progressively open, separate and straighten the fibres. In order to achieve the gradual breaking down of tufts into individual fibres, teeth densities are progressively increased and the worker/cylinder distance is reduced towards the final delivery of the card.

Mixing and randomization of the fibres occur by virtue of the random delay times given to fibres transferred to workers or to fibres which do not transfer immediately to the next cylinder. Thus the length of time taken for an individual fibre to pass through the card may vary greatly depending on the number of times the fibre is transferred to a worker and on any multiple passages around cylinders. At equilibrium a thin layer of fibres will be present on the cylinders and workers, the layer thickness or fibre loadings depending on the proportional split up of fibres at each separation point. Thicker layers indicate a longer average delay time and a greater carding action on the fibres.

Burr removal is an important function of the card and this action is illustrated in Fig. 14. A rapidly revolving bladed roller or burr beater strikes projecting burrs or other foreign matter propelling the particles into a tray for removal. A card will normally have a burr beater soon after the feed to remove the large impurities thus protecting the wire teeth from damage. After some preliminary opening smaller particles are removed by burr beaters operating in conjunction with Morel rollers where opened fibres are

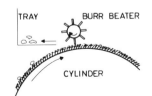

Fig. 14. Burr removal in carding.

pressed below the level of the roller teeth leaving burrs and other particles projecting above the teeth. In some cases an additional device called a Harmel crusher may be used to break up any residual burrs into small pieces.

The removal of fibres from the final cylinder is accomplished by a doffer and doffer comb in conjunction with a fancy roller as illustrated in Fig. 15. The rapidly rotating fancy produces a brushing action on the fibres on the cylinder lifting the fibres onto the points of the teeth on the cylinder facilitating easier removal. The doffer/cylinder action is point to point as for the worker/cylinder action except that the relative surface speeds are considerably reduced. Fibres transferred to the doffer are stripped in web form by the oscillating comb, condensed into a sliver and coiled in a can. In modern cards using metallic wire it is common to dispense with the fancy roller as unnecessary although some recent research appears to indicate that it is beneficial.[17]

The exact machine arrangement for a particular purpose varies between machine makers although each usually offers a range of cards to suit a variety of conditions. Modern practice is for modular design so that a particular arrangement is produced by assembling the appropriate modules. Figure 16 illustrates several arrangements for cards built by Thibeau. The first three arrangements are for wool containing respectively up to 2% vegetable matter, up to 8% vegetable matter and up to 20% vegetable matter. The first two systems differ in burr removal arrangement, i.e. a single Morel roller and two Morel rollers. The arrangement for up to 20% vegetable matter has an initial section identical to the up to 8% arrangement followed by doffing and a further burr removal and carding action. The fourth arrangement is for man-made fibres. All arrangements are multi-cylinder, with the initial cylinders essentially for opening, and the final main cylinder or swift providing parallelization.

For the above cards metallic card clothing is used on the main rollers with satellites being clothed with semi-rigid and/or flexible on the wool cards and

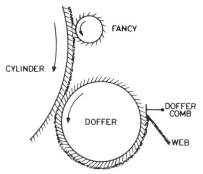

FIG. 15. Action of the fancy, doffer and doffer comb.

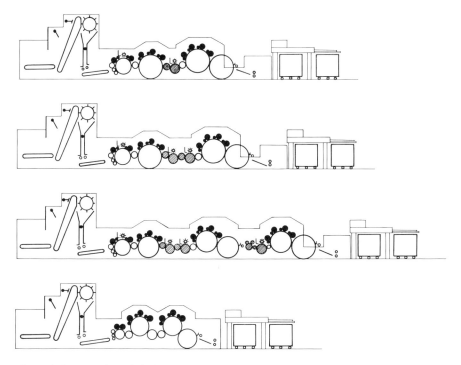

FIG. 16. Range of cards built by Thibeau: from top to bottom—wool with up to 2% vegetable matter; wool with up to 8% vegetable matter; wool with from up to 20% vegetable matter; man-made fibres.

metallic and/or semi-rigid on the man-made fibre card. Working widths are 2.2 or 2.5 m with production rates of around 80 kg/hr for fine Merino wools, 160 kg/hr for crossbred wools and up to 600 kg/hr for man-made fibres, the production rate being smaller for finer fibres. It is interesting to compare these rates with the typical 1.5 m wide flexible wire cards of the early 1960s. Production rates for fine Merino wools and crossbred wools respectively have been given as 20 to 25 kg/hr and 25 to 45 kg/hr.[18a] This indicates that in addition to the trend to increased production by using wider cards the production per unit card width has also shown a dramatic increase. These trends are continuing and the latest machines are up to 3 m wide (3.5 m for woollen carding) with productions up to 1000 kg/hr depending on the raw material.[19]

Improvement of production rates by increased machine widths and roller speeds has been brought about by improved bearings, improved roller and cylinder construction, dynamic balancing of rollers and cylinders, the ability to produce and maintain very accurate roller spacings (necessitating high accuracy in roller and cylinder radii) and sharp wire with constant teeth profile. Metallic wire has been important in achieving these goals, in particular eliminating the need to grind the wire to maintain teeth sharpness and the need to periodically strip fibres caught between the card wire. Improved wire design has reduced fibre breakage and nep production and increased vegetable matter removal.

Accurate feeding of cards is an important feature in maintaining optimum carding efficiency and producing a uniform output sliver. The original type of hopper feed where fibres are fed to a weighing device which delivers constant weight drops to the feed lattice of the card has undergone a range of mechanical design and electrical sensing and control improvements due to the need to improve feed uniformity at increasing throughput rates. With increased speeds the trend has been towards chute feeds where a continuous layer is fed from a vertical chute where a constant volume is maintained using ultrasonic sensors to detect the height of the fibres in the chute. The packing and feed can be assisted by vibrating the chute. The four cards illustrated in Fig. 16 have volumetric chute feeds with ultrasonic sensors. Additional control devices to vary feeding speed to achieve uniformity are also available.[20]

Modern cards are fully enclosed for safety and also to prevent fly. Various stop-motions and braking systems are provided to stop the card quickly in an emergency. The older mechanically operated oscillating doffer combs are limited in speed which in turn limits doffing speeds. One solution to this problem is the electrically operated comb developed by Thibeau[21] which sets up a sustained vibration at speeds of up to 3200 strokes/min as the result of the release of stored energy in a twisted torsion bar. A safety device is incorporated to release the comb to prevent damage if material builds up at the comb.

Other recent developments in assisting increased production have been the development of double doffer systems.[21] Increased throughput can lead to the problem of a too heavy card sliver and one solution is to split the output into multiple slivers. Another is to provide a drafting unit after doffing to reduce mass per unit length prior to coiling of the sliver in a can. Autolevellers can also be included. The draft provided may be beneficial in straightening some of the trailing hooks produced in carding.

## VI. Gilling or Pin-Drafting

The card sliver is subjected to a number of preparatory gilling or pin-drafting operations (usually three) prior to combing in order to straighten and improve the parallelization of fibres, to provide further fibre mixing and to

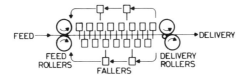

FIG. 17. Principle of operation of a gillbox or pin-drafting machine.

reduce mass per unit length fluctuations in the sliver. Combing affects the randomization of different fibre lengths and reduces sliver coherence. Further finisher gillings (usually two) are provided to mix and randomize fibres, to increase sliver coherence and to even out mass per unit length fluctuations. After topmaking further pin-drafting operations are carried out in the worsted drawing operation prior to spinning.

The basic principle of operation of a gill-box or pin-drafter is illustrated in Fig. 17. A number of slivers (doublings) pass between the feed rollers and are progressively penetrated and held between pins mounted on bars (fallers). Upper and lower sets of intersecting fallers move forward towards the delivery rollers at a slightly greater linear speed than the feed rollers thus tensioning the slivers. At the end of its travel each faller is withdrawn and transported back to the starting point where it gradually repenetrates the fibres as it moves forward. The delivery rollers operate at a faster linear speed than the feed rollers so that the material is drafted (typical drafts of 5 to 15) to produce a sliver of the appropriate mass per unit length.

The pins are used in fibre control in that they assist in maintaining inter-fibre friction and this in conjunction with their restrictive effect minimizes forward, out of turn movement of fibres not gripped by the delivery rollers. In addition fibres being withdrawn have their trailing ends combed by the pins, greatly accelerating the straightening and parallelization action. In particular the pins have an important role in removing the hooked fibre ends produced in carding. In order to provide straightening on both ends of the fibre a reversal of the direction of drafting is required in the next process and this aspect will be discussed later in the section on combing. A progressive straightening and parallelization is provided by increasing the pin densities in later processes, the initial processes having lower pin densities to reduce the severity of the action so that fibre breakage is minimized.

The traditional method for providing the faller motion has been by mounting faller ends in the grooves of rotating screws as illustrated in Fig. 18. At the end of its travel, rotating cams knock the faller on to another pair of rotating screws with coarser pitch and the faller is quickly returned to its starting position where it is mechanically repositioned on the original screws. The traditional screw gill suffers from two major limitations, a

FIG. 18. Faller head of a screw gill (Cognotex).

mechanical speed limitation associated with the screw traversing mechanism and the speed of disengaging the faller bars, and the environmental problem of excessive noise. This has led to a wide range of alternative mechanisms for high production at reduced noise levels.

For many years the upper speed limit for screw gills was considered to be 1500 faller drops/min but pressure from other developments has led to improved machines with speeds up to 2500 faller drops/min or about 22 m/min input. Faller design changes included such innovations as rubber inserts, resin-bonded faller pins and plastic composite faller bars which result in faster speeds and reduced noise levels. A further development recently reported[22] utilizes a single faller bed and penetration and withdrawal of fallers by cams without impacts on the fallers to achieve input speeds of 25 m/min.

Development of other mechanisms has advanced rapidly and input speeds up to 75 m/min and delivery speeds of 400 m/min (independent of draft) are common. These machines have production rates up to 750 kg/hr and only a

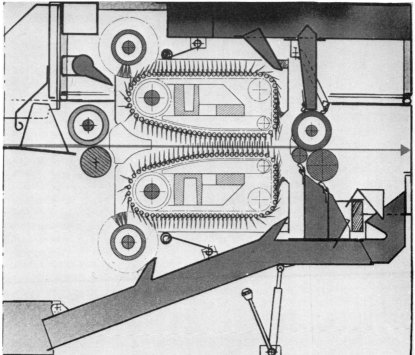

FIG. 19. Chain gill (Schlumberger). (Top) Faller head; (Bottom) cross-sectional view.

few machines are required to handle the entire production of a mill. At least one author[19] has pointed to the potential serious dislocations to a mill's operations if one such machine breaks down.

In the forefront in the development of other mechanisms has been the chain-gill where fallers are attached to a pair of driven chains as in Fig. 19. At the sides of the machine the faller ends are arranged to run in tracks which determine the faller orientation. Impact loadings on fallers are substantially reduced allowing much higher faller speeds. Further developments, suitable with some applications, are the replacement of the upper fallers by flexible rubber lips and the replacement of lower fallers by solid rubber profile strips (see Fig. 20). Figure 21 illustrates a chain-like rotary mechanism without chain drive.

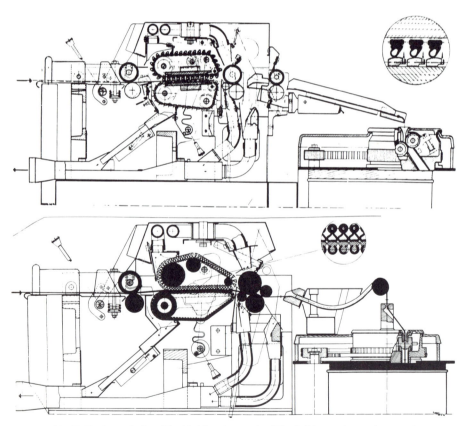

FIG. 20. Fallerless chain gills (Schlumberger). (Top) Pins only on lower element; (Bottom) pins replaced on both upper and lower elements.

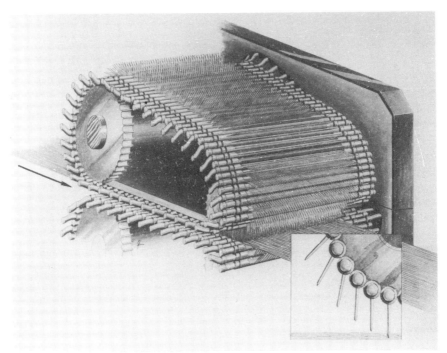

FIG. 21. A rotary drafting mechanism without chain drive (Warner and Swasey).

Rotating head devices where fallers fit in slots in the rotating head and are raised and lowered by cams are also making a significant impact. A system involving two rotating heads which provides the conventional linear pin-drafting zone is illustrated in Fig. 22. A single rotating head system providing a non-conventional curved drafting zone is illustrated in Fig. 23.

Prior to the introduction of autolevellers, reduction in mass per unit length variation was carried out by using a longer processing sequence with many doublings. Improvements in mass per unit length fluctuations by doublings is related to the square root of the number of doublings and this places severe restrictions on the ability of doublings to control irregularity. In addition doublings can only reduce but not eliminate the effects of underweight or overweight slivers. In modern sequences autolevelling, placed at particular stages in the sequence, has an essential role in minimizing sliver non-uniformity. The trend to apply autolevelling as early as carding has been prompted by the knowledge that modern pin-drafters operate closer to their optimum if the input slivers are uniform. Doublings play a lesser role in determining uniformity, their major role being in assisting fibre mixing.

FIG. 22. Rotating head intersecting gillbox (Cognotex).

The traditional mechanical autoleveller compresses the input slivers between tongue and grooved measuring rollers and the thickness variability is transmitted mechanically to movable rods on a memory wheel. After the required delay to allow the slivers to move forward to the feed rollers the position of the rods is mechanically transmitted to a speed variator which alters the position of a belt on cone pulleys thus controlling the feeding speed of the slivers. Variations of this mechanical system are still widely used in modern machines. However control by electronic means is in evidence and such use will probably accelerate in the future as fear about the reliability of non-mechanical devices is reduced.

Modern gill-boxes are generally of modular design being easily adaptable to fit a particular user's requirements. Feed may be from balls or cans, back-draft attachments to allow the addition of small components for blending may be added, autolevelling may be inserted if required, conditioning devices may be employed, the gilling head may have one or two heads, and the delivery may be as a single sliver or multiple slivers into cans or on to

FIG. 23. Rotary gill (Petrie-Sawtri).

balls with or without automatic ball or can changing. Pneumatic devices are employed for removing fly and for cleaning pin-bars and delivery rollers. Various design features including special enclosing covers serve to reduce noise levels.

The new HMG drafter[23,24] gives an indication of future trends in design. This machine is controlled via a microcomputer with a microprocessor as central unit. This memory-programmable control system takes over all control duties, the levelling of the input material and draft adjustment by selector switch. In the autoleveller, variations in the input are transmitted electronically to step motors which apply the appropriate correction. The machine's functions are monitored electronically and an operating panel displays the various machine functions showing immediately the location of a fault.

## VII. Combing

In order to achieve the strong, fine, smooth, uniform yarn required it is necessary to produce a sliver of highly straightened, parallel fibres free of short fibres (which increase yarn bulk and hairiness and reduce strength) and free of entanglements (e.g. neps) and foreign matter (e.g. vegetable particles, dust, etc.). The combing process carried out after a number of preparatory gilling operations is a necessary step in achieving this goal.

For many years wool has been combed using either the Noble comb or the rectilinear comb, the former being employed for longer fibres using high addition of lubricants and the latter using shorter fibres with relatively low oil content. Most wool is now combed using the rectilinear comb and interest in Noble combing has virtually disappeared. Some of the reasons for this change will be discussed later.

The basic principle of action of the Noble comb is indicated in Fig. 24. The machine consists of a large pinned circle (A) and two small pinned circles (B) rotating at a common linear speed. Specially prepared balls (C) of wool are loaded around the large circle, the slivers extending radially inwards. Immediately before the large and small circles meet, the wool fringe which extends beyond the inner edge of the large circle is held clear of the pins. At the tangential point rapidly oscillating dabbing brushes (D) force the fibres into the pins of both circles. As the circles separate, fibres predominantly held by one circle will be withdrawn from the other circle leaving a combed fringe on each circle. The fringe on the small circles is aligned by a rotating star wheel (E) prior to withdrawal. The fibre fringes are withdrawn at points (F) combing the other end of the fibres and are combined into a single sliver. Short fibres trapped in the pins of the small circle are lifted from the pins by stationary blades (G) and removed. Slivers are fed forward by a special mechanism, are lifted clear of the large circle pins and the process is repeated.

Whilst the Noble comb operation is essentially continuous the rectilinear comb has an intermittent action. The principle of action is depicted in Fig. 25. Fibres are moved forward by feed rollers (A) and a feed gill (B) and are

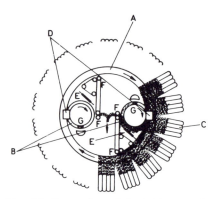

FIG. 24. Operation of the Noble comb.

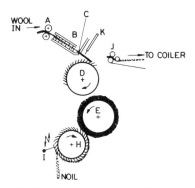

FIG. 25. Operation of the rectilinear comb.

gripped by nipper jaws (C) leaving a fringe of fibres extending beyond the nipper jaws. A partially pinned continuously revolving comb cylinder (D) combs this fringe, removing short fibres and other impurities. Pinning of the cylinder is from coarse to fine giving a progressive combing action. Dust and noil are removed from the comb cylinder by a brush roller (E), noil roller (H) and doffer comb (I). After the fringe is combed the drawing off rollers (J) move to grip the fringe, the top comb (K) is lowered to penetrate the fringe and a tuft of fibres is withdrawn through this comb which holds back short fibres and impurities. This tuft is placed to overlap with preceding tufts to be formed into a sliver. The process is then repeated.

In the mid-1960s production rates for combing were reported as[18b] up to 14 kg/hr for the rectilinear and about 20 kg/hr for the Noble comb when processing Merino wools. However even at that time the latest models of the rectilinear comb gave machine speeds up to 175 nips/min and were reported to be approaching the productivity of Noble combs.[18b] These authors reported the steady increase in popularity of rectilinear combing and the decrease in the number of Noble combs. The advantages cited for rectilinear combing were greater top yield, higher quality top, less waste, greater versatility, less space required, and lower power consumption.

Noble combing has a major speed limitation due to the oscillating dabbing mechanism, which pushes the fibres into a moving bed of pins, and its development in terms of productivity has been slow. Engineering development leading to increased rectilinear comb production rates together with the trend towards using low added oil tops have been reported[25a] as factors contributing to the increase in rectilinear combing at the expense of Noble combing.

Probably the major factor in the demise of Noble combing is the modern trend towards machines which are not fibre specific. The ability to comb only long fibres using the Noble comb provides a positive disincentive to improve the process since to a machine maker this represents only a limited market. On the other hand the rectilinear comb can be used over a wide range of fibre lengths and types and hence the incentive for improvement is high since developments for long staple fibres may flow on to short staple combing and vice versa.

Modern worsted rectilinear combs have production rates up to 200 nips/ min or 20 to 40 kg/hr whilst cotton combs have developed as high as 300 nips/min. Various solutions to increase speed have been applied including such innovations as moving both the feed and delivery sections giving two smaller reciprocating motions rather than one large one,[18c] non-reciprocating detaching rollers with a reciprocating carriage[26] and a stationary detaching device with an oscillating nipper with constant feed roller speed.[27] Combs can be fed from cans or balls, up to two combing heads may be used, mechanical or pneumatic tuft striking devices and pneumatic dust and noil removal systems are available whilst silencing hoods may be used for noise abatement. Figure 26 illustrates the complexity of a modern comber. The numbers one to six indicate respectively the settings for the top comb, nip distance, drawing off, and height of drawing-off rollers, the use of a positive chain drive to the drawing-off rollers and the drive of the nip and of the carriage by cams with two followers and compensating springs.

Production rates of combs are still well below that of gill-boxes and hence combs may be regarded as a bottleneck in the processing sequence. Several attempts to produce continuous high production rectilinear combing actions have been made including feeding in a large circle with continuous fringe combing, top comb operation and drawing-off devices moving around the inside of the circle[28] and combing heads moving in an oval path past continuously operating combing devices.[29] As indicated previously their ultimate success may depend on their applicability to short staple fibres even if they prove very effective for long staples.

The effectiveness of carding and preparatory processes becomes evident in combing when mean fibre length of the output and amount of noil are measured. In first combing fibre breakage rates of 30 to 35% are observed[3b] and are attributed to fibre entanglements behind the top comb and especially to leading hooks broken by the circular comb. The observation that recombing produces breakage rates of between 4 and 10% emphasizes the importance of effective sliver preparation prior to combing.

The number of preparatory gillings affects fibre straightening and the amount of noil produced in combing. The first gilling has a large effect on alignment with decreasing effects for later gillings with four or more gillings having only a very small effect.[3c] A greater number of gillings up to four decreases combing noil.[3d]

FIG. 26. Rectilinear comb (Schlumberger).

Carding produces many fibres with hooked ends the majority of these hooks being on the trailing end of fibres leaving the card and successive gillings in opposite directions are necessary to reduce these hooks on both ends of fibres prior to combing. Reversals are also necessary for uniform drafting since fibres tend to move in groups during drafting and successive drafts without reversals leads to poor regularity whilst reversing tends to break up the fibre groups.[25b]

Studies indicate[3d] that if only one or two gillings are used it is preferable to carry out the first gilling with the majority of hooks trailing, i.e. direct drafting of the card sliver without reversal. However the conventional three gillings gives better performance and the natural reversal between each process is preferable since an initial direct gilling has a more severe effect leading to more noil and a lower mean fibre length top. Thus it is preferable to carry out an initial gilling with the minority of hooks trailing to improve fibre alignment prior to the gilling with the majority of hooks trailing. The conventional three gilling sequence with reversal before each process means that slivers will enter the comb in the same direction as they came from the card, i.e. with the majority of hooks trailing. It has been shown that this direction of combing is preferable irrespective of the number of gillings prior to combing.

## VIII. Tow to Top Conversion

Conversion of filament man-made fibres in the form of tow to staple fibre top may be carried out by stretch-breaking or cutting of the filaments. Only a brief mention of the systems used is possible here.

Stretch-breaking generally produces a highly variable fibre length distribution and rebreaking is often required. Cutting methods involve producing a uniform sheet of aligned fibres, cutting at an angle and redistribution of the fibre ends to give a sliver. Variable length cuts are made to produce the fibre length variability required for efficient drawing and spinning.

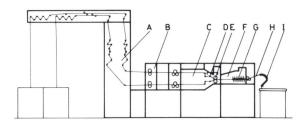

FIG. 27. High performance tow to top convertor (Rieter).

Figure 27 illustrates schematically the components of a modern converter with a production rate of up to 500 kg/hr and a delivery speed up to 400 m/min. Tow is passed over tensioning bars (A), through a threaded filament roll separating device (B), into a relaxation zone (C) and then into a variable cut unit (D) and cutting head (E). The cut fibres are transported by double aprons (F) to a chain intersecting gill-box (G), into a sliver gathering and compressing unit (H) and finally into a can coiler (I).

## IX. Future Developments

Although a range of revolutionary developments have been evident in other textile processing areas, changes in worsted topmaking have been largely evolutionary. The processing sequence and processes are much the same as 50 years ago except that improved materials, design and control have dramatically increased production rates, have improved processing performance and quality and have reduced the length of the processing sequence.

A particular trend taking place is the move towards machinery which is easily adaptable to different fibre lengths and different fibre types. Wool processing is spread widely and is forming a smaller proportion of the increasing fibre market. The incentive by machine makers to develop wool specific machines is low and wool specific machines are less in evidence. The demise of Noble combing as indicated earlier is a good example of this trend. In the future it will become more necessary for wool producers to finance the development of wool specific processes and the adaptation of wool to newly developed non-fibre-specific processes. The recent major developments in scouring and effluent treatment could never have been achieved without major backing from wool producers.

In attempting to speculate about the possibility of bringing about substantial and revolutionary changes in future worsted topmaking a simple picture of what happens to fibres in processes up to carding has provided the main basis for new and radical proposals. Greasy wool essentially consists of groups of parallel, relatively unentangled fibres and the process of scouring effectively removes this parallelism and entangles the fibres. It is then necessary to tear the fibres apart by carding, in the process breaking large numbers of fibres leading to a shorter mean fibre length and to considerable fibre waste. Previous attempts at improvements in these areas have been limited to developing scouring techniques which give less entanglement and devising machines to replace carding whilst still operating with entangled fibres from scouring. As mentioned previously the development of a viable worsted carbonizing process could have important implications for modifying or eliminating the carding process but fibre entanglement from scouring could still be a limiting factor.

Since the original fibres are relatively unentangled one logical proposal might be to produce an aligned array in layer or sliver form prior to scouring; previous attempts to achieve this have not been successful. However, correct mechanical handling in a modified scouring process followed by a worsted carbonizing should preserve this arrangement, giving a raw material requiring only minimal opening. There are numerous other possibilities for dramatic improvements but because of the present structure of the industry the incentive for such developments does not exist. It seems likely, however, that in the long-term future, wool producing countries will increasingly process their raw materials to the topmaking stage and in these circumstances the situation could change dramatically. Nothing short of a massive research and development programme associated with large scale industry could bring about the sort of automated, continuous high production rate worsted topmaking process which will be necessary if wool is to face the 21st century with equanimity.

## References

1. Wood, F. "A Review of Some Factors Affecting the Location of Early Stage Wool Processing". *Proceedings of Textile Institute Annual Conference: Managing Technological Change,* Ashville, USA, October 1980.
2. Barker, G. V. and Stewart, R. G. "The Application of Weighfeeding to Loose Wool Scouring", *Proceedings of 6th Quinquennial International Wool Textile Research Conference,* **III**, 17–29. Pretoria, 1980.
3. Smith, P. A. "Yarn Production and Properties", *Textile Progress,* **1**, No. 2, 1969. (a) p. 35, (b) p. 59, (c) p. 55, (d) p. 52.
4. Lipson, M. and Wood, G. F. "CSIRO Jet Scouring—Industrial Development and Processing Performance of the Product", *Interwool,* **71**, 28–32. Brno, 1971.
5. Samson, A. and Chaikin, M. "The Aqueous Compression Jet Scour", *Cirtel,* Section III, pp. 151–62. Paris, 1965.
6. Chaikin, M. and McCracken, J. R. "A New Method of Wool Scouring—The Hydronamic Scourer", *Interwool,* pp. 8–27. Brno, 1971.
7. McCracken, J. R. "Developments in Wool Textile Wastewater Treatments at the University of New South Wales", *DISC 80 Conference,* pp. 44–49. Australian Wool Corporation, Melbourne, March 1980.
8. Robinson, B., Morgan, W. V. and Gibson, J. D. "Wool Scouring and Effluent Treatments", *DISC 80 Conference,* pp. 15–43. Australian Wool Corporation, Melbourne, March 1980.
9. Stewart, R. G., Barker, G. V., Chisnall, P. E. and Hoare, J. L. *Wool Record, Textile Machinery Supplement,* 1975, **127**, 7 Feb., 8–11; 18 April, 13–17.
10. Stewart, R. G. "WRONZ Scouring Technology", *DISC 80 Conference,* pp. 10–14. Australian Wool Corporation, Melbourne, March 1980.
11. Chisnall, P. E. and Stewart, R. G. "Studies in Wool Scouring, Part IV: Commercial Scouring with Mini-Bowls", *Textile Institute and Industries,* **17**, No. 2, 68–69, 1979.
12. Anderson, C. A., Christoe, J. R., Pearson, A. J. C., Warner, J. J. and Wood, G. F. "Current Status of the Lo-Flo Process, CSIRO Division of Textile Industry", *DISC 80 Conference,* pp. 1–9. Australian Wool Corporation, Melbourne, March 1980.
13. McCracken, J. R. and Chaikin, M. "An Economic Solution to the Treatment of Wastewater from the Scouring Process", *Proceedings of 6th Quinquennial International Wool Textile Research Conference,* **III**, 1–16. Pretoria, 1980.
14. McCracken, J. R. and Chaikin, M. "An Efficient Biological Treatment of Wastewaters from Wool Textile Processes", *Proceedings of 6th Quinquennial International Wool Textile Research Conference,* **III**, 153–68. Pretoria, 1980.

15. Nossar, M. S. "New Developments in Carbonizing and Drying of Wool", *DISC 80 Conference*, pp. 57–78. Australian Wool Corporation, Melbourne, March 1980.

16. Nossar, M. S. and Chaikin, M. "Carbonizing of Burry Wools for Combing", *Proceedings of International Wool Textile Conference*, **IV**, 136–46. Aachen, 1975.

17. Townend, P. P., Hewitt, A. and Chu, C. K. "The Comparative Contribution of Some Card Variables to the Formation of Neps Using Metallic Clothing", *Textile Institute and Industry*, **17**, No. 3, 100–102, 1979.

18. Wetzold, W. W. and von Bergen, W. *Wool Handbook*, **2**, (a) p. 260, (b) pp. 299, 306, (c) p. 310. Edited by W. von Bergen, Interscience Publishers, 1969.

19. "ITMA 79 in Retrospect, Spinning Technology—Desires and Reality", *International Textile Bulletin*, **4**, 509–15. Spinning, 1979.

20. "Card Levelling on the Hopper Feeder", *International Textile Bulletin*, **3**, 391. Spinning, 1978.

21. Iredale, J. A. and Dyson, E. "Long-Staple Yarn Manufacture", *Textile Institute and Industry*, **18**, No. 3, 67–70, 1980.

22. "The Gill in Worsted Spinning", *International Textile Bulletin*, **2**, 244–45. Spinning, 1979.

23. "High-Performance Intersecting Gill", *International Textile Bulletin*, **1**, 39–40. Spinning, 1980.

24. "Chain-Driven Gills Perfected Further", *International Textile Bulletin*, **2**, 155–56. Spinning, 1980.

25. Dyson, E., Iredale, J. A. and Parkin, W. "Yarn Production and Properties", *Textile Progress*, **6**, No. 1, (a) p. 17, (b) p. 22. 1974.

26. Shaw, J. and Chisholm, A. A. "Textile Machinery", *Textile Progress*, **1**, No. 1, 19, 1969.

27. "Long-Staple Comber with Moving Nipper", *International Textile Bulletin*, **1**, 45–46. Spinning, 1976.

28. Unisearch Ltd. BP 1 119 900 (Aust. 9 Dec. 1965).

29. Turpie, D. W. F., Cizek, J. and Klazar, J. "An Introduction to the SAWTRI Comb", *Proceedings of 6th Quinquennial International Wool Textile Research Conference*, **III**, 279–92. Pretoria, 1980.

# Chapter 3

# Long Staple Yarn Manufacture

W. PARKIN and J. A. IREDALE

---

## I. The Post 1945 Period

If one were to consider the most appropriate way to review contemporary worsted yarn manufacture (or long staple yarn manufacture, as it is now more correctly referred to because of the present extensive use of man-made fibres), it would seem necessary to start with the many recent developments that have originated from the significant impetus given to this sector of the textile industry following the application of the work of Raper and Ambler in the mid-1950s. These two, working independently of each other and on different aspects of processing, developed ideas which together probably had the greatest impact on long staple processing since the advent of mechanical combing one hundred years earlier.

## A. The Raper Autoleveller[1]

For several years Raper had been considering some method of automatically controlling the drafting unit to ensure that the output sliver had a uniform weight per unit length irrespective of those variations present in the slivers entering the drafting zone. Prior to this time doublings were used to promote uniformity, the levelling value being given by $1/\sqrt{(n)}$, where $n$ is the number of doublings employed. This was obviously contrary to the main purpose of drafting as an increase in the number of doublings, while improving uniformity, would increase rather than decrease the weight per unit length of the material and consequently increase the number of drafting operations needed to produce a yarn.

Although Raper's work was originally directed towards topmaking, his

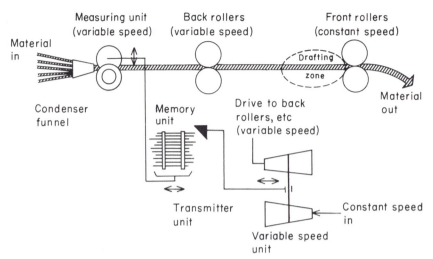

FIG. 1. Diagrammatic representation of the basic components of a Raper Autoleveller Unit.

ultimate aim was to reduce the number of doublings used and, if this could be achieved, to reduce significantly the number of operations required to produce a yarn.

The original Raper unit, or autoleveller as it became known, was initially fitted to a gill box, although it was subsequently shown that the mechanism could be successfully applied on other machines (see Figs 1 and 2).

The unit was constructed so that the input slivers were fed through a

FIG. 2. Feed regulator of an Autoleveller Gill Box. (N. Schlumberger et Cie., France.)

condenser funnel into a pair of sensitive measuring rollers, before passing through the back rollers of the drafting zone. These measuring rollers determined the cross-sectional area of the material passing through them and automatically recorded the variations in thickness as a "pattern line" in a memory unit which retained the recorded pattern until that particular cross-section of material had reached the drafting point. At that instant the pattern of the variation was fed, via a transmitter unit, to a variable speed unit which effected the necessary corrections to the draft by varying the speed of the back roller. The autoleveller comprises four essential component units, viz. measuring, memory, transmitter and variable speed, which were, initially, mechanical devices. With the widespread adoption of open-loop autolevellers, several machinery makers have developed alternative methods for operating the component units.

The use of Raper Autoleveller units succeeded in making possible a significant increase in the uniformity of the weight per unit length of the material being processed, together with a substantial decrease in the number of doublings required. For most applications these advantages were of considerable importance although, when doublings are used to ensure adequate blending of different fibre types or colours, a sufficient number still must be employed.

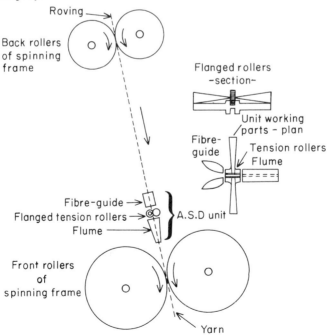

FIG. 3. Diagrammatic representation of the Ambler Superdraft Unit.

## B. Ambler Superdraft[2]

At approximately the same time, Air Vice-Marshall Geoffrey Ambler was considering the problems associated with the size of draft that could be imposed on long staple fibres. In general terms, the greater the amount of draft applied, the greater the degree of irregularity created by the drafting wave. The size of the draft was traditionally limited to an amount equal to the mean fibre length in inches + 1, normally within the range of 5–12. This meant that the smaller the amount of the permissible draft the greater the number of operations necessary to convert a top into a yarn. Ambler developed a drafting unit which gave an increased degree of control over fibre assemblies containing relatively few fibres in the cross-section, i.e. at the spinning stage. This unit, known as the Ambler Superdraft (ASD) permitted, in theory, drafts of up to 300 to be used, although in practice a limit in the region of 150 was more usual (Fig. 3).

A combination of the Raper Autoleveller and the Ambler Superdraft units within the same processing set gave a considerable saving, viz.:

|  | Conventional | New Bradford System (Raper/ASD) |
|---|---|---|
| No. of operations | 8 | 4 |
| No. of operatives | 10 | 6 |
| Floor space (m²) | 446 | 334 |
| Waste (%) | 1.93 | 0.89 |

These figures are typical for producing a 28 tex 100% wool yarn.

This type of drawing set first became available in 1958 and was known as the New Bradford System; it was intended for processing oil-combed materials, but is no longer (1981) manufactured.

The complementary nature of the work of Raper and Ambler is worthy of further discussion. The autoleveller unit was most effective when used on thick fibre assemblies; attempts were made to use it at later stages in the drawing set but as each drafting unit required an individual control this became unacceptable where a large number of units was employed. Conversely, the ASD unit performed best on thin fibre assemblies; an Ambler-draft Finisher unit[3] was developed but did not achieve popularity and thus the high-draft unit was restricted to use at the spinning frame.

The present situation in the worsted industry is that a much smaller weight of oil-combed material is now processed[4] and, as a result, relatively few of the New Bradford System machines are currently in use. This change in demand has arisen from several causes including the current fashion for a more casual form of dress and the much greater use of knitting to produce apparel fabrics. Also, at a more technical level, is the fact that the New Bradford System required twist to be inserted into the material at an early stage, usually the second operation in the drawing process. This was accomplished by means of a bobbin and flyer which imposed a limit on the linear speed at which the material could be processed. Secondly, as has been explained by Chaikin in the previous chapter, the increased popularity in the use of the rectilinear comb has increased the supply of dry-combed material.

## II. Post 1960

Thus, since the late 1960s there has been a steady decline in the use of drawing and spinning machinery suitable for processing oil-combed wools with a parallel increase in that for processing dry-combed materials. This has, in turn, meant a much greater demand for drawing sets of the Continental type. Modern Continental drawing sets include the use of the autoleveller technique, usually at the first gill box, and now rarely incorporate the traditional method of fibre control, the porcupine roller.

## A. Apron drafting[5]

During the early 1960s the makers of machinery for processing dry combed wools, i.e. Continental type machinery, found it necessary to attempt to match the high drafts available for oil-combed wool when using the ASD spinning system. This they did by adapting for wool the apron drafting technique of spinning cotton fibres. This method is now widely used for long staple spinning and the units are manufactured by several machine makers (Fig. 4). Whilst the double-apron method of fibre control does not match the high drafts of the ASD, drafts of up to 30 are quite normal. It is, of course, important that the oil content be limited to approximately 1½% to prevent contamination of the aprons.

Unlike the ASD system, the double-apron method of fibre control has been successfully applied to the drawing operations preceding spinning and, today, fibre control in a Continental set is restricted to the use of either pins set in fallers or some form of double-apron unit.

The Continental system does, of course, process twistless material and one particular advantage the double-apron unit has over pinned fallers is the higher linear speed at which it is possible to operate. Consequently, modified double-apron units may be used at the second or third drawing

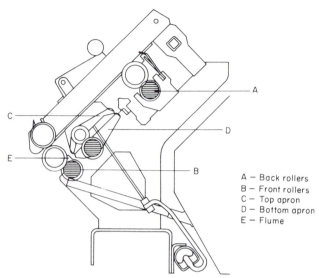

A — Back rollers
B — Front rollers
C — Top apron
D — Bottom apron
E — Flume

Fig. 4. Diagram of typical double-apron drafting zone (Rieter Machine Works Ltd., Winterthur).

operation.[6] The modifications became necessary because of the problem of achieving adequate fibre control for material having a high total input weight when using two flat aprons. Typical of these modifications are aprons of the caterpillar type used in conjunction with either a plain or pinned surface (Fig. 5).

Fig. 5. Fibre control by means of a plain bottom apron and a top "caterpillar" apron. (Upper apron shown in the raised position.) (N. Schlumberger et Cie., France.)

## III. Spinning

In terms of the speed at which a yarn may be spun, the governing factor is clearly the rate at which twist can be inserted to the drafted strand. It is not surprising, therefore, that considerable attention has always been paid to increasing the rate of twist insertion. These attempts have, to date, achieved a considerable degree of success. The production of one operative using a dangling spindle would have been such that it would have taken 12 500 hours to produce 100 kg of yarn (HOK); by the use of a spinning wheel this could be reduced to about 3100 hours. The introduction of power spinning in the early 19th century enabled an operative to produce 100 kg in 390 hours and, by the late 1960s, the time required to produce this weight was down to 0.63 hours. These figures will, of course, vary widely according to the yarn count being spun.[7]

Although both Raper and Ambler contributed to significant advancements in worsted yarn manufacture, neither was directly concerned with the problem of twist insertion. At that time three methods of twisting and winding-on during spinning were available, viz. flyer, cap and ring, all of which inserted real or true twist.

### A. Flyer spinning

Flyer spinning was developed from Arkwright's water frame (patented 1769) and was, perhaps, the most popular system for spinning coarse counts, but it was subject to limitations both in the speed at which it could operate and in the size of yarn package which could be produced. The maximum speed at which the flyer could be rotated depended upon several factors but, in general terms, it was limited to approximately 3200 rpm because at higher speeds, there was a tendency for the flyer legs to be thrown outwards by centrifugal force. Furthermore, as the bobbin was not positively driven but pulled around by the length of yarn between the bottom of the flyer leg and the bobbin, there was a limit to the size of the delivery package. This limit was largely determined by the weight of the package and the strength of the yarn being spun. However, in spite of these restrictions, flyer spinning had one advantage over both cap and ring spinning in that the twist inserted was uniform throughout the package. The bobbin was free to rotate around the spindle, and the tension or drag necessary to retard the bobbin and so permit winding-on to take place was provided by means of a washer located between the lifter plate and the base of the bobbin. One turn of twist was inserted for every revolution of the flyer, viz.:

$$\text{Turns/m} = \frac{\text{flyer speed (rpm)}}{\text{front roller delivery (m/min)}}$$

## B. Baxter flyer

Because of the popularity of flyer spinning, particularly for coarse yarns, an attempt was made to modify the unit to prevent the flyer legs moving outward at higher speeds and to permit more uniform tension being applied to the spun thread. This was known as the Baxter flyer, shown in Fig. 6, and found some application during the 1960s for the spinning of coarse count knitting yarns and carpet yarns but, like traditional flyer spinning, is now only rarely used in industry.

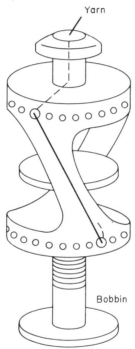

FIG. 6. Diagrammatic representation of the Baxter Flyer.

## C. Cap spinning[8]

Cap spinning was invented about 1829 and became popular for fine count spinning with spindle speeds limited to about 7000 rpm. In addition to this limitation in speed, the delivery package was also restricted to a size that could be accommodated inside the cap. Attempts to increase the cap size demanded reduced spindle speeds and, for this reason, were not readily acceptable. In cap spinning the bobbin is placed on a tube which is rotated by means of a tape drive and winding-on is achieved by a combination of air-drag on the yarn balloon and friction between the yarn and the base of the cap. However, this method of twisting and winding-on results in the spun yarn containing some small twist variation from the inside to the outside of the package. The actual twist may be calculated by:

$$\text{Turns/m} = \frac{\text{tube speed (rpm)} - \left[\dfrac{\text{front roller delivery (m/min)}}{\text{bobbin circumference (m)}}\right]}{\text{front roller delivery (m/min)}}$$

The variable factor in this formula is the continuous variation of the bobbin circumference.

Cap spinning enjoyed considerable popularity amongst fine count spinners and, to increase further its attractiveness, an attempt was made in the 1960s to increase the size of the delivery package. This modification was known as "Featherflex" and consisted essentially of employing a telescopic package which extended as it was withdrawn from the cap, thus increasing the length of yarn it was possible to spin before it became necessary to stop the machine for doffing (Fig. 7). However, this improvement was overtaken

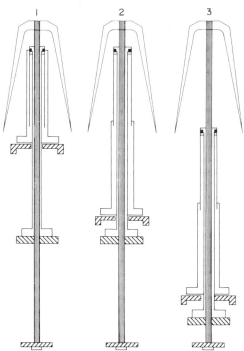

Fig. 7. "Featherflex" Cap Spinning. Diagram to show basic cycle of the telescopic spool.

by other developments in ring spinning and "Featherflex", although an interesting idea, was never used commercially. Fine yarns, like most others, were subsequently spun using the ring system.

## D. Ring spinning[9]

Ring spinning was invented about 1832 and, at the present time, this method is employed to produce about 95% of all long-staple yarns. Its popularity can readily be understood by reference to its spindle speed which can now, under appropriate circumstances, be as high as 10–12 000 rpm when spinning fine yarns. This high spindle speed, as compared to flyer and cap spinning, has only been achieved gradually through a series of relatively minor improvements which have occurred over the past 50 years. It is not appropriate here to consider the detailed history of the development of ring spinning but, because of the overall importance of the system, some brief mention of the various improvements must be included, many of which were originally introduced on ring frames intended for cotton spinning.

Probably the most important single factor was the use of variable-speed motors.[10] These permit the various changes in winding-on tension to be accommodated by variations in spindle speed. Changes in tension are primarily associated with changes in the winding-on diameter, particularly when commencing spinning with a set of empty packages. At this stage the tension is greatest and in order to compensate for this the spindle speed needs to be reduced.

A second important factor contributing to yarn tension during twisting and winding-on is that of air drag on the yarn balloon. The use of balloon control rings[11] limits the diameter of the balloon and thus reduces the tension on the yarn. In certain circumstances a collapsed balloon[12] may be used in order to minimize yarn tension. This technique is employed by means of an extension or modification to the top of the spindle and is often adopted when spinning coarser count yarns.

Higher spindle speeds create greater friction between the ring and traveller and improvements in the design of both components[13] have minimized this, as have improvements in the lubrication of the ring.[14] In appropriate cases either nylon rings or travellers may be used in order to reduce friction.

All the foregoing improvements have contributed towards the application of increased spindle speeds but, additionally, changes have been made which have made possible an increase in the size of the package that can be produced. As in cap spinning, the yarn balloon is retarded and therefore there is a small difference in twist between the yarn wound on to the inside and the outside of the package. The minimum and maximum twist values may be calculated from the same formula as that for cap spinning. These variations are maintained within acceptable limits by ensuring that only relatively small differences exist between the inner and outer winding-on diameters. Because of this, any increase in package size must largely be accomplished by increasing package length, but this would normally result in significant differences in balloon length when winding-on at the upper and lower extremities of the bobbin. To accommodate this, arrangements which allow both the ring plate and the spindle rail to rise and fall were introduced. These have not proved popular and balloon lengths are now usually control-led by the use of a rising lappet rail.[15] Increased spinning speeds increase the skill operatives require to piecen-up, so the introduction of aids such as "Pneumafil" which removes the waste created by broken ends facilitate machine-minding and thus improve efficiency. These improvements are now almost standard equipment on modern ring spinning frames and show why the ring frame has now largely superseded the flyer and cap systems. Amongst recent improvements to be employed industrially are automatic doffing devices.[16] These devices have been available to spinners for several years, but it is only recently that they have achieved general acceptance in an industrial environment. Following their introduction at textile machinery exhibitions, units which patrol the spinning frame to detect and piecen-up broken ends are also gradually finding favour in industry.

The ring spinning frame has progressively been developed to a high degree of sophistication and in some quarters it has been suggested that it may have reached its limit in terms of both package size and spindle speed, thereby encouraging serious consideration of the various alternative methods of spinning that have been developed in recent years and which will be discussed later. However, the textile industry of the world is thoroughly familiar not only with the advantages and disadvantages of frame spinning but also the characteristic properties of the yarns produced, factors which could well militate against a ready acceptance of alternative methods of yarn production. Suggestions that the development of the ring frame has reached its ultimate are belied by the recent revival of interest in the revolving ring technique.

## E. Rotating ring spinning[17]

The floating ring, or rotating ring as it is sometimes known, is yet another device, the purpose of which is to permit an increase in spindle speed, and thereby increase the productivity of the ring frame. Reference has pre-viously been made to the restrictive influence of the friction between the ring and the traveller. If the ring itself can be rotated in the same direction as the traveller, then this source of friction will be reduced. If the ring is rotated at the same speed as the traveller, then there will be no relative movement between the two elements and, thus, no frictional drag. One method of achieving this is to employ a light-weight ring which is free to rotate in a low friction bearing mounted on the ring rail. From the evidence of current developments it would seem that some form of air bearing for the ring is proving to be the popular choice. Different types of air bearings have been used, e.g. one type where a cushion of air is pumped between the ring and bearing, and another design where the air cushion is created by the movement of one surface relative to the other. If an air pump is used, it is essential that the air supply is adequately filtered. If particles of foreign matter are permitted to lodge between the ring and its supporting surface, then the smooth rotation of the ring will be interrupted and result in variable tensions being applied to the yarn. Serious obstruction within the bearing may prevent any rotational movement of the ring and create such large changes in tension that the thread will break. The use of a ring without an air pump is claimed to be successful provided that the two surfaces forming the bearing are machined uniformly. There is a danger with such bearings that, should the ring and its support touch each other, then the heat generated may cause damage. As an alternative to the use of metal bearing surfaces, the introduction of ceramics has also been considered.

Other alternatives to air bearings, including ball bearings,[18] have been used but it has been found that these may considerably increase starting-up difficulties. In addition, the collection of lint, etc., will cause similar problems to those associated with the use of air bearings. Spinning frames incorporating rotating rings are currently being offered by at least five machinery makers. Their industrial application has achieved greatest success when spinning man-made fibres, partly because of the smaller amounts of dust liberated. To date only limited details of performance have been published by the machinery makers, but SACM claim increases in spindle speeds of at least 25%, together with increases in ring diameters where these are desirable. Clearly, this innovation suggests that the development of the ring frame has not reached its end.

Rotating rings are more expensive to produce which means an increase in capital outlay of about 20–25%. It has been suggested that, because of this, it is essential that this type of machine be operated at its highest speed in order to gain the maximum economic benefit.[19] With this type of frame the ring assumes an even more important role and ring maintenance costs can be expected to be greater than with conventional rings. Air bearings require little energy, but the cost of providing a filtered air-supply must be taken into account. One problem associated with higher spindle speeds is that of increased noise levels, although the machinery makers have indicated that these can be kept within the recommended upper limits.

The possibility of employing higher spindle speeds incurs greater total power consumption, and thus an efficient use of all the other aids to ring spinning, as discussed previously, would appear to be essential for the viable use of this type of spinning frame. Nevertheless, current opinion suggests that the rotating ring offers a further opportunity for the ring frame to maintain its position against the alternative spinning systems.

Traditionally, the spinning frame has produced a single yarn containing uni-directional twist, the amounts of which can be changed according to the various end-uses of the yarn. As will be shown later in this chapter, certain of the most recent methods of yarn manufacture no longer maintain these criteria. However, before considering these, brief reference should be made to two other methods of spinning, viz. pot and open end.

## F. Pot spinning

Attempts were made in the 1950s to introduce pot spinning to the worsted industry. This system utilized a rotating container, similar in some respects to a Topham Box, but operating in the inverted position. The roving was drafted by conventional means and twist was inserted into the drafted strand by the rotation of the pot inside which the yarn was held by centrifugal force. When an appropriate amount of yarn had been spun it was then necessary to insert a bobbin into the pot in order to off-wind the material situated on the inside of the pot before doffing could occur. Although machines of this type were used by a number of processors, the system failed to achieve any significant degree of popularity. Difficulties were experienced in piecing-up broken ends and thus, at the end of each doff, the frame was running at a very low efficiency. In addition, there was always the danger of a power failure causing the pots to stop rotating which, with the consequential loss of centrifugal force, would allow the yarn contained in the pots to drop out and effectively be lost.

An experimental machine incorporating the "pot" spinning principle, the "Axispinner", was shown at the ITMA Paris Exhibition in 1971 and was designed for the production of two-ply carpet yarns. This method of spinning was subsequently abandoned by the machinery makers.[20]

## G. Open-end spinning[21]

Without doubt the most revolutionary method of spinning introduced to the industry during this century has been open-end or break spinning. Although, at the present time, this method of spinning is rarely used for spinning wool fibres, machines are available which will spin man-made fibres possessing certain wool-like characteristics, i.e. long fibres. Whilst the resultant yarn could not be claimed to closely resemble a worsted yarn, it being more semi-worsted in character, an open-end spun wool yarn could well find use in certain worsted-type products. The reason for the difference in yarn appearance is that in a conventionally spun yarn the fibres tend to lie very closely to each other which produces a smooth, lean yarn, whereas in an open-end spun yarn the fibres on the surface are much more loosely arranged and will not be parallel to the yarn axis.[22]

Open-end spinning is now quite popular in certain areas of cotton spinning, and research and development work at WIRA and other similar organizations has resulted in some of the problems associated with the spinning of wool fibres being overcome (e.g. self-cleaning rotors).[23]

In order to produce a yarn that has what are generally considered to be normal characteristics, it is necessary to insert twist. To date this has been done on almost every machine by rotating the delivery package. It could, under certain conditions, also be achieved by rotating the supply package, but this would be most unusual. The method used for inserting twist in open-end or break spinning is to introduce a break in the material between the supply and delivery packages. It is then possible to rotate this free end without the need to rotate either the supply or delivery packages (Fig. 8). When the requisite amount of twist has been inserted then the break must be rejoined and a new break formed further along the material. This process must be repeated at frequent intervals and, to produce acceptable results, it should virtually be continuous.

The original concept of break spinning is not new. In a paper in the Journal of the Textile Institute in 1936, Prof A. F. Barker described a machine using this principle.[24] However, most of the significant work on the subject has been carried out since the mid-1960s. There are now many techniques available for open-end (OE) spinning, the early ones originating in Czechoslovakia, and today OE machines for spinning cotton fibres are manufactured in many countries.

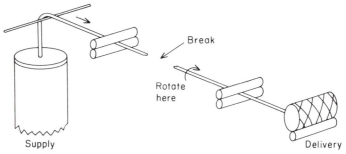

Fig. 8. Diagram to show basic principle of open-end or break spinning.

Turning now to a consideration of the actual methods used, the many hundreds of patents now covering OE spinning may be divided into several groups. Of these only one, usually described as "continuous circumferential assembly", is at present used commercially. This can be described in simple terms with the aid of Fig. 9. Fibres are fed into the system in a discontinuous flow, thus ensuring the break between the feed and the delivery. The method of feeding of the material to the rotor is an extremely important part of the operation and several techniques are available. The fibres enter a rapidly rotating pot which is suitably shaped so that the centrifugal force throws the fibres to the outer edge of the pot, thus forming a ribbon of fibres. If an end of yarn is then inserted into the rotating pot it will also be thrown by centrifugal force to the outer edge of the pot. If this yarn is then slowly withdrawn it will bring with it the other fibres which will have attached themselves to its end and, because of the rotation of the pot, these will assume a twisted configuration. As both feeding and withdrawal continue other fibres will be twisted on to the end and thus a yarn is formed. The pot may be rotated at speeds of up to 100 000 rpm, depending on its size. Thus, it becomes evident that this is a much quicker way of introducing twist into a yarn than by means of a rotating spindle.

One of several machines which are suitable for spinning long-staple material is the RL 10 (Schubert and Sulzer A.G., Ingolstadt). Production rates may be up to 3 kg (6½ lbs) per hour per pot, when spinning 5″ staple acrylic fibres at 30 000 rpm. More recently the French firm of SACM have

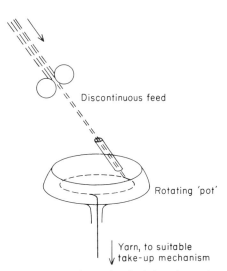

Discontinuous feed

Rotating 'pot'

Yarn, to suitable take-up mechanism

FIG. 9. Diagrammatic representation of principle of continuous circumferential assembly method of open-end spinning.

launched their ITG 300 machine which they uniquely claim is suitable for spinning wool.[25] Several industrial organizations, as well as the IWS at Ilkley and Bradford University, now have experimental machines and evaluation programmes are currently being undertaken (see Chapter 7).

## H. Coverspun yarn

A method of two-fold yarn production, sometimes described as wrap spinning, has been developed recently. This is a process for combining an untwisted yarn of staple fibre with a fine denier filament, the staple material forming the core with the continuous filament acting as the binder on the outside of the yarn (Fig. 10). Commercial yarns of this type produced on machinery manufactured by Leesona are described as "coverspun". The yarn is produced by first passing a conventional roving through a suitable drafting zone. As this strand emerges from the front roller nip it is combined with a filament of man-made fibre. The two components then pass through the centre of the rotating package on which the continuous filament is wound. The rotation of the continuous filament around the core of the drafted material produces the wrapped yarn. Theoretically the core may be composed of any staple fibre. Because of the restricted speed at which the wrap yarn package may be rotated, there is a limit to the degree of cover that may be obtained.[26]

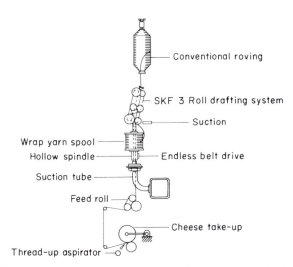

FIG. 10. "Coverspun". Diagrammatic process description.

## I. Twin spun yarn

One of the most recent of the new spinning methods also produces a two-fold yarn but, in this case, both components are spun simultaneously. At the time of writing no details have been published but it is understood that there are various methods available for producing the "twin-spun" yarn. The concept is said to have originated in Australia at CSIRO and is currently being developed in the UK by IWS at Ilkley. Work has reached the stage where the equipment required to modify a conventional spinning frame to one able to spin a "twin-spun" yarn is being manufactured by Messrs Zinser.[27] The method has had extensive industrial trials and several thousands of spindles are currently in use in both Germany and the UK. The machinery produced to the IWS specification is known as Siro-spun, but it is understood that at least one other company is experimenting with similar equipment.

The twin-spun yarn is produced using a conventional ring spinning frame. Two rovings are passed through each drafting zone, and are kept separate by guides. The drafted strands are brought together as they emerge from the

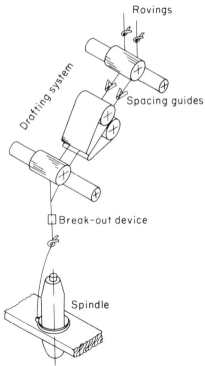

FIG. 11. Diagrammatic representation of the "Sirospun" system.

front rollers and are twisted and wound using the conventional ring and traveller (Fig. 11). To achieve an acceptable two-fold yarn each individual thread must largely retain its own identity within the two-fold structure. There must also be an effective system of detection should one of the component threads break, as the machine would obviously continue to function when spinning a singles yarn. To date two methods of detection have been utilized, viz. Siro I, using a pair of positively driven rollers located between the front roller nip and the spindle tip, and Siro II, which employs a much simpler swinging detector arm (Fig. 12). The equipment being produced by Zinser will incorporate the Siro II technique.

An advantage of this system is the ability to produce a two-fold yarn from identical threads without the need for an additional twisting operation. It is suggested that this itself can create a saving of up to 40% on yarn costs. A further advantage is that the twist is unidirectional, thus removing one of the disadvantages associated with self-twist yarns for certain end-uses—see later in this chapter.

Unlike many of the newer methods of spinning described, the Siro method attempts to produce a two-fold yarn having very similar characteristics to one produced by conventional methods, but certain disadvantages exist. Neither of the component single yarns will contain twist and, secondly, breaks in the yarn will result in a knot in the two-fold yarn, which may have an implication on mending costs or in the end-uses of the yarn. It should perhaps be emphasized that, at the time of writing, there has been no published account of this method of spinning, but it would appear that, because of the cost advantages and the generally satisfactory performance of the yarn, the industry will see much more of this type of spinning in the future.

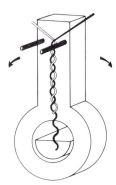

FIG. 12. Diagram of swinging arm for detection of yarn breakage in Siro II system of two fold spinning. If an end breaks the tension in the remaining end swings the detector to the right or left, past its centre of gravity, the detector then swings over and breaks the remaining end.

## J. Self twist spinning[28]

Without doubt the system of spinning which has had the greatest impact on the long-staple sector of yarn manufacture in the last two decades has been self-twist, or, as it was originally known, Repco spinning. The method is now widely known throughout the world and machines incorporating this principle were shown at the Hanover ITMA Exhibition in 1979 by both UK and USSR machinery makers. Self-twist is a method of producing a two-fold yarn, but rather than using unidirectional twist as in the methods previously described, the twist is inserted in alternating directions, i.e. a length of S-twist followed by a length of Z-twist, etc. This can be achieved without the use of a spindle by employing one of several methods, but reciprocating rollers have proved to be the most successful method to date. It must be emphasized that these single yarns contain only *false* twist and that such yarns would have little or no strength and would have limited practical application in that form. However, if *two* such yarns are laid side-by-side it is found that the torque forces in the two yarns are such that they will cause the yarns to twist about each other to form a two-fold yarn. In this latter yarn there are still zones of S and Z twist and, because some of the original false twist is used to fold the yarns together, the amount of residual twist in the single yarns will be reduced.

When twist is inserted in alternating directions, there must be a point in the yarn where the twist direction changes and where there is no twist at all, thus creating the weakest point in the yarn. Should two yarns having alternating twist be brought together in such a way that the points with zero twist coincide then the resultant yarn will be very weak. This is shown in Fig. 13(i). However, if the two yarns can be brought together so that the two points having zero twist are separated by a short distance, then the resultant two-fold yarn will be much stronger as there will be no point in the two-fold yarn without some twist. This is shown in Fig. 13(ii). In practice the condition shown in Fig. 13(i) where the points of twist change-over coincide is

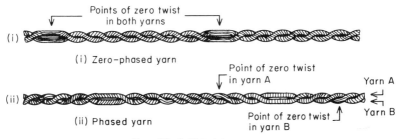

FIG. 13. Self-twist yarns.

known as "zero phase" whereas the second and more desirable condition, shown in Fig. 13(ii), is known as "phased" yarn. In practice the latter condition can be achieved without too much difficulty.

Thus, the basic form of spinning machine to produce this type of yarn differs in many respects from that of a traditional spinning frame. The most outstanding difference is that it produces a two-fold yarn; in fact, it is impossible to produce a truly single yarn. The other essential difference is that a self-twist machine does not require a conventional rotating spindle. Probably the only feature which remains similar to that of a traditional spinning machine is the drafting zone where the double-apron method of fibre control is usually employed. The essential form of a commercial self-twist spinner is shown in Fig. 14.

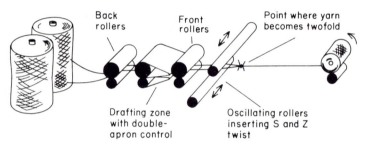

FIG. 14. Diagrammatic basic layout of a self-twist spinner.

The machine at present available, described as the Mark II (Fig. 15), has a delivery speed of 300 m/min (328 yards/min) which, considering that five ends of two-fold yarn are produced simultaneously means that one machine has a production approximately equal to that of 170 ring spinning spindles when spinning 2/27$^s$ worsted (R66 tex/2), or 250 ring spindles when spinning 2/20$^s$ worsted (R88 tex/2) counts. This saving can be expected to be reflected in yarn costings.

The yarn produced by the self-twist spinner is conventionally referred to as ST yarn and in this form is claimed to be suitable for use in both woven and knitted fabrics. However, in practice the ST yarn may be subjected to a further twisting process and is then described as STT.

A modification of the self-twist machine was offered a few years ago, namely the Selfil spinner.[29] This machine was not accepted by industry and is no longer manufactured. The Selfil spinner was an attempt to produce a yarn which, in appearance, more nearly corresponded to a singles worsted yarn. The machine utilized the same self-twist technique as previously described but, in this instance, only one strand of staple fibre was used. After drafting in the usual double-apron drafting zone, the single strand was false-twisted

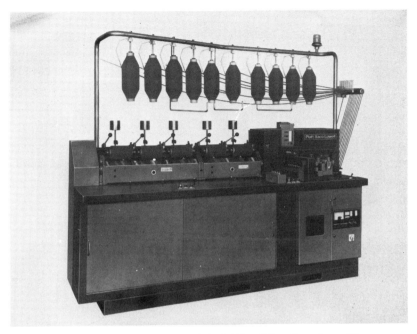

FIG. 15. Self Twist Spinner (Mark II)—(Platt Saco Lowell).

with a continuous filament synthetic yarn as shown in Fig. 16(i). At this stage, a further self-twisting operation wrapped a second continuous filament around the product of the first stage as shown in Fig. 16(ii). It should be noted that the second self-twisting operation was so arranged that the change-over points from S to Z twist do not coincide. The resultant yarn was said not to require further twisting and was claimed to be suitable for use in knitted structures; it was not recommended for weaving applications except as weft.

A large number of alternative self-twist structures have been proposed[30] none of which has been produced commercially. However, these suggested

( i )

( ii )

FIG. 16. Selfil yarn showing the two stages of wrapping the continuous filaments around the staple strand.

structures have demonstrated the wide range of yarn types which can be produced using the self-twist technique. Because this method has been available for several years a library of research work exists which includes reports on the production of worsted yarns at both ends of the count range[31] and other topics.

## K. Twistless spinning

The ultimate in yarn manufacture could well be one produced without twist, thus obviating the need for an expensive spindle or other rotating element. Over the past years several novel methods have been proposed but, to date, none has achieved commercial acceptability. An early idea was the use of adhesives to hold the fibres together and in this way produce a twistless yarn. A range of adhesives was investigated and in the late 1960s the TNO system, using starch based adhesives, showed considerable promise, although the resultant yarn was more irregular than its ring-spun counterpart. Later the Signaal-Twilo system using a PVA fibre was developed.[32] The preliminary work was carried out using cotton fibres and the system was not considered to be suitable for wool because of the oil content and the difficulty in finding a suitable adhesive.

## L. Bobtex yarn[33]

An interesting Canadian idea was that for producing a yarn by a system described as Integrated Composite Spinning (ICS) now more popularly referred to as Bobtex. An outer layer of fibres is coated onto a core yarn as

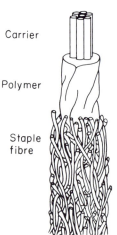

Carrier

Polymer

Staple
fibre

FIG. 17. Construction of Bobtex composite yarn.

shown in Fig. 17. Some twist may be added so that any fibres not actually held in the matrix can be retained. The core yarn is usually a continuous filament of man-made fibre and may thus have considerable strength whilst the outer cover can, in theory, be composed of any type of staple fibre. The proportions of staple fibre and polymer can be varied over a wide range between 85–15% although values approaching 50:50 are more usual. The character of the resultant yarn is quite different from that of a conventional spun yarn and, at the present time, has only enjoyed limited industrial acceptance.[34]

## M. Periloc yarn[35]

A further method of producing a twistless yarn was shown for the first time at the ITMA Exhibition at Hanover in 1979. This was the Vibraire machine produced by the Dutch firm of Signaal who have been active in the general area of unconventional yarns for many years. The machine produces "Periloc" yarn by a felting process for wool or wool-rich slivers. The yarns, which are of coarse counts, are said to be suitable for many types of carpets and upholstery fabrics. It is claimed that fabrics produced from felted yarns combine very small levels of fluff formation with a high resistance to wear, and pile definition, even after intensive use, remains clear. High yarn strengths of up to 4 g/tex are possible. The structural design of the felting unit has not been revealed but delivery speeds of up to 30 m/min are claimed. The machine processes ten ends, each of which is delivered into a net capable of holding between 8–15 kg of yarn, which means that considerable lengths of knotless yarn are produced. The method eliminates the need for any subsequent twisting operation. At the present time research and development work is being jointly carried out by IWS in the UK and TNO in Holland. Twist is inserted not only in the preparation of the single yarn but also in the folding together of two or more yarns to produce a ply yarn, in which form almost all long-staple yarns are used.

## IV. Twisting or Folding

Over the past 25 years the traditional twisting or folding operation has undergone considerable change and development. One technique still employing the ring spindle is that of stage twisting. This method is unusual in that, unlike many other developments, the main object has been to improve yarn quality rather than the more common aim of reducing costs.

Stage twisting replaces a single folding operation with two separate operations. The *first* stage is to combine together, on one package, the component single yarns. The emphasis at this stage of the process is to ensure that the singles yarns are fed forward under uniform tension and thus ensure that equal lengths of yarn are combined. These single ends are held together with small amounts of retaining twist, e.g. 20 turns/m. The *second* stage is to take the combined yarns and then insert the amount of folding twist finally required. This latter operation is frequently carried out on an up-twister type of machine. A further advantage of stage twisting is that it facilitates the yarn clearing operation. The yarn produced at the spinning frame is rewound onto a cone or cheese and during this operation any faults are removed. Thus, a longer length of single yarn than can be accommodated on one spinning package is made available. These cleared yarns are more valuable and are often in demand for those end-uses which require "knotless yarn", i.e. no knots in the doubled yarn.

## A. Two-for-one twisting

The introduction of stage twisting has provided an opportunity for the use of "two-for-one" twisting for long-staple yarns. The two-for-one principle is not new, having previously been used extensively in the production of tyre-cords. Because of the route which the yarn follows during the twisting process (Fig. 18) *one* revolution of the spindle inserts *two* turns of twist into

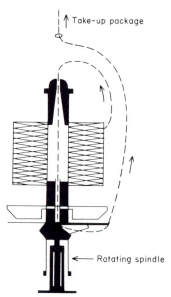

FIG. 18. Diagrammatic representation of a typical two-for-one spindle assembly showing yarn route.

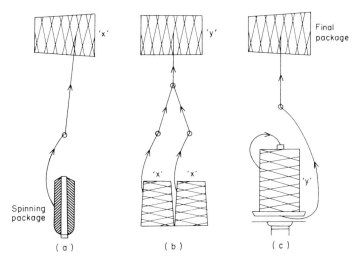

Fig. 19. Winding routine for two-for-one twisting. (a) Rewinding spinning package; may include yarn clearing. (b) Assembly winding. (c) Two-for-one twisting.

the yarn. Two-for-one twisting is possible only when an assembly-wound package containing the required component yarns has previously been prepared (Fig. 19). In theory, three-for-one or four-for-one twisting is possible but such units become too complicated for normal standards of operative skill and are not used commercially.[36]

## B. Fancy yarns

Brief comment should be made concerning the production of fancy yarns. These continue to be a very fashion-sensitive area of the industry with periods of very limited production followed by periods of considerable demand. Clearly many of the newer methods of spinning produce yarns with characteristics which differ, in varying degrees, from those spun by conventional methods and these ideas have been adapted by manufacturers of fancy yarns. In particular, the use of the hollow spindle which can permit a binding thread to be wound round a fancy yarn without the need for an additional twisting operation, is proving popular.[37]

The period since 1945 has been of particular significance for the long-staple industry with the introduction of many more new methods of spinning. Whilst it must be emphasized that a far greater proportion of yarn is still produced using the traditional ring spindle, it must be acknowledged that techniques such as self-twist, open-end (see Chapter 7) and Siro are, in certain areas of application, beginning to be more widely used. Although these yarns possess different characteristics when compared with those conventionally spun, it may be confidently predicted that, largely because of economic advantages, these newer techniques will continue to gain in popularity.

## References

1. Grosberg, P. (1968). *An Introduction to Textile Mechanisms*, p. 9, Benn.
2. Ambler, G. H. and Hannah, M. (1950). *J. Text. Inst.*, **41**, 115.
3. *WIRA Bulletin* (Spinning) (1971). **2**, 239.
4. Iredale, J. A. (1974). "Oil—versus Dry-Combing, the Present Situation in Worsted Combing". *Tl & I*, Jan. p. 16.
5. *Inst. Text. Bull.* (Spinning) (1972). **1**, 19.
6. *Inst. Text. Bull.* (Spinning) (1971). **2**, 239.
7. Dyson, E., Iredale, J. A. and Parkin, W. (1974). *Textile Progress*, **6**, No. 1, 1, The Textile Institute, Manchester.
8. *Wool Science Review* (1955). **14**, 13. International Wool Secretariat.
9. *Manual of Cotton Spinning* (1965). **5**, The Textile Institute, Manchester, and Butterworths.
10. "Variable Speed Spinning" (1949). *Tex. Manuf.*, Dec., p. 593.
11. Charnley, F. (1958). *J. Text. Inst.* **49**, 107.
12. *Platt Bulletin*, **10**, 9.
13. Fuch, H. (1970). *Int. Text. Bull.*, **3**, (Spinning).
14. Fuch. H. (1973). *Int. Text. Bull.*, **3**, (Spinning).
15. Wool Research (1949). *Drawing and Spinning*, **6**, c. 11, WIRA, Leeds, UK.
16. Lord, P. R. (1974). *The Economics, Science and Technology of Yarn Production*, pp. 168–69, North Carolina State University.
17. Igel, W. (1979). *Int. Text. Bull.*, **3**, 486, (Spinning).
18. Wolf, B. (1980). *Int. Text. Bull.*, **3**, 270, (Spinning).
19. Lord, P. R. (1980). *Textile Month*, June, pp. 73–74.
20. Hunter, L. (1978). *Textile Progress*, **10**, No. 1/2, p. 9, The Textile Institute, Manchester, UK.
21. *Break Spinning* (1968). Shirley Institute, Manchester, UK.
22. Cygan, W. (1975) In "Rotor Spinning" (E. Dyson, Ed.), p. 213, *Textile Trade Press*.
23. Hunter, L. (1978). *Textile Progress*, **10**, No. 1/2. c. 12, The Textile Institute, Manchester, UK.
24. Barker, A. F. (1936). *J. Text. Inst.*, **27**, 98.
25. Lord, P. R., Chapter 4, this volume.
26. Caban, J. C. (1974). "Wrap Spinning"—paper at *49th Annual Research and Technology Conference*, Textile Research Institute, USA.
27. *CSIRO News Letter*, No. 10, August 1981, Geelong, Victoria, Australia.
28. Henshaw, D. E. (1971). *Self Twist Yarn*, Merrow.
29. Walls, G. W. (1975). "Selfil, A New Spinning System for Yarns for Knitwear", *CSIRO*, Geelong, Victoria, Australia.

30. Walls, G. W. (1970). "Recent Research and Industrial Applications of Self-Twist Spinning and Related Techniques". *In The Yarn Revolution, Annual Conference,* The Textile Institute, Manchester, UK.
31. Gruoner, S. (1973). "Shortened Worsted Yarn Production Process with Repco Spinning Units". *Mell. Text.,* Oct., pp. 1013–16.
32. van Dort, J. M. (1975). *Industrie Text.,* 11.
33. Bobkowicz, A. J. (1975). *Canada Text. J.,* **92**, No. 7, p. 55.
34. Anon (1981). *Textile Month,* Feb. p. 38.
35. Pitts, J. M. D. (1980). "Felted Wool Yarns", *Wool Science Review,* **56**, 61, International Wool Secretariat, Ilkley, UK.
36. Teijin Ltd., BP 1 215 877 (Japan, 16 Dec. 1970).
37. Iredale, J. A. and Dyson, E. (1980). *Tl and I,* Manchester, p. 67.

# Chapter 4

# Sliver Production

## P. R. LORD

---

## I. Economic Motivation for Machinery Development

Engineering developments in textile machinery usually arise because of economic pressures. The textile industry is international and highly competitive. Countries with low labour costs and low levels of capital investment compete with others that have intermediate and high labour costs and high levels of capital investment. Usually, the higher the capital investment, the lower is the requirement for labour (i.e. the lower is the HOK value) (see Fig. 1). As shown in Fig. 2, there has been an almost logarithmic decline in HOK over the years. This improvement in worker productivity has been

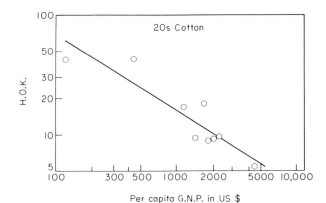

Per capita G.N.P. in US $

FIG. 1. Productivity as a function of investment. Base year 1968; each point is a different country.

P. R. LORD

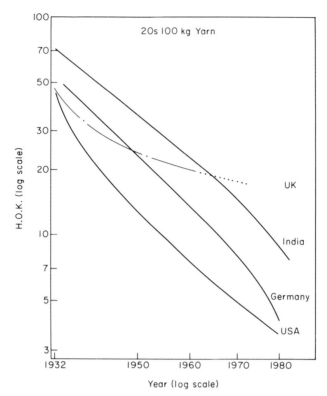

Fig. 2. Changes in spinning productivity. 20 s cotton count; HOK = number of operator hours required to produce 100 kg yarn.

brought about by changes and improvements in the design of the equipment. Such changes are mandatory if the high wage countries are to be able to compete.

For a ring frame, the productivity

$$(P) = \frac{U\eta}{K(TM)N_e^{3/2}} \ \text{kg/h}$$

where $U$ = spindle speed in rpm, $\eta$ = efficiency, $TM$ = twist multiple or twist factor, $N_e$ = English cotton count, $K$ = constant.

If the labour assignment is $(a)$ spindles per operator, then the HOK = $1/(100/ap)$ operator hours/100 kg.

Since both $(a)$ and $(P)$ are functions of $(N_e)$, then it is not surprising to find that there is a roughly logarithmic relationship between (HOK) and $(N_e)$ as

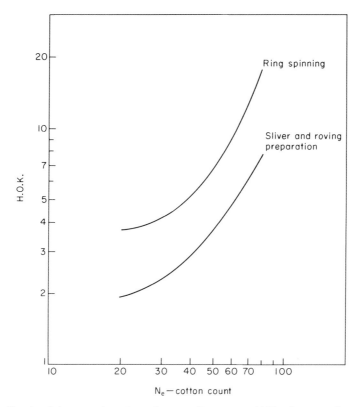

FIG. 3. Productivity as a function of count. Base year 1980; short staple systems.

shown in Fig. 3. The slope of these curves depends on the fibre purchasing policies and this will be discussed later.

A textile mill is originally designed with a balance in both space and equipment to ensure that the outputs of the sliver preparation and roving production sections are almost the same as that of the yarn production section. As the productivity of the means of yarn production increases, there is a pressure to increase the productivity of the preceding sections. There is also a greater need for more extensive sliver preparation for the finer counts (e.g. combing). Thus, in fact, one finds that the (HOK) of the sliver and roving production systems in a series of well-designed mills is related to the (HOK) of the corresponding spinning sections (see also Fig. 3). The curve does not apply to any given single mill.

Yarn production costs can be divided into: (i) fixed costs per unit time; (ii) variable costs (except handling); (iii) handling costs per unit mass. The variable costs under item (ii) consist of the costs of power, piecing, machine cleaning and any other items that vary with machine speed. The handling costs under item (iii) consist of those related to doffing, creeling, transport and storage and do *not* vary with machine speed. The design and development of new machinery is costly and new types of equipment are nearly always more expensive than their predecessors. As shown in Fig. 4, an increase in fixed costs makes it necessary to increase the running speed to give the lowest cost per unit mass for the new machine.[7] Hence not only *can* the new machine usually run faster, it *must* if it is to be economic. The more expensive the machine, the faster must it run.

In sliver and roving production, the handling costs can be strongly affected by the design of the machinery. To reduce the handling costs, modern trends have been to: (i) increase sliver can sizes; (ii) increase roving package sizes; (iii) reduce the number of package or can transfer points; (iv) use automatic handling; (v) use autolevelling; (vi) use better blending methods; (vii) increase machine productivities. Inasmuch that it is necessary to increase the cost of the machine to achieve some of these aims, then the increases in

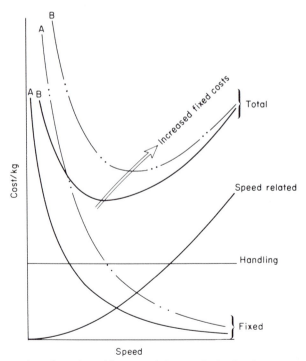

FIG. 4. Cost as a function of machine speed. In case B the fixed costs are double those in case A.

machine productivity become necessary on purely economic grounds. Also the increase in automation implies dispensing with lap handling, and the like, in favour of continuous flow systems such as chute-feed carding systems.

As the machine speeds rise, the stresses acting on the textile material rise and the incidence of end-breakage rises also. This is particularly so in spinning and the end-breakage rate here is highly dependent on the evenness and homogeneity of the material being processed. The evenness and homogeneity are largely determined at the sliver and roving preparation stages. It thus becomes increasingly necessary to use autolevelling and superior blending for purely economic reasons related to the spinning process.

## II. Environmental Motivations for Machinery Development

Over the last decade there has been a growing awareness of the dangers to

FIG. 5. Automatic fibre feeding (Automatic Bale Handling).

health and well-being of both the working and surrounding populations due to pollution of air and water. In the case of cotton spinning, the incidence of air-borne particles from the non-lint portions of the raw material have been related to byssinosis and other lung-related health problems. This has led not only to efforts to prevent dust escaping into the atmosphere, but also to efforts to reduce the population of workers exposed to these dangers. Hence, there has been increased attention to automatic material handling (see Figs 5 and 6) in mills processing cotton. The need for planning flexibility has caused such equipment to be used for short-staple blends and also (by extension) for wholly man-made fibres in the same fibre length range.

In the case of wool spinning, the concern has been more with water treatment[9] and this has had less direct effect on the handling of the cleaned fibres. However, the earlier experiences of having to scour and then oil the fibres to control electrification and lubrication were not lost on the man-made fibre industry when they came to produce staple fibres (which had to be carded and otherwise converted into yarn). The art of applying and controlling the application of fibre finish has become very important. So has the control of fibre crimp.

Fig. 6. Chute feeding (Fibre Controls).

## III. The Traditional Technological Background

In early times, the great bulk of yarn was made from natural staple fibres. Of these, the most important were cotton and wool. Cotton is a relatively short fine fibre with convolutions which tend to give fibre assemblies a degree of coherence. Wool is usually longer and coarser; the fibre surface is scaly and the fibre is naturally crimped. This gives an enhanced degree of fibre coherence. Furthermore, the greater average fibre length makes it possible to use relatively low twist levels in the yarn, and, despite the somewhat coarser fibre, to produce a soft yarn. The use of a relatively high-twist in wool yarn makes a durable product. Both fibres are variable in fineness and length; this especially applies to wool.

The mechanical processing of all such fibres involves both drafting (or drawing) and twisting, but in sliver production we are concerned only with the former process. In this drafting process, fibres are caused to slip past one another as the strand or sheet of fibres is elongated. The regularity of fibre flow under these circumstances is dependent on the fibre length and the distance apart of the grip points. (In roller drafting this distance is the ratch setting, but in carding it is the distance between the fibre grip points.) Consequently, the length of the fibre and its variability are matters of great importance. Furthermore, the greater fibre coherence and springiness of the wool fibre, as compared to cotton, require that different drafting techniques be used. Thus, in carding wool one uses roller-top rather than flat-topped or fixed-top cards. In the drawing process one uses pin drafters (gill boxes) rather than simple roller drafting. It follows that there must be two separate streams of development. Historically this has been so, and one example of each of these main streams of development is sketched out in Fig. 7. There are, of course, many variants. For example, in short-staple spinning it is possible to introduce combing, or to replace one passage of drawing, roving and ring spinning by open-end spinning. In long-staple spinning it is possible to replace the worsted system by the woollen system (which has a very much shorter flow chart), or to introduce self-twist spinning. Generally speaking, there is a greater variation in plant layout and equipment in long-staple spinning than in short-staple spinning.

Since the natural fibres are inherently variable, the matter of selection and blending becomes extremely important. In a given mill, the ratch settings are fixed and it is a costly and time-consuming task to change them. Hence it is desirable to maintain a fibre supply in which the fibre length is within the range that can be accommodated by the equipment at the particular setting. It is necessary also to control the fibre fineness and other properties to ensure that the quality of yarn desired is, in fact, produced. The most desirable fibres are usually the most expensive and, consequently, it is necessary to

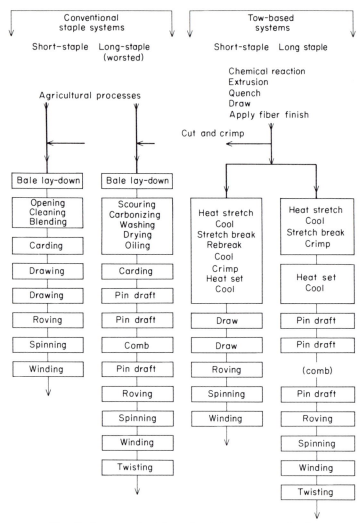

FIG. 7. Some typical flow charts for yarn production.

blend fibres to give the best mill performance and yarn properties at the minimum fibre (and processing) cost. A typical blending machine for short staple is shown in Fig. 8. It is one thing to make a laboratory blend to achieve these ends, and quite another to achieve it consistently in industry. The homogeneity of the blend throughout the whole mix is of the greatest importance. Consider the consequences of an inhomogeneous blend. The stream of fibres going forward might contain separate pockets of long and

FIG. 8. Intimate fibre blending (Trutzschler).

short fibres. The long fibres get broken and become short fibres and, in breaking, they disturb the flow to cause unevenness in the product. The short fibres cause drafting waves which produce periodic unevenness each time they pass through a drafting system.[4] With an inhomogeneous blend one finds bursts of unevenness which are difficult to detect and yet become very visible in the final product. The matter is complicated by the multiplicity of error wavelengths produced. The same pocket of short fibres produces a band of error wavelengths of some twice the fibre length at each stage, but these errors are elongated by the following drafting processes. The mean wavelength of the error for each stage as it appears in the final yarn is ($\lambda D$), where $\lambda \approx 2 \times$ fibre length and $D$ is the total draft experienced by the fibre from the point of error production to the point of measurement. Since there is drafting at the cards, draw-frames, combing machines, roving frames and ring frames, each with several drafting zones, then there is a spectrum of error wavelengths produced for each pocket of short fibres. To these must be added the inevitable mechanical errors. Also the grip point is always at least a short distance from the ends of the fibre. Thus, the effective fibre length is always less than the actual length and is variable. A further complication is that the effective ratch setting is not the simple distance between the contact points of fibre and roll (or tooth). There is a pressure distribution when a bundle of fibres is nipped between two elements as in Fig. 9. The effective ratch distance for fibres at the centre of the bundle is different from that applying to fibres on the outside of the bundle. The differences such as $l_1$ and $l_2$ in Fig. 9 are dependent on the weighting and sizes of the rolls, the stiffness of the fibres and the linear density of the bundle. Fibre lubrication and crimp level strongly affect the effective fibre length. If the linear density or the stiffness or other properties of the fibres vary along the length of the strand, the ratio $l_1/l_2$ varies and there are varying amounts of drafting waves. Thus, errors in blending or evenness beget further errors in later processing stages.

Several possibilities exist to control, or partially control, these errors. These are to use: (i) superior blending techniques; (ii) combing to reduce the short fibre content; (iii) doubling to reduce long-term error; (iv) autolevelling; (v) aprons, trumpets or like devices to control the fibre flow in the drafting zones.

The experiences with short-staple equipment modified to meet the clean-air requirements have indicated that reduction in tuft size fed to the card not

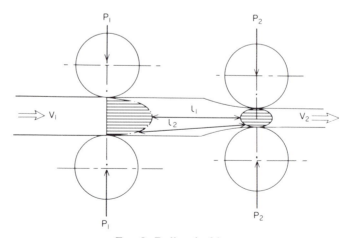

FIG. 9. Roller drafting.

only enables better cleaning to be achieved, but also gives a better blend homogeneity with the beneficial results that can be deduced from the foregoing. It is clear that the selection of a multiplicity of small samples from an adequate number of bales produces a more uniform product than would be achieved by hand-feeding large samples from a small number of bales.[12] Bearing in mind the need for automatic handling at this point, it becomes understandable why the bale plucker and automatic fibre feeding systems have developed so rapidly in the last few years. In addition to this, blending machines have been developed to homogenize the fibre supply. Such machines have a large but finite capacity and can integrate the blend variances over a certain mass of fibre. They are not capable of integrating over extended periods and variations over days, weeks or years must be controlled, as always, by careful purchasing and mill supply policies. These policies dictate the variance in the fibre length and properties, which, in turn, dictate the amount of blending required. The policies also dictate, to a large extent, the performance of the mill. A policy of buying superior fibre for fine counts tends to reduce the slope of the curves in Fig. 3. The extra cost of the fibre is offset, to some extent, by a reduction in processing costs. Conversely, a policy of buying low-grade fibre makes fine-count spinning impracticable and the extra costs of processing can absorb much of the saving in fibre costs. Similar considerations apply to long-staple spinning. In both cases, combing can be used to up-grade the fibre but not only is this expensive, it also requires that the fibre purchasing policies be adjusted accordingly.

Doubling is an effective means of reducing long-term error.[1] If a sliver has a single enlarged portion which is $x$ cms long and $y\%$ oversized, and this sliver is placed with $(z - 1)$ other slivers which are perfect, then the total feed material has a portion $x$ cms long which is $(y/z)\%$ oversize. If this is drawn with a draft ratio of $D$, the output contains a portion of $(xD)$ cm in length and $(y/z)\%$ oversize. Where there are $(n)$ passages of drawing, each of the same doubling factor, the error is reduced to $(y/z^n)\%$. In this case, there are considerable advantages in using several passages of drawing. In practice, all slivers have many errors, not just one, and consequently the improvement in evenness depends on the probability that the thick spots coincide with sufficient thin spots at the particular cross-sections to give a balance in net error which is better than that obtained in a single sliver. Multiple passages of drawing still give appreciable benefit. As shown earlier, the draw-frame dilutes the error by doubling and elongates the error by drawing. The drawing is imperfect and tends to introduce error. These new errors are mostly short-term, and therefore the draw-frame tends to reduce long-term error and then insert new short-term errors. If a sliver blend of two different fibres is being used and reliance is placed on doubling alone, there will be periodic changes in blend proportions. If the original error is from carding or from the earlier processes, the input error wavelength can be very long. If the error wavelength is greater than the length of sliver stored in the sliver can, then it is possible that the draw-frame doubling process will not correct the error. In such a case, it is necessary to use autolevelling to adjust the linear density by changing the draw ratio in the card[2] or draw-frame or by controlling the earlier processes. If there is a store of full sliver cans prior to drawing, then it is possible to sliver-blend from different bale lay-downs and to even out differences in fibre lots as well as differences between card sets. However, such a store ties up space and capital; for these reasons the mass stored is usually limited.

Autolevelling is a form of automatic control in which the linear density is measured at one point and corrected at another. As shown in Fig. 10, there are two basic forms. In the first, the feed-forward system, the error ($E$) is measured and the transducer generates a signal which is proportional to the

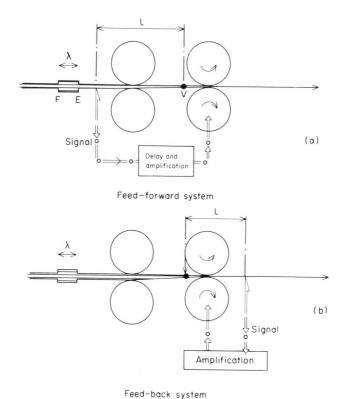

FIG. 10. Autolevelling.

error. This signal is stored until the material in error reaches the vicinity of the velocity change point ($V$); at this time the amplified signal is applied sufficiently to cause the draft ratio to change by such an amount that restores the linear density to its proper level. In other words, $A_1 = A_0$, where both amplitudes are expressed in terms of output linear density. As point ($F$) reaches the measurement point, there is another signal change generated which ultimately causes another change in the draft ratio. The changes in draft are not instantaneous because of the inertia of the system, and a typical response is shown in Fig. 11(a). Unless the delay time is accurately set for the prevailing conditions, there will be some residual errors in the output material as also shown in Fig. 11(a). The delay time is a function of the input velocity of the material, the transit response and the length ($l$) in Fig. 10(a). The fibre velocity change point ($V$) varies in position and alters ($l$), especially

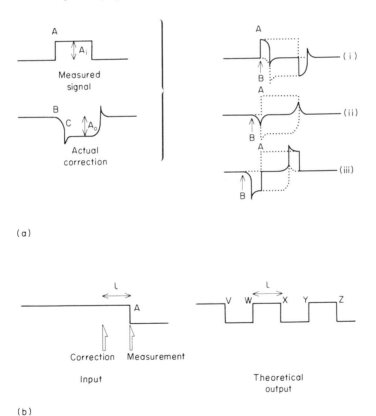

FIG. 11. Control loop responses. (a) Effects of altering delay times with a feed-forward system. (b) Behaviour of a feed-back system when a single step change occurs in the feed material.

if there is a varying blend of staple-lengths and other fibre properties. Apart from movements of the velocity change point, the response time of the system [(C) in Fig. 11(a)] affects the appropriate setting of the delay time as illustrated in Fig. 11(a) at (i), (ii) and (iii). Thus, the system has to be tuned to give the best average performance for the given blend in any given machine, but there can never be a perfect elimination of the input error. Furthermore, the feed-forward system can never correct drafting errors in the machine of which it forms part. This is because it never measures such errors.

With the feed-forward system, it is necessary to set exactly both the amplification and the delay time for the given conditions, otherwise there will be improper control. With the feed-back system, such accurate setting is not so essential because of the regenerative nature of the system. If the distance ($l$) in Fig. 10(b) were zero, then the errors would produce a signal tending to reduce the error to zero; and the error signal would persist until the error is reduced to zero. The practical problem is that if too high an amplification is used, the system goes into oscillation and is uncontrollable. When the distance ($l$) is not zero, the same thing happens as shown in Fig. 11(b). A step change in the input material at (A) has no effect until it reaches the measurement point. In the simple case, the signal then causes the output material to be reduced in linear density at (Y). This stage persists until (Y) reaches the measurement point, at which time the linear density is changed again at (X). This is because the portion (XY) shows no error and the draft ratio returns to normal but the input material is oversized. Consequently, there is an alternation in linear density as at (VWXYZ) in Fig. 11(b). To overcome this problem, it is necessary to integrate the signal over at least ($2l$) and this reduces the range of wavelengths that can be corrected; it is helpful to reduce ($l$) to the minimum amount by the design of the drafting system.

It can thus be seen that blending, doubling and autolevelling are complementary rather than competitive. Poor blending can cause drafting waves in a draw-frame which the feed-forward system could not control. If the fibre length < $l$, then the feed-back system could not control it either. Many modern control systems use both feed-forward and feed-back systems to control both long- and short-term errors. But even if these work perfectly, excellent blending is still required because it is not practical to autolevel at the ring frame, and the most severe drafting waves are generated at this last critical stage.

It is possible to introduce various devices into the drafting zones to control the fibre flow and to stabilize the velocity change point. A floating fibre passing through this point has two sets of forces acting on it. A relatively few rapidly moving fibres act to accelerate the subject fibre from the linear velocity of the back roll to that of the front roll. A greater number of slowly moving fibres (mostly in contact with the back roller) tend to restrain the fibre from accelerating. There is thus a tug-of-war. When the floating fibre actually accelerates, it tends to do so suddenly because of the differences between static and dynamic coefficients of friction. When the fibre is travelling at the same speed as the fast-moving fibres, the coefficient of friction between these rapidly moving fibres tends to be higher than that between the fast- and slow-moving fibres. Also, when a floating fibre accelerates, it changes the balance of forces acting on its neighbours that are also "floating". Suddenly there is one more fast fibre and one less slow fibre, and the other floating fibres also accelerate. Thus, the velocity change phenomenon tends to be unstable but, if mechanical or chemical forces are brought to bear, the instability can be reduced to acceptable levels. The most well-known way of doing this is to use aprons but these cannot be used at high linear densities. The use of trumpets helps because the trumpet walls force fibres on the outside of the fibre stream into contact with the others and

FIG. 12. Autolevelling at the card (Zelleweger Uster).

E

some of the worst effects of the phenomenon described in Fig. 9 can be ameliorated. An interesting way of controlling the fibres is to lightly glue them together as in the Pavena system. This restrains the fibres from accelerating until the leading end of the fibre reaches the mechanical nip at the front roll. A similar effect can be obtained with wet drafting. In this case it is the surface tension of the water which acts partly as a trumpet and partly as the binder (as in the Pavena system). These latter devices have not yet become commercial and one reason for this is the cost and trouble of drying the yarn or roving.

The surges of floating fibres cause local concentrations of short fibres which cause even further problems in later drafting stages. Generally speaking, the problems with drafting waves increase with the draft ratio. Thus, the main draft in ring spinning is one place where great unevenness can be produced if all the machines in the line are not correctly set and the proper fibre is not used.[3] Also, the break draft in ring spinning can give trouble, particularly if the roving twist or linear density or break draft level is too high. In an uneven yarn, the twist runs to the thin spots and tends to give intermittent bursts of drafting waves and slubs. In addition, twist can be

FIG. 13. Autolevelling at the drawframe (Schubert and Salzer).

forced back along the roving and then suddenly flow into the draft zone to give similar difficulties. Both of these phenomena are troublesome if the twist level is near the critical value beyond which the fibres will not draft properly. The critical value is a function of the fibre length and fineness. With man-made fibres, the crimp level and fibre finish can also play a part. Thus, it is seen that the choice, the blending and the production of an even homogeneous roving is of the greatest importance if a high quality yarn is to be produced.[4] Since the evenness of the roving also depends on the sliver evenness and homogeneity, then it can be seen that various problems associated with blending and evenness control are cumulative. Recognition of the importance of controlling the early processes, coupled with the economic advantage of controlling the sliver rather than roving or yarn, has led to the development of autolevellers at both the card and the drawframe (see Figs 12 and 13). Overall management of the quality control system is very important.

## IV. The Nature of Polymer

The tow-based systems shown in Fig. 7 all involve synthetic polymers; it is necessary to understand the nature of these materials before the problems can be properly discussed. These polymeric materials are assemblies of long-chain molecules, and the structure and form of these are of the utmost importance in the final product.[8] Also, these materials are nearly all thermo-plastic, which means that they have a softening point (glass transition temperature $= T_g$) as well as a melting point (phase transition temperature $= T_m$). Above $T_m$, the molecules move independently and the material exists as a viscous fluid. As the temperature drops below $T_m$, portions of some molecules become united to form solid crystallites and these tiny crystals grow until they start to interfere with one another. The crystals are surrounded by amorphous material which has different properties from the crystals. The molecules, or portions of molecules, in such amorphous zones are capable of segmental motion when the temperature is between $T_g$ and $T_m$. This is associated with the rubbery, plastic nature of the material in this temperature range. It is in this state that the material is set during normal heat treatment. Below $T_g$, the molecules become frozen and the material behaves as a glassy solid. Both $T_g$ and $T_m$ are affected by the local stress conditions within the fluid/solid material. Consequently, it is possible to find variations in the properties of the polymer according to the variation in local stress levels during solidification and drawing as well as the variations in viscosity. All of these are affected by changes in temperature but, because the fluid is thixotropic and the rate of crystallization is stress sensitive, there is a complex relationship, and variations in local temperatures and stress levels become important as far as the final product is concerned.

Undrawn polymer changes its characteristics considerably when it is drawn. In this context, drawing describes the movement of molecules over one another rather than the movement of fibres over one another as described previously. The elongation of the polymeric material causes the molecules to become better oriented, and may cause some changes in the crystallinity. The outcome of drawing is that the mechanical strength is much improved as the molecules become better aligned. Also, the density of the material changes and so does dye affinity. The more closely packed the atoms become, the more difficult is it for relatively large dye molecules to penetrate and find sites where they can become attached or entrapped. Also, crystals have their atoms closely packed in precise order with the result that dye absorption is much different from that found with the open and disorderly arrangement of atoms in the amorphous zones.

When the polymeric material is heated above $T_g$, changes in morphology can occur. The degree of crystallinity and the size of the crystals can change; also there can be changes in orientation. Thus, heat affects the mechanical properties and the dye affinity. In fact, both the mechanical and thermal histories of the material are of great importance. Variations can cause differences which might not appear until after the dyeing process is completed. Streakiness, barré and moiré patterning can then emerge and the problems may have been created in extrusion, drawing, heat-setting, or any of the thermal or mechanical processes that follow. A new form of quality control is thus required.

## V. Melt Extrusion

The great majority of man-made fibres are melt-extruded because such extrusion can be carried out at very high speeds and this enables the enormous investment in capital equipment to be spread over a very large mass of fibre. In other words, the fixed cost/kg can be brought to a reasonable value. (The arguments in Section I can also be applied.)

A typical tow-producing extrusion plant consists of a chemical reactor directly coupled to a series of extruders and the associated ancillary equipment.[11] The hot polymer is pumped to the extruders, and, of course, the pipes are designed to keep the polymer temperatures at the correct level. The basic extruder consists of a screw, filter, metering pump, spinnerette, quench, draw and take-up systems (see Figs 14 and 15). There may be an extra (separate) stage of drawing. The amount of working that the polymer receives before it reaches the spinnerette affects its viscosity, which, in turn,

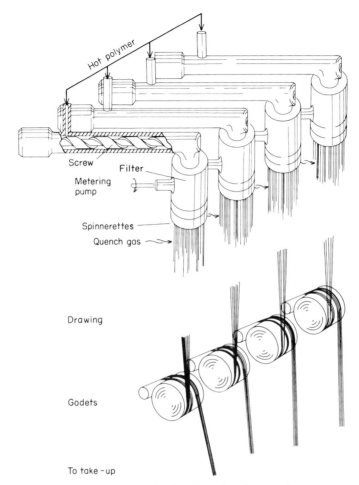

FIG. 14. Schematic drawing of melt extrusion.

affects the extrusion. Consequently, care has to be taken to control accurately the temperatures and speeds in the process.

If solid or unmeltable material is allowed to pass to the spinnerette, it is likely to block (or partly block) one or more spinnerette holes. The flow through that hole (or those holes) breaks down and a drip may form. This drip can get caught up by the other filaments and flow along with the rest of the production. A high incidence of polymer drips causes problems in later processes, especially carding where the solidified polymer drip gets embedded in the wire. Such "plaque" in carding then can cause other problems

FIG. 15. Fibre extrusion line (Barmag).

such as periodic generation of nep. Thus, the use and maintenance of an effective filtration system in extrusion is of considerable importance. A further problem arises because, when there is a break, a length of undrawn filament is produced, and such undrawn material will, if allowed to remain in the flow of fibres, produce a more deeply dyed streak in the final product. The higher linear density of the fibre can also cause some problems. Also the very high extensibility of the undrawn material causes differences in drafting.

The streams of hot polymer emerging from the spinnerette have to be cooled. In high-speed tow production, this involves the use of quench air or gas. If the rate of cooling between one filament and others in the bundle varies, then there is a difference in crystallinity and morphology which also shows up in the final product. The cooling rate during the critical solidification phase is very important and so is the strain level. This latter is related to the draw ratio and the potential contraction due to cooling. The molecules in the melt being extruded are made orderly by the flow process, but changes occur during solidification. At low throughput speeds, the molecules in the amorphous regions have time to become disorderly before they are confined by the surrounding crystals and it is necessary to draw the material to restore some degree of order. The strength (and dyeability) is related to this degree of order. At very high throughput speeds with rapid quenching, it is possible to freeze some of the molecular order caused by the flow process coupled with the elongational strain of take-up. This requires rapid and even quenching of the fibre bundle. Such material has many of the attributes of a normal drawn polymer. The economic advantages of ultra-high-speed extrusion are fairly obvious, and therefore it is to be expected that more of such materials are likely to be found in the marketplace. However, the molecular structure is, in fact, different from that of conventionally drawn materials. Differences in structure due to variations in speed and cooling rates affect the final product.

The denier of the filaments is controlled mainly by the ratio of speeds between the metering pump and the final take-up velocity. The size of the spinnerette hole does not directly affect the mean linear density of the filament, although its shape does affect the cross-sectional shape of the fibre. If the spinnerette hole is of the wrong size, it can affect the stability of the process, but it can normally be expected that this is properly controlled. The amount of drawing employed affects both the linear density of the filaments and the characteristics of them. The godets and other moving elements have to be true, otherwise periodic errors are produced. Build-up of finish, monomer, polymer or any other substances on these active surfaces can produce such periodic errors. These errors not only affect the denier of the fibres but also the dyeability and mechanical properties.

The last stage in the fibre forming process is the procedure for adding finish and, perhaps, crimp. Variations in finish or crimp affect the coherence of the staple fibres eventually produced. Also such variations can contribute to the generation of drafting waves in the staple fibre processes that follow. Such variations are not easy to check and control; hence the variations, especially batch-to-batch variations, are quite likely and need to be taken into account.

The quality requirements for tow to be used in the production of bales of staple fibre differ from those required in the direct production of stretch-broken fibre. In the first case, which is typical of normal short-staple fibre production, the tow is cut and mixed with fibres from many other tows to produce a reasonably homogeneous blend with reasonably consistent characteristics. In the other case, however, there is less opportunity of blending.

There is some doubling at the stretch-breaking machine(s), but there could be difficulty with the consistency of the product if the quality of the tow is allowed to vary. Thus, for the stretch-break market, it is normal to select the best tows and this increases the cost. Also the stretch-breaking equipment has to be designed with the difficulty in mind. Similar remarks also apply to long-staple cutting systems such as the Pacific Converter.

## VI. Stretch-Break Process

The stretch-break process usually involves not only the conversion of tow to staple fibre but also the production of a high-bulk material. At one time, the Pacific type of converter was usually used to process polyester tow and the Turbo type of stretch-break process was usually used to produce long-staple tops. Modern stretch-break equipment (see Fig. 16) can produce long- or short-staple slivers of either polyester or acrylic fibres with varying degrees of relaxation.

If a thermoplastic filament is heated above $T_g$ under tension, and then cooled under this tension, the extension will be set in a similar way that bending and torsional deformations are set during texturing. If these heat-set filaments are then re-heated above $T_g$, they will tend to return to their original length and the shrinkage can be considerable. The exact amount depends on the polymer and the load applied. In a large bundle of filaments, such as tow, not all filaments are similarly stressed, nor are the characteristics the same; consequently some filaments shrink more than others. The differential shrinkage causes some filaments to be put in compression; they buckle and become more bulky. Furthermore, a filament may not shrink the same amount on one side as compared to the other. This makes the filament curl like a bicomponent filament, and this, too, creates bulk. These effects can be heightened by mixing some heat-stretched material with non-heat-stretched material, and/or using bicomponent fibres. Acrylic fibres are very suitable for this sort of treatment and they are usually regarded as high-bulk materials because of this.

A heat-stretched yarn is easier to stretch-break than a non-heat-stretched yarn. For one thing, the breaking extension of the heat-stretched material is relatively low and a proportion of the molecular chains are under a stress that is locked-in by the heat treatment. Consequently, it is normal to find that a stretch-breaking machine consists of a heat-stretch section and a stretch-break section. The first section consists of a feed system, two sets of rolls to grip the tow, a heater, and a means for cooling the tow before the external stress is released. The second section consists of two sets of rolls, breaking zones and a delivery section.

Fig. 16. Stretch breaking machine (Seydel).

Fig. 17. Actual drawstand (ARCT).

The stacks of rolls which grip the tow at the various critical points are key parts of such a machine (see Fig. 17). Nip rolls are unable to provide the required tensions without flattening the filaments. It is better to rely mainly on the logarithmic accumulation of tension in a roll stack such as shown in Fig. 18. Nip rolls give additive increments of tension, whereas the wrap action provides a multiplicative effect, where the output tension is $e^{\mu\theta}$ times that of the input tension. A combination such as shown in Fig. 18 gives a tension distribution such as shown by the dotted line. The figures are for illustration only. The size of the steps in this curve depends on the pressure applied by the large roll. The value of the coefficient of friction affects both sorts of tension mechanisms and quite large variations are possible by changing the surface materials and the state of lubrication of the fibres. Bearing in mind the previous heat treatment, it is not always easy to predict the value of the coefficient of friction ($\mu$). Furthermore, rubber surfaces which would give high values of $\mu$ are more likely to wear than hard surfaces. This is particularly so on the last roll where tensions vary between J and K. The total tension $T_o$ may easily run as high as 2000 kg—it being recalled that the filaments have to be broken in the stretch-breaking zone. In fact the wear on this element is usually high and many users periodically change the rolls around so as to distribute the wear. Also the roll diameters have to be large so as to reduce the surface stress; large diameters are also required for stiffness and to give an adequate cooling surface. Hence it is difficult to make short-staple fibre in a single step.

Final control of the actual breaking is by the use of nip rolls. Usually there are at least two zones so that long fibres from the first zone get broken in the next zone or zones. Once again, the nip rolls have to be large in diameter

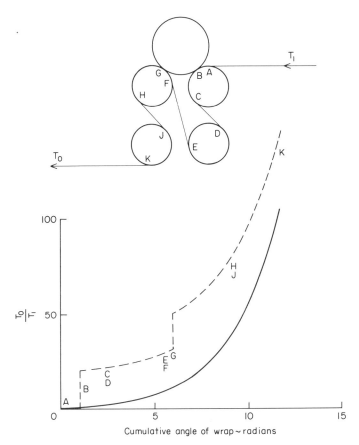

FIG. 18. Theoretical tension accumulation in a roll stand. Assumed value of coefficient of friction = 0.4.

but, in the first zone, the fibres are gripped by the previous roll stand. Of course, when a fibre breaks, its tension immediately drops and stays at a low level until the leading end reaches the nip roll. The tension then rises rapidly until a break occurs at the weakest point between the nip rolls. There is a stress concentration at the feed-roll nip and, if the feed material is even, the fibre will break there. If the material is uneven, then it may break elsewhere in the break zone. Thus, the regularity of fibre length from this first zone depends on the regularity of the material entering. This, in turn, depends, in part, on the regularity of the preceding stress and heat treatment. In theory, short fibres entering subsequent zones will not accumulate tension and will emerge unchanged in length. Long fibres get broken. The resulting staple length is reckoned as the final ratch setting less the relaxation shrinkage. For long-staple products (75–150 mm), it is then only necessary to crimp (usually by stuffer box), relax and, perhaps, oil so as to secure the fibre characteristics desired. The steaming can be done by batch using an autoclave or continuously. The steaming is very important in generating bulk and releasing the locked-in stresses caused by the stretch-breaking process.

For short-staple fibres, it is normal to re-break the already stretch-broken material. The sliver is fed to the re-breaker without steaming or crimping. The re-breaker usually consists of up to four breaking zones in which the length is reduced to the required level (usually in the range 25–60 mm). This process is then followed by the crimping and steaming operations. The sliver produced on a modern machine has a good fibre diagram and good evenness of linear density. If operated at improper conditions, the re-breaker can generate unwanted amounts of fly.

The loads involved in breaking a tow are very large by textile standards. If all the filaments in a 100 ktex tow equally shared in the applied load and the tenacity was, say, 30 g/tex, then the applied load would be 3000 kg. In fact, at a given filament extension, the loads would not be equal and some would have broken and some not. However, the main point is that the loads on the system are high. Apart from the requirements for ruggedness of machine construction, there must also be a limit on the linear density of tow that can be processed. For acrylic fibre, this is about 120 ktex but for the stronger polyester fibres it is only about 70 ktex. In a re-breaking machine, only a proportion of the fibres are loaded at a given instant, and it is thus possible to use higher feed linear densities. As a rough guide these are usually about three times that permissible in the first stretch-break machine.

In stretch-breaking there are changes in linear density of both the fibre and the sliver. A typical delivery sliver has a linear density of 20 ktex. Thus, the feed to a re-breaking machine could contain $(3 \times 120) \div 20 = 18$ slivers. A usual figure is to creel 16 slivers and this gives a useful blending. It is also possible to use a blender, which is a form of drawframe, in which one might blend up to 20 slivers. Thus a total doubling of $20 \times 18 = 360$ is possible. As will have been realized, the processes of extrusion, drawing, heat-stretching, stretch-breaking, fibre relaxing, fibre crimping and oiling all can introduce variables at both the fibre and molecular level. To avoid streaking and shading in the final products, whether they be due to differences in bulk, denier, dye affinity or other cause, requires a sufficient blending from a judiciously chosen range of input tows. These days, very large cans are used (e.g. 1 m diameter × 1.2 m high), and it is quite possible for the creel to hold over 2000 kg and this represents roughly one day's production.

A stretch-breaking system costs as much as a set of opening lines, together with, perhaps, a dozen short-staple cards. Thus, the productivity of the machines must be commensurate with the cost. A modern stretch-break or re-break machine can run at about 250 m/min. Thus, the productivity ranges from 100 kg/hr to 300 kg/hr depending on the permissible linear density. On a cost basis, this is roughly comparable to the traditional system. However, less labour is required and the sliver is normally of good quality. Taking all elements of cost into consideration, it is calculated that there is usually about 6 or 7% gain in the conversion cost per unit mass over traditional methods. However, the cost of tow can be higher and the final cost balances under various world-wide conditions have yet to emerge.

To operate at 250 m/min it is necessary to work with powerful heaters and forced cooling. A typical stretch-breaking machine might require a 20 kW heater (which is 30% of the total installed electrical capacity), as well as a flow of 2 m³/hr of chilled water. Even a re-break machine requires a small amount of chilled water. The steam relaxing stage needs at least 50 kg/hr of steam. Thus, apart from mechanical complexities, the operators have to cope with items that do not form part of the traditional expertise. It is very important that the operators be properly trained to control these variables. Quality control in this sphere is often not so much the checking of things mechanical but of things chemical and physical.

## VII. Fibre Cutting Systems

Fibre cutting is used by both the fibre and primary industries (see Fig. 19). In the fibre industry tow is chopped, baled and sold by the bale. These bales are then blended with like fibres or with dissimilar fibres by the primary industry in the manner already described. In fairly recent history, tow has been supplied to the primary industry where it has been cut using Pacific Converters and then transformed into yarn. This latter route was favoured for a period because of the difficulty in stretch-breaking polyester. It is now less popular. However, interest is rising in the cutting of natural long-staple fibres to give better fibre diagrams. This interest has been sparked because of the needs of "long-staple" OE spinning where over-long fibres cannot be used successfully. It will be interesting to observe progress.

The largest volume of fibre cutting is carried out by the fibre industry. A major problem has been to eliminate (or to reduce to an insignificant level) the incidence of over-length fibres. A damaged sharp cutting edge mis-cuts when the damaged portion of the edge encounters a fibre. Thus, there is a periodic production of over-length fibres which can interfere with the drafting operations in the primary industry. However, this is not a new problem and major fibre producers are able to keep this problem within bounds. The major benefit of cutting by the fibre producer is that he is able to blend

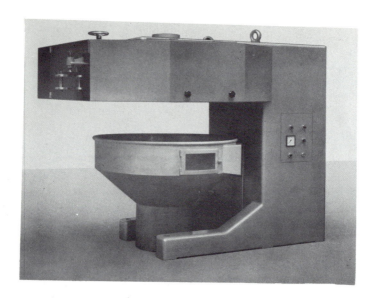

FIG. 19. (a) Fibre cutting machine (Fleissner). (b) Fibre cutter (Fleissner).

massive amounts of fibre which helps to maintain the quality with the increasingly productive equipment being used; this is an important factor.

### References

1. Bowles, A. H. and Davies, I. "Influence of Drawing and Doubling Processes on the Evenness of Spun Yarns. The Characterization and Analysis of Irregularity". *Text. Inst. & Ind.*, **16**, No. 10, 317, 1978.

2. Douglas, K. (Ed.). "Uster Autolevelling Systems for the Spinning Mill". *Uster News Bulletin*, No. 24, Nov. 1976.
3. Douglas, K. (Ed.). "The Uster Statistics, 1975". *Uster News Bulletin*, No. 23, 1975.
4. Douglas, K., Chapter 8, this volume.
5. Lloyd, B. V. "Bale Opening: A Stride Forward". *Textile Industries*, 1979, **143**, No. 4, 1978.
6. Long, H. E. "Protecting the Environment, the Producer, the Consumer". *Shirley Institute*, **30**, 125, 1978.
7. Lord, P. R. "Economics, Science and Technology of Yarn Production". *Textile Institute*, 1981.
8. Mark, H. *et al.*, (Ed.). "Man-made Fibers: Science and Technology". 3 vols., *Interscience*, N.Y. 1967.
9. Stewart, G. G. *et al.* "Studies in Woolscouring—Grease Recovery and Dirt Removed in Optimised Liquor Handling". *Proc. 5th Int. Wool Text. Res. Conf.*, **V**, 647, Aachen, 1976.
10. Wirth, W. "New Ideas in the Preparation and Carding of Man-made Fibres for the Spinning Mill". *Chemiefasern/Textilindustrie* 29/81 (1979), E108.
11. Ziabicki, A. "Fundamentals of Fiber Formation: the Science of Fiber Spinning and Drawing". Wiley, N.Y., 1976.
12. Anon. "Now Justifying Mechanical Blending". *Textile World*, **45**, Feb. 1978.

# Chapter 5

# Staple Fibre Spinning on Condenser (Woollen) System

R. T. D. RICHARDS

This chapter outlines the principles adopted and the practical engineering techniques used in the production of singles yarn from staple fibre by the comparatively short woollen (condenser, cotton waste) processing line. The adjective woollen refers to the process (or type of machine) and does not of necessity imply the presence of wool fibres. Although short fibres can be used, the woollen system also handles long fibres with an average fibre length usually up to 100 mm.

The weight per unit length of the singles woollen yarn is usually between 40 and 2000 mg/m (tex) with the majority of yarns having a count between 50 and 500 tex. The yarns can be used directly in fabric production (woven, tufted and knitted) but often they are used as two-fold and even up to six-fold yarns. The twist of the singles yarn is usually between 60 and 600 turns/m with the majority of values between 100 and 500 turns/m.

Note will also be made of processes using carding machines (cards), some of them conventional woollen ones, which are used to form layers of fibrous webs for use in non-woven fabrics (with or without a lightweight scrim). Card webs can be used also as a fibrous layer(s) (cover) to be needled or otherwise affixed to a knitted or woven fabric.

Note should also be made of cards (some of them woollen) which are used to produce slivers for direct use in sliver knitting or in open-end spinning. That is, the final web is gathered up into a sliver as in a worsted card and this sliver is wrapped into a "ball" or coiled into a can. This sliver is one of a number (up to 12) which, usually with filament yarn, are fed to a knitting machine. Note however that the fibres in the sliver are well separated by a fast moving card-clothed roller and that it is the individual fibres which are used to give a pile surface on the fabric.

For open-end spinning, two, three or four slivers are formed at the end of the card by a sliver forming unit in place of a condenser. The slivers have been spun very successfully on open-end spinners without any gilling operation in some instances. Sometimes the fibres in the sliver have been stretch broken prior to open-end spinning using rotor diameters up to 80 mm, i.e. the maximum diameter available on some commercial models.

In almost all these other applications the specifications for effective blending and carding are at least as stringent as those for the processing of woollen yarns.

## I. Blend of Fibre Components

For technical and commercial reasons it is usually advisable for a woollen blend (stock) to have at least four components.

Commercially there is the obvious direct aspect of price of a blend component but as important is the scarcity aspect if a fibre component is unavailable. This applies particularly to wool which may not be easily available at all seasons of the year. It is easier to maintain uniformity both of cost and technical characteristics of the yarn over a period if only one component of the blend has to be changed.

One popular component is termed waste. This term is unfortunate as it is usually most valuable. It can be one or more of the following types of fibres as examples:

(1) Those which have dropped out of the woollen machinery or have been discarded as "soft" waste during the processing of a previous blend (e.g. slubbing cut off condenser bobbins).

(2) Those which cannot be used on other yarn processing lines (e.g. noil).

(3) Those reclaimed by rag grinding and/or garnetting from yarns, fabrics, garments, etc. (e.g. shoddy).

(4) Those known to be not up to specification (e.g. man-made fibre of variable denier).

The fibre components can differ in many ways such as lustre, smoothness, crimp, length, average diameter (denier or decitex), colour (even of undyed fibres), surface finish, fibre strength and amounts of fibre dust, dirt and other impurities.

Naturally, differing properties are needed for the fibre components as well as differing yarn specifications to cover the vast range of final products containing woollen yarns. In fact it is often necessary for the fibre component to have different properties to give all the necessary parameters to the final product. The handle of an all-wool knitted garment is changed by changing the average fibre diameter from 29 to 34 microns or by changing the twist level in the singles yarn from 180 to 240 turns/m if a blend of shorter fibres is used.

## II. Blending

The prime objective of the process of blending (mixing) of a number of fibre components is to ensure the relative (proportionate) amounts of components (and processing additives) remain the same between different parts of the blend. Naturally, the proportionate amounts should be equal to those for the blend as a whole. The minimum part weights necessary for this objective to be realized will be discussed later.

In practical terms this objective is a stringent one for the comparatively short woollen system. For example, a blend of 2 tonnes of fibres (with 6% of oil added and 8% of the initial total weight lost in processing) yields over 8500 km of singles yarn of count 228 tex (mg/m). The aim is to obtain the same properties (e.g. colour) for every short length of yarn. Thus, reliable and adequate blending is an essential pre-requisite for quality woollen yarn, particularly as the process is a short one and there is a limit to the degree of fibre mixing in a woollen card.

## III. Blend Cleaning

It is assumed that a good proportion of dirt and fibre dust (length less than about 0.1 mm) has been eliminated before the fibres reach the main blending line. Further, with wool it is assumed that any fleeces or portions of fleeces have been partially separated and scoured. This can be done by a powerful machine, a bale opener (breaker) up to two metres wide and with production rates up to two tonnes per hour.

In some instances blend cleaning is carried out by a shake willey machine in which the fibres are kept for a short but variable time within the housing of a revolving spiked cylinder. Part of the housing is perforated and the dirt and fibre dust released is extracted through this part.

In an alternative machine for extracting impurities and partial opening of fibrous masses the main cylinder with spikes is revolving within an enclosed drum. Between the cylinder and drum there are baffle plates which constrain the fibres to move from one side of the cylinder to the other along the direction of the cylinder shaft. The impurities which are rejected are continuously discharged into a collecting chamber below the machine. This chamber can be emptied automatically. Production rates up to 1200 kg/hr are achieved.

For some blends, before the processing additive (e.g. oil) is added to the fibres in the blending line it is still desirable, after further fibre tuft separation, to get rid of the extra dirt and fibre dust released. This can be done when fibre tufts are conveyed between bins or when the tufts are fed to an opening (separating) machine such as a fearnought.

In principle the usual means adopted is to arrange that fibre tufts are blown into a perforated chamber and the dust temporarily deposited in an outer sealed periphery chamber from where it continues its journey to the collecting dust bags. The cleaner fibre tufts are contained within the bottomless perforated chamber and fall down on to, for example, a conveying lattice or into a hopper of a fearnought.

It is essential to empty the dust bags regularly and also to see that at all times they allow enough air to be released at a sufficiently slow speed. Otherwise any dust or dirt released from one component is spread over all components.

In yet another efficient cleaning machine, some five spiked rollers of the same diameter and rotating in the same direction are arranged with their parallel axes on an incline. Grids are arranged under each roller and suction hoods can be placed over each roller. The fibre tufts are dropped out at the uppermost roller.

## IV. Blending Lines

There are a large number of types of blending lines. An outline will be given of one which illustrates the principles of operation and the textile engineering problems involved (Fig. 1). In many firms the same principles will be applied but the operation will be carried out manually or mechanically. Further, the bins are often of a different shape (e.g. circular in horizontal section).

A number of bales (mostly pressure packed) are placed on the floor between and alongside up to six horizontal parallel and comparatively long lattices. Usually fibres from these bales are fed manually onto these lattices.

Each of the lattices moves slowly (intermittently or continuously) towards a nearly vertical spiked lattice. This lattice then drops fibre tufts into a weigh pan or a chute where a monitor and control mechanism for the flow of fibres can be introduced. Examples of monitors are weighing hoppers, pressure gauges on lattices or pans receiving the fibres from the chute and a gamma ray device beamed across the width of the fibre mass.

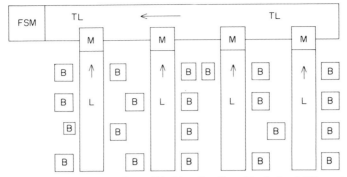

FIG. 1. Plan view of initial part of possible blending line. B—bales (different components); L—lattice; M—monitor and control feeder; TL—transverse lattice; FSM—fibre separating machine. Arrows show direction of motion.

Each of the lattices can drop fibres on a horizontal lattice moving at right angles to the six lattices. Sometimes the fibres can be dropped directly without any weighing, etc., on this other lattice.

This traverse lattice can have the fibres from the other six lattices placed on it in varying forms. The fibres can be in six strips alongside one another, they can be in six layers on top of one another or they can be in masses so that the feed from any one of the six lattices is every sixth mass on the traverse lattice. Alternatively the fibres are fed into pneumatic chutes. The fibres are transmitted to a bin or fibre tuft separating machine such as a fearnought (picker).

This lattice moves and carries the fibres towards a fibre separating machine such as a cotton picker, a fearnought or even a blend cleaning machine as noted above. This separating operation can also be used to clean the fibres rather than in an earlier separate operation.

The fearnought is a popular machine for separating (opening) fibre tufts (Fig. 2). A typical working width is 1200 mm and for many blends containing a good proportion of wool a maximum production rate of 750 kg/hr is suggested for efficient separation and cleaning of the larger fibre tufts.

The fearnought gives the classical carding or fibre working action in which the tips of the teeth (spikes) of some rollers (workers, strippers) are set usually between 5 and 25 mm away from the tips of the teeth of the main cylinder (swift). The partial opening of the tufts is achieved by the tips of the teeth of the cylinder and workers pointing towards each other and by the great difference in the surface speeds of the fast moving cylinder and the slow moving workers. There are often four workers set to a cylinder. Some fibre cleaning is carried out at the fearnought. Grids of horizontal round bars underneath and across the cylinder allow the released impurities to drop out.

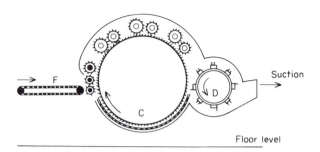

FIG. 2. Typical fearnought machine with cover (three pairs of workers and strippers set to cylinder). F—feed sheet; C—cylinder; D—doffer. Arrows show direction of rotation of cylinder and doffer and of feed sheet movement. The doffer is a "fan" doffer consisting of a few rows of spikes and also ones of leather strips.

The doffer is often made up of six rows of spikes (pointing radially) and leather flaps to get the fibre tufts off the fearnought cylinder. From here they can be sucked away. An alternative doffer covered with straight and tapered pins seems to give better fibre tuft separation.

After the opening machine the fibres are then conveyed pneumatically to a rectangular bin. Such bins have varying dimensions with a capacity of up to ten tonnes of fibre tufts. The overall density of the bulk of fibres depends on the fibre density, the degree of fibre separation and the height of the bin. Many when loaded have an average mass density between 10 and 30 kg/m$^3$. Such bins are usually up to 25 m long, 6 m high and 4 m wide. The bin is filled through the open top by a traversing condensing cage (or rotary spreader) which reciprocates centrally some 20 m and 8 m high above and along the length of the bin. Thus the fibre tufts from the opening machine are layered in this bin. The cage is fed by tubes of up to 0.5 m in diameter inside one another and which extend and contract as the cage traverses along the length of the bin. Similar tubes are used for conveying the fibre tufts away from the bin.

One of the 6 by 4 m walls of the bin consists essentially of an almost vertical endless pinned lattice. When emptying the bin, this lattice and its heavy superstructure moves very slowly (e.g. 50 mm/min) on rails set into the floor, into the fibre mass. The lattice "rotates" so that the pins in contact with the fibre mass moves upwards at a comparatively much faster speed. In some bins, the slow movement of the vertical lattice has a safety control mechanism preventing the movement causing excessive pressure of the fibre mass on the vertical lattice. If the blend size is two-thirds or less of the bin capacity then it is advisable to fill the bin reduced in floor area by moving in one of the smaller walls (one wall is the lattice). The aim is to have the height of the pile of fibres as high as practicable.

It is usual to have the rails in the floor, level to within 3 mm along their length. If the floor is unsuitable, e.g. not sturdy enough, than an alternative is to have a very strong, well supported and movable horizontal lattice as a floor. When emptying, the shorter vertical wall (opposite the pinned lattice) moves very slowly towards the pinned lattice. That is, the pins on the lattice move upwards comparatively fast but the lattice and its superstructure does not move into the fibre mass. Instead, the fibre mass moves slowly towards the lattice.

In either type of bin the fibres are taken off the pins at the upper edge of the lattice and then sucked away. Quite high rates of fibre tuft transfer (e.g. 1.5 tonnes/hr) are achieved if the tufts are being transferred to another bin.

To ensure adequate fibre mixing most woollen spinners prefer at least two operations of placing fibre tufts in a bin in horizontal layers and taking them out by vertical sections. During the transfer between the two bins some spinners apply a spinning additive (oil, emulsion, etc.) to the fibres. However this can be troublesome as the tubes and second bin, lattices, etc., tend to have greasy deposits. Thus many spinners prefer applying the additive when the fibre tufts are conveyed to bins behind the cards or to baling machines.

This later stage of application is adequate since the spread of additive can be made more uniform in the card than can the fibres themselves. The teeth, all the surfaces and the fettlings at the bottom of the card teeth of a carding machine have a layer of the additive from previous blends. The fettlings on the cylinder often have three times the percentage of additive than that on the blend, so the additive has plenty of opportunity to be spread.

## V. Bale Packing

The packing in modern baling machines is carried out in a variety of ways. In one, the fibre tufts in a high column are compressed under pressure. In another, the fibre tufts are blown into a cage, fall onto a pair of horizontal lattices moving towards a nip and the tufts are compressed into the bale situated below the nip. In a third, the tufts are blown and compressed into a bale by a pulsating system of air conveyance. In yet another form the inside of the bale is lined with a thin plastic bag. The fibre tufts are condensed in a tall sealed column above the bale and air is extracted through a hole at the base of the plastic sheet and bale.

A bale of fibre tufts of average density of 70 to 100 $kg/m^3$ is usually formed. That is, a typical bale of dimensions 1 m by 1 m by 1½ m would contain 100 to 150 kg. It is neater at the card hopper for the bales not to be too tightly packed especially in the top third of a bale. Also the fibre tufts should remain separated enough for the carding machine.

Bales are usually conveyed manually to near the card hoppers and the hoppers are fed manually at intervals from these bales. It is preferable to feed in turn from at least two to four bales to ensure adequate fibre blending up to the last point before the carding machine.

## VI. Card Bins

If the blend weighs are large enough (e.g. carpet blends) then card bins near the card hoppers are used for the temporary storage of fibre tufts. Card bins are also used at some firms with blends of smaller weights. Sometimes, a card bin is considered as the second bin in the blending line. Thus the utmost care is necessary to withdraw the fibre tufts in as vertical slices as possible to complete the fibre mixing in the blending line. This care should also be exercised even if there are two bin operations in the blending line.

It is advisable to divide the floor area into four to six regions by two or three straight lines painted on the floor and continued up the bin sides. This helps the staff to empty the bins with the maximum possible fibre mixing as they should aim to clear one floor region before another.

Sometimes it is convenient and useful to have the bins constructed on a mezzanine floor with the floor level a little above the top of the card hoppers. The card hoppers can then be filled from above through a trap-door, often with a chute leading down into the hopper bin. For safety, the trap-door opening should be guarded. Usually the hopper filling from the bin is done manually through the trap-door.

If the card bin floors are at the same level as the card room floor then it can be a somewhat laborious task to fill the card hoppers manually. This is particularly so if the backs of the hoppers are rather high. One alternative is to have pneumatic suction openings at convenient positions in the card bins to convey the fibre tufts from them to above the card hopper and to drop them into a hopper by means of a condensing cage or a neat delivery wheel.

A delivery wheel is a rotating perforated disc bigger in diameter than the cylindrical fibre-conveying trunking, whose axis is parallel to the disc's axis of rotation. The perforations allow the air (and dust) to pass forward while the fibre tufts are arrested and drop down into the card hopper.

The bins are usually constructed in pairs so that at a change of blend on a card there is no delay when the one bin has been emptied before refilling with another blend. With two bins, precautions are necessary. Firstly, staff must empty one bin before starting on the other even when the same blend is involved. This avoids leaving parts of a blend to settle for a comparatively long time at the back of one or both bins. Secondly with two bins to each card width and the need to have a good bin capacity there is a tendency to have each bin with a rectangular shaped floor and with the longer side of the rectangle twice or even six times the length of the shorter side. Unless some extra precautions are taken in filling such a bin, fibre separation of blend components can take place. For example, larger fibre clumps of dyed wool fibres could become separated from "candy-floss" masses of well separated man-made fibres. This separation tends to take place more if the bin is filled by a vent (opening) immediately above the door or bin opening. One precaution is to have a hardboard sheet about one metre square at an angle of about 45° to the vertical and hung from the ceiling about a quarter to a third of the length from the back wall. The stream of fibre tufts strikes the centre of this sheet before falling down into the bin.

## VII. Oiling

Processing oils, emulsions and additives are added to the fibres for various reasons. The amount added can affect the blend yield, that is it affects the weight of fibres dropped out or blown out in carding as well as in spinning. In one trial on a card with four carding units and no grids under the cylinders a 10% addition of oil reduced the fibre loss by a third compared to the loss with a 5% addition. There has been a tendency in recent years for the amount of oil added to be reduced but not the amount of water.

In one example, an additive with a silica suspension increased the yarn strength of an all-wool blend by 15% by increasing the inter-fibre friction. In another example, the additive has excellent lubricating properties to enable the "oiled" hosiery yarn to perform better on the knitting machine.

Additives of many types are added to woollen blends and in percentages usually between 0.5 and 10. This excludes any water added as a separate liquid or contained in the emulsion or additive mixture. Usually the total amount of liquid does not exceed 15% and in many instances does not exceed 12%.

There is no evidence that the addition of oil affects fibre breakage in blending or carding as measured by the estimated fibre lengths after carding. The degree of fibre entanglement does affect these lengths. That is, any partial felting of wools during scouring or dyeing does affect the resultant length after carding.

There is a tendency with all oiling devices to concentrate on the constancy of the rate of application of oil (liquid) without realizing how variable is the rate of flow of fibres. With manual emptying of bins, rates of flow of fibres can vary four to one, when measured between periods of 30 minutes—not just minute to minute—and yet the flow of additive remains unchanged. The important factor is that the *total* liquid added to a given weight of fibres gives the desired percentage. That is, the desired total weight (or volume) of liquid is monitored closely by pre-arranging that this amount and only this amount is available for the blend weight. The modern bins with one "moving wall" to empty the bin does provide an excellent facility to control the rate of flow of fibres and thereby give a more reliable percentage addition of additive when the flow of liquid is also monitored.

Let the percentage $p$ (by weight) of additive be added to circular fibres of diameter $d$ which have a density $r$ times that of the additive. Then if the liquid additive is layered uniformly to a thickness $t$ over all the fibre surface then $400t = prd$. Taking a fibre diameter of 25 microns, $p$ as 6 and $r = 1.5$, then $t$ is just over half a micron. This emphasizes the difficulty of achieving a uniform additive layer on the fibres in the blending operation. One has to await the carding operation to spread the additive thoroughly and much more uniformly over all the fibres.

The oiling application is often in the form of a spray of droplets on to the fibre masses conveyed pneumatically and quickly past the spray. This application takes place where the fibres are constrained to pass through a constriction or through the spray. One example is a loop in the conveying tube such as a circular loop some three-quarters of a complete circle. Another example is a narrowing of the tube. In yet another example, the fibres are swirled around through the spray. One essential requirement is that the fibre tufts keep wiping the sides of the tubes, etc. This prevents an accumulation of additive since the wiping of the sides by fibre tufts maintains an essential self-cleaning action. In another form of application, the spray is applied to layers of fibres as they are deposited in a bin.


## VIII. Functions of Woollen Carding

The two main functions of woollen carding are to separate fibre tufts into individual fibres and also to mix these individual fibres as much as possible. In the most popular basic design of woollen card the two main functions are performed simultaneously. The ultimate aim is to have a final web which has a uniform weight per unit area (across and along) throughout the carding of the blend and for the fibre components in this web to be always in the desired proportions.

A third function is the elimination as much as possible of impurities. With some blends during carding there is a differential loss of impurities from the various components. Thus if the specification of the yarn or other product is strict on proportions, then in making up a blend an allowance has to be made for this differential loss. Thus a blend of 82% of fibre A (e.g. wool) and 18% of fibre B (e.g. nylon) may be required to give a web of 80% of fibre A and 20% of fibre B. Note however that there can also be a differential loss of fibres even if the fibres are of the same type, e.g. two wool components with different fibre stiffness.

## IX. Woollen Card

A typical card consists of four individual units each with a cylinder (swift) of diameter 1 m to 2 m (a popular diameter is 1.3 m) and with a doffer set to it of diameter 0.5 m to 1.5 m (again, a popular diameter is 1 m) (Fig. 3). Also set to the cylinder are four pairs of workers (rollers) and stripper rollers (strippers, clearers). About half-way along the card is situated an intermediate feed (see later) (Fig. 4).

The first two carding units are called the scribbler (breaker) and the last two, carder (finisher). Quite often at the beginning of the scribbler there may be an extra smaller cylinder with one or two workers set to it to separate the larger tufts. Less frequently a similar small cylinder is situated at the start of the carder.

The working widths of different cards, i.e. widths of cylinders and other rollers covered with card clothing, are between 1 m and 3.5 m—popular widths are 1, 1.3 and 2.5 m, especially when a condenser is positioned at the end of the card.

The cylinders rotate at speeds of 50–200 rpm but for a cylinder of diameter 1.3 m, the usual angular speed is between 75 and 125 rpm. At an angular speed of 100 rpm such a cylinder has a surface speed of 400 m/min (Fig. 5).

The cylinder and the rollers set to it are covered in card clothing. This card

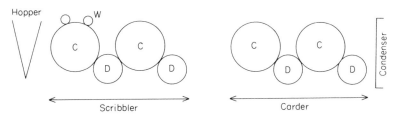

FIG. 3. Sketch outline of a typical woollen card. C—cylinder; D—doffer; W—worker. Cylinder diameter can be taken as 1.3 m. Just two of the workers set to the first cylinder are shown. In modern cards the scribbler is made up of one carding unit or two or three such units. The carder is made up of one carding unit or two carding units.

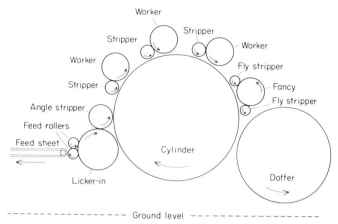

FIG. 4. Typical woollen carding unit with feed sheet. The diameter of the cylinder is 1270 mm. The fancy roller wire is set into the wire of the cylinder. All other rollers except the licker-in and feed rollers are set to the cylinder with clearances typically between a third and two-thirds of a millimetre. A further carding unit following immediately after the unit shown would be fed via an angle stripper set to the doffer shown. The arrows show the usual directions of rotation and the web of fibres travels from left to right on the underside of the doffer.

clothing can be flexible with steel wire (pins, teeth) set into a textile cloth and synthetic rubber backing or rigid (metallic, garnett) which is a metallic wire with teeth formed on it. Usually the number of teeth/m$^2$ is between $10^5$ and $7 \times 10^5$. A few cards have a few rollers in the feed region covered with tapered steel pins.

It is fascinating to see a machine of over 80 rollers extending nearly 25 m

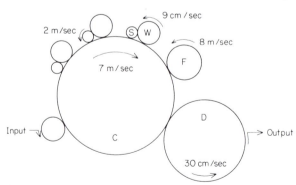

FIG. 5. Typical surface speeds in a carding unit with three pairs of workers (rollers) and strippers (clearers). D—doffer; S—stripper; C—cylinder; F—fancy; W—worker.

and width up to 2.5 m producing almost a perfect, very thin and comparatively weak web of fibres at one end and doing this automatically from a mass of fibre tufts fed in at the other end. The whole operation is carried out at production rates which are usually between 5 and 150 kg/hr.

Many of the other rollers are set to the cylinders to give clearances between them of 0.2 mm to 3 mm. The clearances are usually between 0.25 and 1 mm. These must be maintained reliably otherwise the card delivers an inferior product. Further the clearances must not vanish and the teeth touch since this causes, at best, a loss of sharpness of the teeth and, at worst, irreplaceable damage to the teeth putting the card out of action until the card clothing is renewed.

In addition, each clearance should be constant across the width of the card and be maintained often over a period of weeks. Sometimes the card will be actually running for a total of 140 hours each week.

The bare cylinders are ground to a tolerance of 0.05 mm around a circumferential circle. This tolerance is also the target when they are covered with card clothing. Similar targets apply to the doffers.

## X. Hoppers

There is some fibre mixing in the bin(s) of the hopper itself especially if the bin is large or there are two bins. Cards are fed from a hopper containing fibre tufts and the objective is uniform feeding of the mass of fibres.

The most popular type of control for the flow of fibres from the hopper to the card is to weigh the tufts. The reliability of this weighing at high production rates can pose problems. For example, a mechanical hopper should deliver a weight of 0.75 kg each cycle of 18 seconds to achieve a production rate of 150 kg/hr. In addition, the weighing operation weighs oil, impurities and water as well as fibres. If the water content is high and variable over long periods then the slubbing count at the condenser will change, since a good deal of the added moisture leaves the fibre mass during the carding operation.

Another means of control is to have a chute feed from the hopper and to measure the mass of the bed of fibres on the scribbler feed sheet. This is done by a gamma-ray beamed across the card width. If the mass is insufficient then the speed of the delivery from the chute, the scribbler feed sheet and the feed rollers is increased. If the mass is too much the speed is decreased.

In other control methods the fibres from the chute fall onto an inclined plane or a lattice and the weights of these and the fibres are measured by strain gauges. The signals can be used for control purposes, e.g. any excess weight in any weigh can be compensated for in the next or following weighs.

After the blending operation the objective should be that any 10 kg of the blend contains the fibre components in the desired proportions. The card (of four carding units) with its reservoir of fibres can mix the fibres within 5 kg but not so easily between such lots.

Almost all hoppers in the woollen system mix the fibre tufts so that each 5 kg of tufts presented to the first feed rollers of the card has the blend components in the correct proportion. This is the desired degree of uniformity to enable a card (of four carding units) to cope with the remaining non-uniformity. Naturally, if the number of carding units is less than four a tighter specification may well be required depending on the final web specification. It is assumed that each carding unit has a cylinder, doffer and four workers. If it is a different kind of carding unit (good at opening fibre tufts but not so good at fibre mixing) then the specification may have to be tighter.

As is emphasized later the reliability of the hopper in delivering a constant total weight over a period of time is a very important property (e.g. 5 kg/5 min).

## XI. Carding Action

The fibre tufts are usually well separated into individual fibres when they have reached halfway in their passage through a typical card. The fibre separation is achieved by the relative motion (one slow and the other fast) of two card-clothed surfaces, set closely to each other and with the points (tips) of the teeth of the clothing pointing towards each other. This action happens an increasing number of times to each portion of a tuft which is formed as the portions proceed through the card. It is often called working action to distinguish it from a stripping action when a faster roller with tips leading (in direction of motion) takes all the fibres off the backs of the teeth of a slower roller.

## XII. Reservoir of Fibres

Fibres are found on much of the card clothed surfaces of the card. That is, there is a reservoir of fibres on the clothing. These fibres are ones which sooner or later pass along the machine to the output web. With flexible clothing we are not considering the fibres and impurities (fettlings) which settle to the bottom of the teeth. Periodically, with a loss in production, these have to be cleaned out (fettled) by a wire brush, comb or a very powerful pneumatic suction head.

If a constant feed rate of fibres (e.g. a lap feed) to a carding unit of cylinder, doffer and workers is suddenly changed (i.e. a step change in the feed rate), then the output rate of fibres will not change at all for a time $M$. When the output rate begins to change it will only do so gradually until it ultimately reaches the input rate. The comparatively small loss of fibres as fettlings or droppings can be ignored with no loss of generality.

The minimum time $M$ is the sum of the times for the fibre to travel from the input region over the upper part of the swift surface and the underside of the comparatively slow moving doffer. This latter is the longer time and, for example, if the doffer revolves once every 15 seconds, then the usual time a fibre spends on the doffer before the take-off line is 10 seconds.

## XIII. Electrical Analogue of Fibre Flow

The gradual change in the output rate is, under normal operating conditions, approximately exponential in mathematical terms. That is, the output rate response is similar to the voltage response $V$ from a resistor–capacitor circuit when a sudden voltage change $U$ is applied across the input terminals (Fig. 6). That is:

$$V = U(1 - \exp(-t/T))$$

where $t$ is the time from the start of the change and $T$ is the time constant of the circuit. $T$ is the product of the resistor value and capacitor value measured in suitable units (e.g. megohms and microfarads).

This analogy has proved most useful in enabling experimental verification of fundamental work on carding units and their responses to be demonstrated without recourse to expensive and time-consuming practical experiments with cards and fibres.

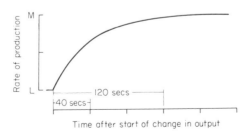

FIG. 6. "Smoothed" response curve from one carding unit subjected to a sudden and sustained change (L to M) in the input. The average time a fibre takes to travel from the input to the doffer is 40 sec. Approximately two-thirds of the change LM is completed after 40 sec. The whole change is practically completed at the output after three times this period of 40 sec.

For a carding unit, the change in output rate $S$ is related to a sudden and sustained change in input $R$ by:

$$S = R(1 - \exp(-t/T))$$

where $t$ and $T$ are measured sometimes in number of swift revolutions and $t$ is measured from the moment the change shows up in the output. $T$ has been defined as the delay factor of a carding unit.[3]

The average time a fibre spends in a carding unit is $M + T$. The time $T$ changes with the number of workers set to the cylinder. For example $T$ might have a value equivalent to the time for 10 swift revolutions when no workers are present. It might have the value 40 with four workers set to the cylinder.

The origin of the reservoir of fibres is that the doffer and each worker take only a small fraction (collecting fraction) of the workable fibres presented to them by the cylinder. A possible value for the collecting fraction of a doffer is 0.1 and for a worker 0.06.

The formula for the delay factor $T$ for a carding unit with four workers is:

$$T = \frac{1}{f}\left(1 + \frac{4np}{1 - p}\right)$$

if each of the four workers are considered to have the same collecting fraction $p$. There is no great difficulty in dealing with unequal values of $p$ and $n$. The collecting fraction of the doffer is $f$ and the formula emphasizes the profound effect of its value on the value of $T$. This fact is amply verified by practical experiments. If $n$, the time (expressed in terms of cylinder revolutions) a fibre, captured by the worker, spends off the cylinder before it returns to it via the stripper is taken as 12 then $T$ equals 40.

## XIV. Transfer of Fibres

The small values for the fraction of fibres collected by a doffer or worker from a cylinder is explained when considering the mechanism of transfer of fibres when the teeth of the two lots of card clothing are pointed towards each other.

It is recalled that the surface speed of the cylinder is much faster than that of the worker or doffer. Thus if two teeth, one on the cylinder and one on the doffer are at the setting line (line of smallest clearance between the two rollers) at the same instant, then a circumferential movement of only 2 mm by the tooth on the surface of the slower roller (doffer) usually means a movement of say 60 mm of the other tooth on the surface of the cylinder. Thus once a fibre is held in the clothing of the doffer, its other end on the cylinder is moved smartly through the setting line. The fibres on the doffer are layered like tiles on a roof. Once a region of the clothing on the doffer is covered with fibres, that region cannot capture and hold more fibres.

Thus the fraction of fibres transferred from cylinder to doffer varies directly as the ratio of the surface speed of the doffer to that of the cylinder. This last ratio is usually small since for example a cylinder of diameter 1390 mm (over the card wire) rotating at 100 rpm and a doffer of 930 mm rotating at 5 rpm means a ratio of 30 to 1 in surface speeds.

On the above hypothesis of fibre transfer, if the doffer is changed to one of smaller diameter but the speed of rotation is increased so that the two doffers had the same surface speeds then there should be no difference in the rate of transfer of fibres. This has been confirmed by experiment. Sometimes a smaller doffer has to be used in industry because of the lack of floor space.

If the direction of angular rotation of a doffer is changed then the transfer of fibres should be unaffected. This also has been confirmed by experiment and the result is not so surprising when one considers how slow is the surface speed of the doffer compared to that of the cylinder. This change of direction of rotation is used sometimes at the doffer of the first carding unit if there is difficulty in conveying and holding the web on the underside of the doffer.

If the two surface speeds are equal (i.e. very slow cylinder and very fast doffer) then no fibres should be transferred and the cylinder would become choked with fibres. This also has been confirmed experimentally.

## XV. Variations in Fibre Flow

The time constant $T$ of a given carding unit has been found to depend mainly on two features. These are the number of workers and the speed of the doffer. Other features can affect the time constant but experimental evidence suggests that they do so to a lesser degree.

The other features include the type of fibre processed, the type of card clothing, the speed of the cylinder, the lead of the fancy roller (brushing roller) over the cylinder surface speed and the speed of the stripper rollers.

Thus for a given design and construction of a card and the processing procedures adopted (e.g. speeds of doffers) it is not easy to change the overall value of the time constant.

For a given time constant $T$, the output response of a carding unit to a given input variation can be calculated. Bearing in mind the resistor–capacitor electric circuit analogue for a single carding unit, it is then possible to give the output response after two or more carding units (two or more resistor–capacitor stages) (Figs 7 and 8).

An oscillatory sinusoidal input variation of amplitude $V$ volts appears

F

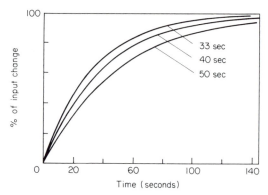

FIG. 7. Smoothed output responses from a single carding unit with three different average times of fibre delay. The input production change was a sudden and sustained one of magnitude equal to the 100% level. Note the small differences between the responses for the three average time delays (analogous to time constants of a resistor–capacitor electrical circuit). Thus it is only necessary to have an approximate value for the average time delay. If the higher frequency variations present are included then the differences are difficult to discern (see Fig. 8).

after one stage as an output variation with a lesser amplitude of $V/(1 + W^2T^2)^{1/2}$ where period of oscillation is $2\pi/W$ seconds. That is, in the analogous carding problem if the time constant $T$ is 40 seconds and the period of oscillation $(2\pi/W)$ is 40 seconds, then the output amplitude is some 15% of the input amplitude.

That is, the swing in the input from maximum to minimum is reduced to 15% after one carding unit and to 2¼% after two carding units. The mixing ability of each carding unit smooths out many irregularities in the feed by the time they reach the output of a card.

Note from the above formula that the carding unit and a carding machine are better at smoothing out the higher frequency irregularities. This can be verified by experiments such as the following.

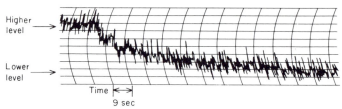

FIG. 8. Trace of weight change of output web from one carding unit when the unit is subject to a sudden and sustained change in input level (higher value to a lower one). Note the presence of higher frequency weight changes and the difficulty of pinpointing exactly when the change starts in the output.

A card of four carding units was fed by a weighing hopper with (a) alternate weighs of different colours, (b) alternate two weighs of different colours and finally (c) alternate four weighs of different colours. The final web was examined closely for uniformity of shade and non-uniformity was only definitely established for hopper feed conditions (c), that is, the lowest frequency of change.

A further very important practical carding implication of this smoothing is in deciding on the real criteria to use for card hopper feeds—weighing or non-weighing types (Fig. 9). For a weighing hopper it is more important to have reliability between the total weights of groups of ten consecutive weighs than between individual weights of weighs. That is, the first objective is the constancy of the total weight of ten consecutive weighs. A more stringent objective is for the constancy of the total weight of five consecutive weighs.

For a non-weighing hopper the total weight of fibre feed to a typical card over each period of five minutes or less should be constant. That is, for all hoppers good control of medium-term and long-term irregularities is very important; the card can smooth out the short-term irregularities.

Put in another way, the final web at the end of a typical card of four carding units contains fibres which have been fed to the card over a period of at most six minutes.

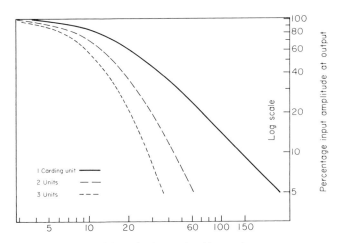

Frequency of input (cycles per hour) log scale

FIG. 9. Output amplitude of variation from a carding part with one carding unit and two or three units. Input to the part is a sinusoidal one of different frequencies. Note smoothing effect of two units for frequencies higher than 20 cycles/hr (period of three minutes).

## XVI. Tracing Source of Some Yarn Count Irregularities

The woollen spinning system does not affect the pattern of count (mass per unit length) irregularities when converting slubbing into yarn. A record of the weight irregularities of the yarn can give an important clue to the origin of the irregularities.

It will be recalled that a sinusoidal input irregularity of a known frequency to a carding input appears in the output with a diminished amplitude but with the *same* frequency. Thus the frequency of the irregularity in the carded web gives the frequency of the original fault in a previous carding unit or at the hopper. Care must be taken in interpreting results if the cause of the irregularity arises in the parts of the card before the intermediate feed. Otherwise the irregularity may be virtually eliminated at the feed if the period of the irregularity is short compared with the repeating mechanics of the feed.

Thus a powerful method of tracing a periodic yarn count irregularity (variation) is to calculate from an instrument chart the periodic length in slubbing by using the spinning draft. Then with the knowledge of the speed at which slubbings are condensed, this periodic length can be converted into a frequency. The final objective is to isolate a mechanism or roller (often in the last two carding units) with the same frequency of repeated action or rotation.

## XVII. Length of a Card

A typical woollen card is a comparatively long machine and naturally there is a temptation to shorten it and also to use what is meant to be alternative machinery but with a smaller floor area.

The specifications for card hopper feeders given above are meant to apply to typical cards which have four carding units with a doffer and four workers set to each cylinder. If the card has fewer units and the output specifications (e.g. web uniformity in weight and colour) remain the same, then the input specifications would have to be tighter. This is because the card has less fibre mixing potential.

However the problem may be more serious than this since there may be insufficient capacity for fibre tuft separating. This highlights the problem with some non-woven plants. The card used may be too short to carry out adequate fibre mixing or even fibre tuft separating.

The potential for fibre mixing and separating are inter-related as will now be explained. The fibre mixing ability of a carding unit is related to the value of $T$, i.e. average time in swift revolutions a fibre spends in the unit (omitting time spent on the doffer):

$$T = \frac{1}{f}\left(1 + \frac{4np}{1-p}\right)$$

for a unit with four workers each with a collecting fraction $p$.

The collecting fraction of the doffer is $f$ and $n$ is the time (expressed by the number of cylinder revolutions in that time) that a fibre, captured by a worker, spends off the surface of the cylinder before it is returned to that cylinder via the stripper.

The definition for $C$, the degree of fibre tuft separation, is taken as the average number of times a fibre passes through a setting line, i.e. between cylinder and a worker or between cylinder and a doffer. The formula for $C$ is:

$$C = \frac{1}{f}\left(1 + \frac{4}{1-p}\right)$$

As $T$ increases so does $C$, that is, the greater the opportunity for fibre mixing, the greater the opportunity of fibre tuft separation. In fact, the two formulae are more strongly related than at first appears. If $n$ decreases (e.g. faster speed of the workers) then $p$ increases and vice-versa. That is, $np$ tends to be approximately constant.

The above briefly underlines the difficulty of trying to design a practical, alternative to a series of conventional carding units. There are units of woollen cards which are not conventional cylinder and doffer units. These other units are usually good for fibre tuft separation but have a comparatively small capacity for fibre mixing. An example is a small cylinder (up to 760 mm in diameter) which is often covered with rigid wire and with one or two workers set to it but with no doffer. In place of a doffer, there is another roller, sometimes called a transfer roller. It rotates at a much faster speed than a doffer and with a surface speed faster even than that of the cylinder. The direction of the teeth of the card clothing is different from that of a doffer, and the teeth of this roller, running point (tip) first towards the backs of the teeth of the cylinder, take all the fibre tufts off the cylinder. That is, the transfer roller strips all the fibre tufts off the cylinder and there is no reservoir of fibres on the cylinder.

Such a unit is most useful at the beginning of the card if there is a necessity to separate tufts from the larger fibre masses. However if such a unit is used at the beginning of a carder, it cannot be the equal of a conventional carding unit as far as fibre mixing is concerned.

Another type of carding unit consisted of five cylinders (each 410 mm diameter) with axes in the same horizontal plane, the last two acting as the cylinder and doffer of a conventional unit. There were six rollers set in pairs to each of two adjacent cylinders. Three of the rollers were above the setting regions between two adjacent cylinders and three below these regions. This unit provided good fibre tuft separation but not too much fibre mixing. In one calculation it was estimated that the average time delay of such a unit was less than 15 seconds.

## XVIII. Intermediate Feed

A full length card of four carding units or more almost invariably has an intermediate feed between the second and third carding units. The most popular type of feed is the simple and reliable Scotch Feed. The adjective intermediate is used since the hopper feed is regarded as the primary feed.

With this feed, the web at the end of the first two carding units is gathered together into a comparatively flat sliver (rope). This is conveyed upwards, over and downwards to the feed sheet for the next two carding units. This feed provides the opportunity of a floor space so that staff can move from one side of the long card to the other.

At the end of the downward path the sliver enters the nip of a pair of rollers fitted into a carriage. This traverses backwards and forwards to lay the sliver across the carder feed sheet (Fig. 10). This feed sheet (table, board) feeds the feed rollers of the carder part.

It will be appreciated that the comparatively slow forward speed of the feed sheet, with the layered sliver moving sideways towards the feed rollers, must be matched to the speed of the sliver delivered by the part of the intermediate feed forming the sliver.

Note that an apparently ideal layering of the sliver on the feed sheet is not necessarily the way the sliver is fed by the feed rollers. Often these rollers have a higher lead (up to 25% has been measured) over the feed sheet than is claimed. Thus towards the sides of the card, the feed rate is more non-uniform since lengths of sliver can be snatched by the licker-in (taker-in) roller.

The main function of a Scotch Feed is to eliminate any persistent differences in weight per unit area across the web coming off at the second doffer. The differences arise at the hopper, whether of the weighing or non-weighing types, which can feed non-uniformly across the width to the carding units. The *total* amount fed across the full width could, of course, be constant. For example, in one test, when splitting the web into five equal widths from the second doffer of a 1.6 metre-wide card, the percentage range of weights was 20.

There is no real mixing of fibres across the width of a carding unit—fibres rarely travel across a card except by air currents. Consequently, if there is

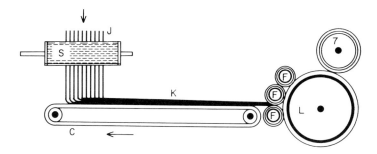

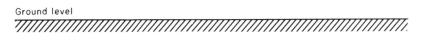

Ground level

FIG. 10. Vertical section (parallel to sides of the card) through feed sheet of carder (finisher) part. Horizontal arrow shows direction of motion of endless feed sheet C. F—feed rollers (bottom one rotating clockwise, other two rotating anti-clockwise); L—licker-in (taker-in) roller (diameter of 350 mm and rotating clockwise); S—one of the two rollers of Scotch Feed carriage (shuttle) traversing the full width and above the endless carder feed sheet C; J—Scotch Feed sliver moving downwards into nip of carriage rollers; K—Scotch Feed sliver laid on carder feed sheet C with the upper portion of sheet moving towards feed rollers F.

some persistent pattern in the variation of weights fed across the hopper width, then this variation will manifest itself across the fibre loading across the two carding units and the final web from them at the end of the scribbler.

A few cards have been built with a lap (layers of web) as an intermediate feed and with the scribbler and carder units at right angles to each other. They are not popular as the lay-out uses a great deal of floor space and also the drives are fairly complex.

There is little fibre mixing (along or with time) arising from the Scotch Feed. For example, if the sliver is too light for a time, then so will be the feed to the following two carding units.

With a minority of intermediate feeds it is possible to have fibre mixing at the intermediate feed by forming a much heavier and broader sliver. More correctly, this sliver is formed by having a number from 6 to 30 web layers. Naturally, it is an advantage to have this extra fibre mixing.

However it is much more difficult to layer these heavier slivers without causing variations in the weight per unit area of the final web. For example, it is not always easy to arrange that one edge of one part of a feed is in contact with the edge of another part of the feed on a carder feed sheet so that there is no comparatively big change in the rate of fibre feed at the feed rollers. Further, the change can last for a longer time than with similar but smaller changes with Scotch Feed.

Since after the intermediate feed there are only two carding units, there is less chance of smoothing out variations of feed. That is, the variations arising at the intermediate feed run a greater risk of appearing in the final web than equivalent variations arising at the hopper end. This risk applies to variations arising from any cause and it becomes greater the nearer the source is to the output web. This emphasizes the need for high standards of mechanical engineering and uniformity of drives at the final carding unit (Fig. 11).

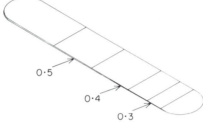

0·5

0·4

0·3

Fig. 11. Taper gauge (thickness of gauge in millimetres is shown at three positions). The use of such a gauge helps in getting more uniform clearances between two rollers at the two sides of the machine. It also helps if two people are attempting to set rollers to the same clearance. The clearance is estimated by the wiping off of chalk dust or a slight smear of grease put on the gauge.

## XIX. Crushing Rollers

The vast majority of woollen cards have a pair of heavy crushing rollers fitted full width at the end of the scribbler before the web is arranged as an intermediate feed. The function of these rollers is to crush seed and burrs, and also any hard fine cotton and worsted threads if there is a waste component in the blend.

These rollers, introduced as Peralta rollers in the late 1930s, have been the means of increasing the versatility of the woollen card in its capacity to process and clean blends containing impurities. Each roller (diameter usually between 250 and 400 mm) weighs up to 1.5 tonnes for a card 2.5 m wide. If necessary the crushing power can be increased by applying extra force (transmitted by hydraulic means) on each shaft. This extra force is distributed uniformly across the width of the rollers by having a very small angle in vertical planes between the axes of the two rollers.

Although some of the impurities drop out of the fibre blend in the region of the crushing rollers, more drop out at the feed region of the carder part.

Further drop out takes place if a straight edge is set close to the first stripper roller of the carder part and over its whole width. Some impurities are separated from the blend at this edge, drop down into a full-width channel which is continuously cleared to one side of the machine by blades mounted on an endless chain. Alternatively the base of the channel is an endless moving belt carrying the impurities to a container at one side of the machine.

## XX. Production Rate of a Card

The constant objective in carding is to produce a quality web, uniform in weight per unit area and clear without cloudiness. That is, a web which is an open structure of separated fibres. Note that a web which is of poor quality can lead not only to a poor spinning performance and poor yarn appearance but also to inefficient consolidation of web layers in non-woven fabrics.

Trials confirmed that with less fibre loading on the cylinder there was better quality of web at the doffer (Fig. 12).[3] However this fibre loading increases if the production rate of a carding unit increases. Thus to maintain a quality web there may well be a limitation of the production rate. However there is one aspect which helps and which deserves further attention.

It was noted above that the transfer of fibres from the cylinder to the doffer was very sensitive to the speed of the doffer (Fig. 13). The physical process of transfer of fibres depends on the area of a doffer's card clothing not covered with fibres which cross the setting line between cylinder and doffer. That is, the collecting fraction $f$ of the doffer is dependent on the

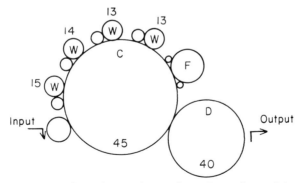

FIG. 12. Weights of fibres (in gm) on various rollers of a carding unit in one example. Production rate 18 kg/hr (5 gm/sec). C—cylinder; D—doffer; F—fancy; W—worker. The total weight on the cylinder and workers is 100 gm (i.e. equal to weight of 20 seconds of input). Weights on workers are reduced a little as their position moves from input to output region of cylinder. Speed of cylinder taken as 100 rpm and doffer as 5 rpm.

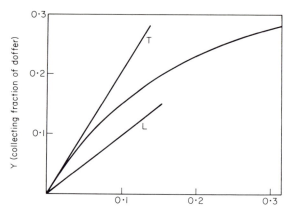

FIG. 13. Relationship between Y (collecting fraction of a doffer) and X (ratio of *surface* speed of the doffer relative to that of the swift). Note that for a typical cylinder speed of 100 rpm, the doffer speed must be as high as 15 rpm for X to reach the value 0.1. The curve is a typical relationship between the collecting fraction and X. Straight line T is the tangent to the curve at X = 0. Straight line L is Y = X. The figure illustrates that for the vast majority of practical values (X less than 0.07), Y is approximately proportional to X.

speed of the doffer. If *P* is the production rate in the time of one cylinder revolution then the fibre loading on the cylinder is $P/f$. Hence the greater the doffer speed relative to the swift, the greater is $f$ and the smaller is $P/f$.

Note particularly that with $f$ being a comparatively small fraction, the production rate of fibres moved by the cylinder is much greater than the production rate measured by the output from a carding unit.

For many carding units, the actual speeds of cylinder and swift are such that as the doffer speed increases, the collecting fraction $f$ increases proportionally. Consequently the fibre loading on the cylinder decreases inversely in proportion to doffer speed.

Thus an increase in the production rate of a card can often be achieved without a consequent loss of web quality by increasing the doffer speed. It has proved possible to increase some carding production rates by as much as 100% by concentrating attention on the doffer speed.

The overall aspect of increasing carding production rates by increasing doffer speeds can be considered as analogous to maintaining the reservoir of fibres at a constant level whilst changing the input and output rates. For example, the input and output rates of fibres could be 30 kg/hr while the total loading of fibres in the carding units, on the doffers and on the feed sheets could be 3 kg or less. If the production rate was doubled and the speeds of the doffers were doubled then the total loading would still be approximately the same.

Note that only a kilogramme of this total fibre loading could be on the cylinders and workers. It is this lower figure which is available for fibre mixing. The remaining weight of fibre is being transported on the feed sheets, intermediate feed and on the doffers.

Tackling the practical problems of increasing the carding production rates highlights other aspects. These are:

(1) The increased versatility of a card if the drives to the hopper, two feed sheets, intermediate feed, crushing rollers, condenser and all doffers come from a number of lay shafts not from the nearest cylinder shaft. That is, the drives to all rollers and mechanisms concerned with feeding fibres to a card, conveying fibres within a card (e.g. on the doffers), and receiving the final fibrous web can be changed fairly easily without necessarily changing the speeds of the cylinders and workers within each part.

(2) The need to have sufficient power to drive the various carding units, hopper, intermediate feed and condenser and to maintain synchronism between the units. A very popular drive is to have one motor for the first two carding units, hopper, first part of the intermediate feed and the crushing rollers and another motor for the last two carding units, second part of the intermediate feed and the condenser. This second motor, and also the driving means from it, particularly to the eccentric shaft of the condenser, should also be sufficiently powerful. The condenser eccentric shaft requires more power than is often imagined. For high production cards (even for card widths of 1500 mm) two vertical shafts are used for the double set of rubbing aprons (rubbers). Such shafts have been known to consume up to 15 kW of power even for widths of 1500 mm. Thus for many cards with tandem sets of rubbing aprons (two eccentric shafts), a third motor is introduced to drive these shafts. This must be synchronized with the other two motors.

(a) It has been found that following an overnight stop in the running of a card (but with the temperature remaining constant) the power required on re-starting can be a fifth greater. The power can drop to the previous day's value after 20 minutes running.

(b) In one trial, part of a card (1500 mm wide) consisting of a double-bin hopper, two carding units, crushing rollers and the first part of the intermediate feed consumed nearly 3½ kW when run empty. The consumption doubled when fibres were carded at 55 kg/hr and trebled at 110 kg/hr.

(3) The reliability of a hopper even at higher production rates must be maintained. The practical problem for a conventional weighing hopper has been noted earlier. Since the time to achieve the weigh may be some

two-thirds of the hopper cycle, then fibre tufts have to be delivered into the weigh-pan at an average rate of 63 gm/sec.

(4) The speed of formation and layering of the Scotch Feed sliver at higher production rate requires some thought as it is inadvisable to have too thick a sliver. Thus the doffer of the second carding unit and the comb stripping the web have to be run at comparatively high speeds. The practical problem is illustrated by a last doffer (diameter 910 mm) rotating at 10 rpm. This means the sliver (formed at 475 mm/sec) has to be layered backwards and forwards at least at this speed across a feed sheet of width 1500 to 2500 mm. That is, a traverse time from side to side for the carriage of three to six seconds. The stripping of the web by an oscillating comb from the doffer can also be a problem. If the traverse (stroke) length of the comb is 20 mm then to strip reliably the above web the comb oscillations must be 1800/min. For some cards the web speeds are 50 or 75 m/min. For this last speed the comb with a stroke of 30 mm would have to maintain a speed of 3000 strokes/min. If it is necessary to keep the web disturbance (vertical movement) to a minimum, then it is desirable to set to the doffer a fast rotating special stripping roller (not covered with card clothing). Such rollers have proved extremely successful in many applications.

(5) In carding, particularly if there are no grids fixed underneath the cylinders, there is a collection of fibres and impurities (droppings) underneath each carding unit and around both sets of feed rollers and the crushing rollers. The amount dropped underneath each carding unit tends to decrease as one moves from hopper to condenser. Thus in one example, twice the amount was found in total under the two units of the scribbler and the crushing rollers as was under the remaining two units forming the carder part. In carding at a higher production rate by the procedure suggested above there is the same fibre loading on the cylinders, doffers and feed sheets. The cylinders are the fastest rollers carrying fibres, and as these speeds remain unchanged, so to a first approximation one could surmise that the droppings in the same time would weigh the same for the different production rates. However, expressed as a fraction of the weight carded, the droppings would be reduced as the production rate increased. In one trial, on a card without undergrids, the production rate was doubled, the actual total weights of droppings were approximately constant so the percentage loss was halved.

## XXI. Fancy Roller

Set into the teeth of each cylinder there is usually a fancy (brush) roller. The teeth of the card clothing on this roller are between 20 and 30 mm long—

typically 25 mm. They are set at an angle to the radius of the fancy roller.

The teeth are set into the cylinder at a depth of from 0.75 to 3 mm with 1.5 mm as a typical value. In addition, the surface speed of the tips of the fancy wire run between 5 and 35% faster than the tips of the teeth of the cylinder. This relative speed (lead) enables the backs of the fancy teeth to brush through the backs of the teeth of the cylinder. This and the little flick of the longer pliable fancy teeth causes some of the fibres on the cylinder to stand proud on the surface of the teeth of the cylinder. Naturally, this flicking action is dependent on the tooth diameter and the time it has been in use.

As the fancy roller is between the last worker and the doffer, the action of the fancy teeth is an aid to maintaining a higher value of the collecting fraction of the doffer. On a few cards with six pairs of workers and strippers set to the cylinder in the first and second carding unit, the third pair has been removed and replaced by a second fancy roller. This improves the fibre separating action of the units by the fourth, fifth and sixth workers taking a greater fraction of fibres.

It is most desirable to monitor the lead of the fancy roller surface speed over that of the cylinder (expressed as a percentage of the surface speed of the cylinder) and also the depth of setting (penetration). If the cylinder is covered with metallic wire then the fancy lead should be kept low and the depth of setting kept shallow and not exceed half the height of the tooth.

The lead can be calculated immediately from the recorded angular speeds and the diameters, including wire, of the fancy roller and the cylinder ($d$ and $D$ respectively). The depth of setting ($s$) can be calculated by measuring the cylinder circumferential distance $L$ swept clean of chalk dust (placed on the tips of the cylinder teeth) after turning the fancy roller by hand. Twice the product of $s$ and the harmonic mean of $d$ and $D$ equals the square of $L$ approximately.

Some cards have covers which fit over the workers and strippers. These covers prevent fibres floating (fly) into the atmosphere and falling on other cards or collecting in the ceiling of the card room. The fancy roller itself usually has a separate cover and it is important that the two types of covers do not create ledges and crevices where fly (and even oil droplets) can collect. The inside of the covers should be made as smooth as possible.

## XXII. Variation Between Slubbings

In most spinning systems, except woollen or cotton condenser systems, a number (doublings) of slivers or tops (or rovings) are brought together and attenuated (drafted) to form one or two other slivers (or rovings). This process is carried out once, twice or more times and at the final stage, yarn is spun. However in the woollen system, the final web is split at the condenser into a number of web strips, which are rubbed into ends of slubbings (rovings) (Fig. 14).

These slubbings are converted directly into yarns without further mixing. Actually at the condenser the slubbings are wound individually into cheeses for conveyance to the spinning machine. These cheeses are formed on condenser bobbins, each containing 5 to 30 cheeses. The average weight per unit length (length per unit weight, count) of these slubbings are different. The range of the values expressed as a percentage of the overall average value can be as high as 40. A practical target for many blends is to keep this percentage below 12½, and for longer fibre blends below 17½. The

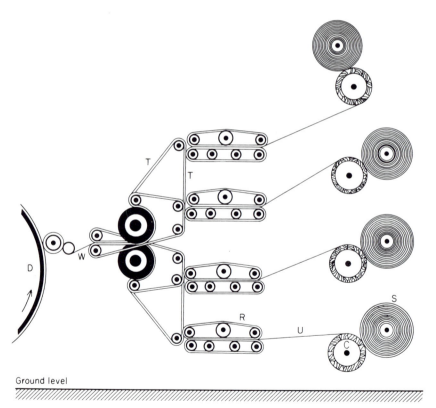

Ground level

Fig. 14. Vertical section of a series tape condenser with a single bank of rubbing aprons. D—tape doffer (rotating in anti-clockwise direction); W—final web from the card; T—tape; S—cheese of slubbing (roving); C—condenser drum; U—slubbing end; R—rubbing apron.

lowest reliable percentage range figure encountered has been 5¾ when considering all the slubbing ends on a condenser.

There are two main causes of these differences. The first is the differences in web weight and the second is the differences at the division of the web by the tapes of the condenser. Perhaps web separation should be used in place of web division as the separation must be carried out without fibre breakage.

The differences in web weights arise mainly at the intermediate feed—the formation of the sliver and its layering on the feed sheet of the carder part.

Examples of things which can give rise to differences are:

(1) A small difference in the speed of the feed shuttle (carriage) rollers traversing the width of carder feed sheet relative to the speed of formation of the sliver. This can arise insiduously by small but sustained changes in the speed levels of one or both of the motors driving the scribbler part (at the end of which the sliver is formed) and the carder part (the shuttle is driven by this part). The speed differences can cause a change in tension of the sliver and affect its degree of relaxation on the feed sheet.

(2) Slight speed variations as the shuttle traverses the width of the feed sheet. Practical trials with a variable speed gear in this drive showed how much this speed affected the slubbing end-to-end weight difference. Speed variation can be caused by too slack a chain drive and also by the size of the two chain sprockets usually driving the shuttle.

(3) Differences in the sliver tension caused by the extension of any spring attached to the overhead lattice conveying the sliver. The differences cause differential stretching of lengths of sliver which cause repeated weight differences on the carder feed sheet at different portions across the width of the carder feed sheet.

(4)(a) Layering of the sliver wide enough on the carder feed sheet. The margins by which the widths of clothing and feed sheet exceed the total width of all the condenser tapes are smaller than in similar situations in other industries. (b) The width of layering is affected by the speed of layering—the faster the speed the less the width.

(5) Incorrect layering of sliver on the feed sheet. For instance the feed sheet and feed rollers could move too fast in relation to the sliver speed. The result is that the feed is not completely covered with sliver particularly near the two sides. In fact, there are instances of greater count variation in the slubbings condensed on the two side regions of the condenser (about a third each of the width) as compared to the remaining third of the width around the centre of the condenser.

The fibres in a woollen carded web are not parallel. Indeed from a visual examination of a comparatively small square web sample of short fibres it can be difficult sometimes to decide on the direction of web movement (direction of carding). In one estimate the projections of the ends of fibres on a line parallel to the direction of carding was 60 units (100 if all fibres were parallel to the direction of carding). The estimate for the projections on a line at right angles to the previous direction was 40 units (zero if all fibres were wholly aligned in the direction of carding). Undoubtedly the longer fibres in the web have a bias to be in the direction of carding.

Thus for a condenser card web separation into web strips can be a tricky job with the comparatively narrow tapes, each with a width usually between 12 and 25 mm and with the majority between 15 and 20 mm. There are two main types of tape condensers, one called endless, in which a long continuous tape travels into every tape position. The other called series, in which there are a number of tapes corresponding to the number of tape positions (that is, number of ends of slubbing on the condenser) (Fig. 15).

Examples of the features which can cause weight differences between the web strips, and hence the corresponding slubbings, are given below. These are additional to the causes given above.

(1) Web more hairy on its lower face than its upper face—with endless tapes this can cause the upper banks of slubbing to be heavier. On one or two cards a rotating plain roller (up to 75 mm in diameter) has been set close to the underside of the tape doffer to assist with this problem.

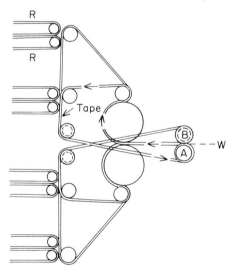

FIG. 15. Vertical section through the tape section of a series condenser. W—web; R—one pair of rubbing aprons; A, B—taker-in rollers. The complicated path of one of the many tapes (each endless) is traced using the arrows. The division of the full-width web into the numerous web strips is completed at the nip of the two larger rollers on the left of the taker-in rollers.

(2) Plane of web not horizontal when entering the first rollers of the condenser—with endless condenser this can cause the upper banks to be heavier. Some firms make the web entry horizontal by passing the web over a rotating plain roller (diameter usually between 100 and 150 mm).

(3) Markedly different tensions between banks of tapes—this is especially so with series tapes.

(4) Differential wear at the edges of some tapes or lengths of an endless tape. This leads to lack of control in web splitting operation.

(5) Waste or impurities nipped between tape(s) and roller(s) where the web first enters the condenser.

## XXIII. Tapes

Wholly synthetic tapes or sandwich tapes with a film of synthetic material sandwiched between leather facings are the most popular types now in use. With the wholly synthetic tapes there have been instances where it has been difficult for the web strips to cling to the tape in the region where the web strip is conveyed on the underside of the tape. A few firms have found it beneficial to have the tape faces embossed or slightly scratched to overcome this difficulty.

As noted previously the tensions in individual tapes (series) and in different sections of tapes (endless) in the web splitting region can be an important factor in deciding on the degree of difference between the web strips and corresponding slubbings. Thus it is extremely important that all the many rollers in a condenser are parallel (distances between any pair of corresponding shafts at the two sides of the condenser should be within 3 mm). Further these rollers should be parallel to the tape doffer.

## XXIV. Monitoring and Controlling Web Weight

The uniformity of the weight per unit area of the final web is all-important in determining the uniformity of the slubbings along their lengths and also that of the ultimate yarns. Even if there is no condenser and the web is used directly in a product then less variation in the web weight is still a desirable feature.

In view of the possible variation in web weight across the web width and the complexity of measuring at a number of positions across this web (see above), one instrument measures the web thickness in only two circular areas (each about 45 mm diameter) fairly near to each other and about a third of the way across from one edge of the web. That is, the instrument

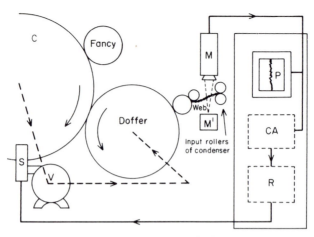

Fig. 16. Autocount control system (closed loop). C—cylinder; V—variable speed unit (VSU); S—servomotor positioning VSU setting; M—measuring head; P—pen recorder; R—regulator; CA—controller/amplifier; M¹—mirror unit to reflect the light beam from M back to M by different path; – – – – – mechanical linkage; ——— electrical linkage.

concentrates entirely on controlling the weight variation (along with time) of the web (Fig. 16).

The measuring of differences in web weight is achieved by noting the differences in the shadows of the totality of fibres in an area of the web when a light beam is transmitted through this area. That is, the greater the number of fibres of the same diameter, the more the shadow and the more the measuring instrument's response (the transducer is a photo-electric cell). For a given fibre type and density the instrument is comparatively insensitive to any features (e.g. fibre colour) other than the weight per unit area of the web and the average fibre diameter. In deriving the formula for the total shadow to light of a region of the web it was necessary to take account of the intersection of fibres. That is, two fibres intersecting do not create as much shadow as when they are not intersecting.

Many of the differences in web weight used (with time) can be controlled automatically (Fig. 17). The control loop is a closed one. That is, any change in the web weight induced by the action of the control mechanism is checked by the monitoring device. The response of the control loop must of necessity be comparatively slow.

Firstly, it should not respond to the higher frequency deviations in web weight that can arise from the incorrect layering on the carder feed sheet of the Scotch Feed sliver. Secondly the process itself (that is, the last carding unit of the swift) has a transfer lag of 20 to 40 seconds as a typical value.

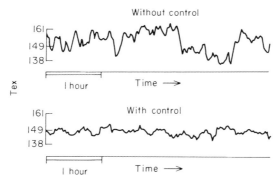

FIG. 17. Slubbing count at a card condenser. The controller of the autocount was operating for the slubbing count shown in the lower trace. The upper trace shows the slubbing count for the few hours previously from the same card. The card had a weighing hopper, three carding units forming the scribbler part and two units forming the carder part. The three marks on the two vertical scales (showing count differences) correspond to 138, 149 and 161 (top) tex.

Thirdly, in the control loop there is a distance–velocity lag (dead time) of 5 to 15 seconds as a typical value. This time is the time when the web, conveyed on the card clothing of the doffer, is on the underside of the doffer and has not yet reached the position for it to be monitored. In control work, a distance–velocity lag is known to cause "hunting" (oscillatory responses of the control mechanism) if an attempt is made to have too rapid a response.

Fourthly, in addition to changing just the speed of the final (tape) doffer, the forward speed of the condenser with all its parts must be changed in exactly the same proportion. Thus the speed of a comparatively powerful drive must be changed. The speed change is usually carried out by changing the setting of an infinitely variable mechanical gear box. More recently, the regulation is a thyristor control of a DC motor which drives the tape doffer and condenser. Strictly the speed of the eccentric shaft driving the rubbing apron of the condenser should also be changed. This would ensure that the degree of rubbing of the web strips into slubbing would be the same. However this is rarely done. It must be noted however that for low-twist singles yarn (less than 250 turns/m) the yarn strength depends on the degree of slubbing consolidation.

## XXV. Non-Woven Applications

The web from the carder part of a woollen card or even from a different type of card can be used directly. It may cover other webs, woven fabrics, knitted fabrics, etc., or it may be cuttled backwards and forwards to form a number of layers as in some non-woven fabrics (e.g. felt). The ultimate weights will usually be between 10 and 2000 gm/m$^2$. Usually the web is conveyed at speeds up to 100 m/min by a cross-lapper or more rarely by a "camel back" lapper, both web conveying mechanisms.

The initial delicate web is conveyed at speeds up to 100 m/min—a good speed would be a web off a one-metre diameter doffer rotating at 25 rpm. To eliminate any variations in weight across the web, the web is usually layered across a lattice moving at right angles. However if only a few (as low as two) thicknesses of web are required then special care is required to minimize the effect of too much overlapping by the edge of the web.

In a few applications (for example, in blanket making) more than one card may be used. The various webs are fed onto one lattice. This method enhances the uniformity of weight, the directional strengths and fibre distribution in the final product.

The greater the attention that has been given to the carding aspect of non-woven fabrics the better has been the product and the more versatile the plant. At one firm a full-length woollen card (four carding units) produced an improved quality of final product from cheaper fibre blends than a card of one carding unit (that is, one cylinder).

## XXVI. Spinning Draft in the Woollen System

In all spinning systems the attenuation given to the final roving, sliver and slubbing before the thinner (finer) singles yarn is spun is called the spinning draft. For example a sliver of 6 kilotex (blend of man-made fibres, 150 mm in length and 17 decitex in weight/unit length) could be given a spinning draft of 20 to 1. This would give a 300 tex single yarn in the semi-worsted spinning system.

However in woollen spinning the spinning draft is comparatively small (low) with values between 1.07:1 to 1.65:1 and with the majority of values between 1.25:1 and 1.45:1. Such small values of draft are used in other systems at the feed regions of spinning frames, roving frames and gill boxes so that the slivers (or rovings) are conveyed under a slight tension and with fibres as straight as possible.

Thus in woollen spinning it would be better to consider the spinning draft as more of a fibre straightening operation. In fact, a spinning draft of about 2:1 imposed on woollen slubbing leads to a very irregular yarn. Clusters of shorter fibres occur at intervals along the yarn with a few long fibres between the clusters. In woollen spinning there is very little chance of improving the levelness of the slubbing delivered from the card and condenser. However the levelness can be made worse unless proper attention is given to the spinning. In general the main features of the slubbing weights per unit lengths are reproduced in the woollen yarn.

## XXVII. Mule Spinning

In Italy and the UK, mules are still used in woollen spinning to a reasonable degree. For example, in 1979, the UK had three times as many mule spindles as ring frame spindles. The total number of mule spindles is decreasing much more rapidly than the increase in the number of frame spindles.

Since 1825 (following the work of the engineer Richard Roberts) most of the mule mechanisms have performed automatically so the mule is often termed self-acting. Excellent summarized accounts of the complicated operation are given in References 2 and 1.

Mules work intermittently and each cycle of operation is very involved. Most mules are entirely mechanical in their operation with usually one electric motor as a power source. However there are some "electrical" mules in which some of the mechanical mechanisms are replaced by electro-mechanical switching devices and in which also the drive to the spindles is modified.

In most mules the long carriage, in two usually approximately equal lengths on each side of the main gearing, moves outwards during the delivery and drafting of the slubbing. The movement is away from the line of condenser drums supporting the condenser bobbins (e.g. 12) each containing a number of cheeses (e.g. 25) of slubbing. Each part of the carriage has 75 to 225 spindles with a pitch of 45 to 120 mm, with 60 as a typical figure.

The carriage remains almost stationary or is stationary while most of the yarn twist is inserted. It then moves in, back towards the condenser drums, while the new yarn is wound on the yarn cop (yarn package) which is on each spindle. Note that if any of the yarn from the previous draw has slipped off (sloughed-off, roved-off) the spindle tip then this length of yarn (up to 200 mm) will be twisted further, and there are many instances of such yarn lengths having twists of 50% or more greater than normal twist.

In a few electrical mules the carriage remains stationary and the lighter structure supporting the condenser drums, as long as the carriage but with fewer drives, moves out and in. In yet another mule, both carriage with spindles and the drums with condenser bobbins move apart from one another for part of the delivery operation.

In any type of mule the drafting takes place while there is a small amount of the final twist in the thread. This twist gives the control necessary over the fibres and care is needed not to have too much twist during this stage.

## XXVIII. Ring Frame Spinning

In woollen frame spinning the operation is carried out continuously. Slubbing from each of the cheeses (totalling up to 250 per condenser doff but usually between 100 and 200) is unwound by friction contact between the cheeses and drums. At some mills the full condenser bobbins are conveyed on a slow-moving overhead rail from the condensers to the frames and the empty bobbins returned to the condensers.

Each slubbing end is taken away through the nip of the back rollers. The thread then passes through a "false" twisting device which is close to the nip or nips of the front rollers (Fig. 18). The word "false" can be misleading since the twist is real enough although only temporary. The purpose is to

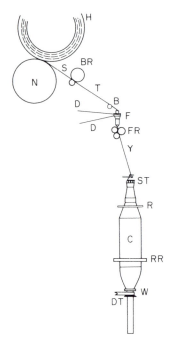

Fig. 18. Vertical section through axis of one spindle of woollen frame and section at right angles to condenser drums. BR—back rollers; N—condenser drum; S—slubbing; FR—front rollers (double nip illustrated); T—thread; H—cheese of slubbing; Y—yarn; R—balloon control ring; W—spindle wharve; B—movable deflector rod (changes path of yarn to make rotating false twister device F more or less effective); D—driving belt to false twister device F; DT—driving belt to spindle wharve W; C—cop of yarn (full); ST— spindle top; RR—ring rail (the cop of yarn is full and the rail nearing its lowest position ready for the operation of doffing).

provide twist, and thereby more inter-fibre friction and greater fibre control during drafting. This control is required as the delicate slubbing is attenuated into the lighter thread (and finally yarn) by the faster speed of the front rollers relative to the back rollers.

Most false twister devices are rotating ones. These seem to be much more popular and more versatile than the non-rotating ones. The ratio of the speed of each of the false twisting devices on one frame to that of the spindle is usually between a third and two-thirds. There are a number of types and, in addition to ensuring some grip on the thread to insert temporary twist, it is desirable that they can be "threaded" easily and that they collect as few fibres (particularly short ones) and as little oil and grease as possible.

Theoretically, as the thread moves through the nip(s) of the front rollers, there should be no temporary twist left in the thread. However, the fibres have been twisted and on twist testing yarn by the untwist/retwist method differences in test results of up to 10 turns/m have been recorded. This maximum difference is between yarn lengths spun when the false twister tube is rotated in each of the two different directions but with all other spinning parameters (e.g. spindle speed) kept the same.

There are many variations of the path of the yarn from the nip(s) of the front rollers until the yarn is placed on the yarn package. There can be one or two rings to restrain the size of the yarn balloon and enmeshing with the yarn balloons of the adjoining spindles. Almost invariably there are also separator plates or a ribbed structure to restrain the balloon.

The spindle top may have an extension probe or finger attached to it which entraps the yarn. This reduces considerably the length of yarn which balloons. Alternatively, the spindle top may be shaped (notched, castellated, etc.) so that there is no balloon at all except in the region of the traveller. The yarn coils around the spindle and also around the yarn tube near the top. The yarn then travels out to the ear-shaped traveller on the ring before it is wound onto the yarn package.

With this collapsed-balloon spinning it is most desirable that the pot-eye or pig-tail near the top of the spindle is kept at a certain but constant distance above the spindle top. Further, the "centre" of the yarn gap should be vertically above the axis of the spindle. Slight deviations cause big differences in yarn tension—for example there is a big increase if the pot-eye is too near the spindle top.

With these modified spindles, a particular feature is that the yarn is twisted by the spindle top modification or attachment. The stated objective is that twist can be driven more reliably and quickly up to the nip of the front rollers. This assists greatly in reducing ends down at higher spindle speeds. However, some feel higher spindle speeds are possible because of the reduced yarn tension with smaller balloons. They note that the speed of travel of the twist is in excess of 100 m/sec, i.e. the twist has travelled at least 7 cm in the time of one spindle revolution if the spindle speed is 8500 rpm.

The travellers are ear-shaped and made of steel, nylon, nylon with glass or carbon fibre, and nylon with steel inserts in contact with the yarn. The yarn naturally passes between the traveller and the ring and usually the ring is of sintered steel or steel which should be lubricated regularly.

Most traveller weights vary between 7 and 70 mg and the weight used can vary from firm to firm. They depend on the spindle speed ($n$ rpm), the ring diameter ($d$ mm), the yarn count, the yarn type (e.g. a fancy yarn with neps, knops or nips added) and yarn strength among other things.

## XXIX. Spindle Speeds

The values of the product $nd$ can be used as a guide to the spinning performance of a non-fancy yarn. For steel travellers the maximum value of the product is $6 \times 10^5$ with well-lubricated rings. Excessive heating of the ring and traveller is the limiting factor. For nylon travellers $7 \times 10^5$ is good, $8 \times 10^5$ is very good and $9 \times 10^5$ is extremely good. Note that for $nd$ values exceeding $7.5 \times 10^5$, the balancing of the spindles requires extra attention. With nylon travellers, the speed limitation when spinning heavy count yarn (heavier than 200 tex) made from all-wool or wool-rich blends may arise from the wear of the yarn on the traveller. The yarn wears a deep groove and long before a traveller breaks, this groove causes a much hairier (rougher) yarn and/or causes ends to break.

With some man-made fibre blends, the melting and fusing of the basic materials of the fibres on the travellers and rings can be the limiting factor for spindle speed. This again emphasizes the high temperatures which occur because of the friction between traveller and ring and even between the yarn and traveller.

For some blends the maximum spindle speed is determined by the number of ends down and the number of operatives manning the frames. For some frames the maximum spindle speed is determined by the speed and dexterity of the operatives in making a good piecening after an end breaks. For example, with a spindle speed of 8000 rpm, yarn with 320 turns/m is being spun at just over 400 mm/sec.

A machine utilization of 80% maintained over long periods with a variety of blends is commendable—the creeling of full condenser bobbins of delicate slubbing can be time-consuming. No automatic doffing of full cops and creeling of empty cops is performed. There have been one or two attempts at partial mechanization of the task of doffing empty condenser bobbins (spools) and the bringing of full bobbins into position on the bobbin drums.

## XXX. Spinning of Elastomeric Yarn

On some types of woollen frames it was very practicable to spin a yarn with a light elastomeric filament core. The thin filament was introduced just above the rotating false twister device at a linear speed of 20 to 25% of that of the surface speed of the front rollers. That is, the elastomeric filament was drafted between four to one and five to one by the faster front rollers. It was buried in the woollen thread as both passed through the false twisting device (e.g. a tube with nipping jaws). Rarely did the filament appear on the surface of the composite yarn.

## XXXI. Power Consumed in Ring Spinning

The power consumed by a woollen frame should not be under-estimated. The motors used are rated up to a quarter, and in some cases up to a third kilowatt per spindle depending on the ring diameter and spindle speeds envisaged.

There is the usual power consumed by the gearing in the headstock (main drive region) and also from the headstock to the spindles. Then there is the power consumed by the yarn balloons when rotating through the air. Finally there is the power consumed by the rotation of the yarn package itself through the air. The package is usually between 50 and 125 mm in maximum diameter and when full between 200 and 600 mm in height.

In one trial the power to rotate a frame package of filament yarn was recorded. The package was then covered with one layer of heavy count woollen yarn and the power to rotate at the same speed recorded once more. The second result was five times greater than the first which again emphasizes the power consumed and the windage created by the rotation of the packages on a woollen frame.

The movement of air (draught, windage) caused by the package is such that at some firms precautions have to be taken to prevent excessive loose fibre collecting on the rings, travellers, tape drives to the spindles and, at times, at one end of the frame.

In addition, some firms have had to space their frames further apart because of excessive windage set up by the rotating packages. One or two firms have a fabric barrier between frames. Others have a wooden or metal barrier in sections fixed along the whole length of the frame which is at an approximate angle of 20° to the vertical, and is sloping outwards so as not to interfere with the doffing of cops and donning of tubes. The bottom horizontal edge is near the spindle wharve (which has the spindle tape passing around it) and the top horizontal edge is between two-thirds and three-quarters of the way up the height of the spindle.

## XXXII. Variation of Spindle Speeds

Earlier the maximum spindle speed was discussed. However during the build of a whole cop the motors driving the frames (and naturally the spindles and yarn delivery) are run at variable speeds with the aim of preventing the yarn tension becoming excessive.

During the building of the base of the cop (longer length of yarn balloon and winding of the yarn length from the traveller onto a small package radius) the speed is kept at a lower level. At a few firms, the speed is kept constant at the lower level during the building of the whole of the cop base. Then the speed is changed manually to a higher speed while the main body of the cop is built. More typically, the motor speed is gradually increased automatically so that it reaches the pre-set maximum speed when the base is built.

As the yarn package becomes bigger the speed of the spindle is kept almost constant and then is gradually reduced until the cop is full when the spindle speed is approximately equal to that at the start of the cop build. The onset of this gradual reduction of speed differs at different firms but has been known to start after three-quarters of the total yardage on a cop has been spun. There is less length of yarn balloon during this period of spinning and so less length available to allow for sudden changes in yarn tension and so prevent excessive end-breaks.

In addition, on many frames the motor speed is changed by smaller amounts than above but much more frequently. The frequency is the same as that when yarn is wound on the nose of the cop. The yarn is wound as binding coils of coarser pitch down to the shoulder of the cop and then back up to the nose by winding coils of finer pitch. The speed variation is changed automatically.

## XXXIII. Versatility and Development of the Woollen System

Finally the woollen system of spinning and the uses of woollen blending lines and cards to produce webs for non-woven purposes or slivers for open-end spinning are attracting increasing attention. The versatility of the card, together with the practical application of scientific analysis and modern engineering practices means a bright future for the system. Proposed future work will make one of the achievements of 19th century mechanization into a fully automated modern manufacturing system.

## Acknowledgements

The author acknowledges the assistance and advice of colleagues at WIRA, Leeds and elsewhere; also the staffs of many firms in the textile industry over the past 30 years.

## References

1. Mackereth, L. (1966). "Woollen Carding and Spinning". *Wool Review.*
2. Brearley, A. and Iredale, J. A. (1977). "The Woollen Industry." WIRA, Leeds.
3. Jowett, P. D., Richards, R. T. D. and Thorndike, G. H. (1968). "Woollen Carding." WIRA, Leeds.

# Chapter 6

# Recycling of Fibres and Waste Textiles

## F. HAPPEY

## I. Introduction

The recycling of fibres and waste textiles provides an important contribution to world textile output. It is of interest to consider the changes which have occurred over recent years in the total quantities of new raw materials processed on the basis of world consumption. These were discussed at a conference in New Delhi in 1976 on Man-made Fibres for Developing Countries.[1] They are summarized in Table I of Chapter 1 where it is shown that consumption of fibres doubled between 1950 and 1974, with a general increase in fibre production of all kinds but with an extraordinary rise in the synthetic field. On the basis of these figures the problem of recycling must be considered, as raw materials from natural sources can no longer keep pace with the universal demand for textiles of all kinds. Having drawn a global picture of textile needs it is interesting to consider the broad local needs of different climatic areas as these determine to a great extent the most suitable textile raw materials for use in each region. A broad estimate would suggest heavy fabrics in cold and temperate zones and lighter materials in the tropics but even in temperate regions the introduction of bulked yarns and fabrics has influenced clothing design.[2]

The "heavy woollen" area of Yorkshire and the Prato district of Italy afforded good examples of earlier fibre recycling for the woollen and worsted trades and up to 20–30 years ago the shoddy and mungo trade was well established in the manufacture of woollen goods from fibrous waste and rags. The later introduction of man-made fibres into worsted manufacture, mixture fabrics and knitting wools produced used materials which were difficult to recycle on existing machinery. Man-made fibre production is determined by the raw organic products from which they are synthesized, such as distillates from oil and, to a lesser extent, coal, and this production now far outstrips that of animal hairs, rapidly approaching that of cotton. There had been a declining interest in the field of fibre recovery by recycling of woollen and worsted rags, but with the rapid changes in world economics production of cheap materials has become a matter of urgency. Man-made staple fibres have been introduced into woollen and worsted-type materials and also used alone in such products and the recovery of these should provide a major source of fibres for recycling.

The traditional scheme of fibre recovery from waste in the shoddy and mungo trade can be summarized as follows:[3] (a) rags shaken and cleaned prior to sorting into classes and qualities; (b) sorted rags "ground" to reduce woven and knitted fabrics to fragments; (c) the products of (b) opened by garnetting to break fabric down to a yarn and fibre complex; (d) the product of (c) mixed and blended by carding to produce a fibrous mass to pass on to the processes of yarn manufacture.

A similar scheme was used for cotton rags but recovered cotton fibres are very short and only suitable for the waste cotton trade. This has been of little importance in the past in the shoddy industry but could be re-developed with a view to introducing recycled cotton mixtures on an increased scale, particularly fibres recovered from cotton/man-made fibre waste.

It is in stages (a), (b) and (c) that major modifications were required in order to be able to deal with the recovery of modern synthetic materials and mixtures. These were put forward first in BP 1 684 991[4] and have since been developed commercially as shown later in this chapter.


## II.  Fibre Recovery from Mixed Rags Including Waste Containing Man-Made Fibres

Depending on the market served, the "lower woollen trade" used a variety of raw materials ranging from newly sheared and scoured wool through semi-manufactured products such as combing noils and spinning waste, new fabric in the form of tailors' clippings to cast-off clothing. Whatever raw material was used the early stages of mechanical processing had the same objective as any other sequence of textile processing, i.e. the separation of individual fibres preparatory to re-assembly in the desired form. Where fabrics were the starting point of the operation especially if they had been subjected to felting or milling it was clearly a difficult process and one which subjected the fibres to considerable stress and a high degree of breakage.

Before the introduction of man-made fibres the heavy woollen industry was able to classify its products as shoddy from soft waste with a reasonable fibre length, and mungo which was an inferior short staple product made from material much felted and which required increased mechanical effort in the "grinding" process to separate the fibres. A further by-product was known as "extract". This was produced from fabric containing cellulosic material such as cotton and the latter fibres were removed by acid carbonization of the cellulose by sulphuric acid or other chemical means.

It was pointed out earlier that wool, especially finer wools, is limited to use in the worsted and woollen manufacturing industry. The fineness of the fibres depends on the breed of sheep or goat and these fine fibres were processed mainly in Europe and America, but more recently in Eastern Europe and the Far East. Meanwhile the man-made fibre industry has expanded almost explosively and fibres can be made with filament deniers as fine as the finest wools. They have usually a breaking strength greater than the wool fibre (~4.5 g/denier as compared with ~1.5 g/denier), and they can be cut or produced as broken staple to match the fibre dimensions of any type of wool.[5] Thus the synthetics were introduced in "wool" fabric to form mixtures as distinct yarns or as mixed yarns of wool/man-made fibre. The "worsted" type fabrics produced, especially lightweight, are now very popular, but these fabrics, many undefined in content, caused havoc in the reclaimed fibre industry because of their increased strength. This led to a reappraisal of the technology of fibre recovery.

The first suggestion was to devise a separation technique to obtain wool separately for traditional processing methods.[6] This was not pursued for technical reasons. Also it was more realistic to try to recover the whole complex than to utilize only a proportion of the material. In the traditional method of grinding mixture rags in the fearnought etc., before garnetting and carding the fabric was pulled apart ("ground") by spiked rollers into small pieces of random size and shape. This method broke down with the synthetic mixtures as their strength was great enough to damage the rag grinding machine particularly when 50/50 by weight of wool/terylene was used. It was seen in Table I, Chapter 1, that the global output of wool and synthetic fibre in 1962 were 9% and 7% respectively. Even at this stage wool/nylon and wool/terylene mixtures were being used and therefore it was decided in the textile department of the then Bradford Institute of Technology to consider fibre recycling with a view to recovery of the whole mass of mixture fabrics.[7] These had become the first field into which the polyesters had penetrated, closely followed by nylons and acrylics.

The process of rag "grinding" was not the most satisfactory means of fibre recovery because of the great degree of stressing of the material during transformation from fabric to fibre. When man-made fibres were incorporated in wool fabric the power needed to reduce the material became so great as to damage the machine. It was decided to replace the rag-grinder with a cutter in order to produce materials ready for disintegration by the garnett in the second stage of the standard process, but requiring much less force.[4] As a preliminary experiment cut pieces of a wool/man-made fabric nominally 2″ square were passed into a garnett intended for the opening of "soft waste". Despite the fact that this machine was not designed to handle torn rags the product obtained was acceptable for carding, contained no pieces of unopened fabric, and only a reasonable number of lengths of unopened thread. Carding and spinning of the samples was entirely conventional. Carding was on a modern 60″ wide two part woollen card with Scotch intermediate feed and tape condenser. Yarns were spun both on the mule

FIG. 1a. Platt woollen ring spinning frame showing excess fibre fly collecting on guide-rail, on surface of machine and at thread guide in spinning recycled waste rags previously converted to fibres by traditional "rag grinding" process. This fly can also be seen attached to the condenser sliver on its passage from back to front roller of the machine. It should be noted that excessive fly in the spinning shed and particularly on the machines is not only a health hazard, but on a short term basis it is more importantly a fire risk. This is especially the case when man-made fibrous fly may well generate sufficient static electricity to cause ignition of the material.

FIG. 1b. Ring spinning frame showing spinning of fibres recovered from cut rags (CGC process), from the same batch of material as in Fig. 1a. There is a considerable decrease in fly accumulating on the machine. It also shows a more aligned fibrous sliver and twisted yarn during processing.

and ring frame to counts of 17½ YSW (111 tex). At this stage it was felt that a satisfactory basis had been established on which to carry out further development.[8,9]

A sample of wool worsted tailors' clippings was used to compare fibre breakage during conventional rag pulling and in the new sequence of cutting, garnetting and carding (CGC). Half the material was processed by this new method and half by a firm of commission rag pullers. The fibre length diagrams obtained showed the large reduction in mean fibre length after pulling and that the commercially pulled sample suffered much more fibre breakage than the CGC sample. In particular the latter contained fewer very short fibres. This was shown strikingly during ring spinning of the samples by the difference in the amount of fly produced (Fig. 1a, b). It is thought that this reduction in fibre breakage is achieved because the small cut pieces are more easily reduced to their component yarns and fibres.

G

## III. Technological Development

At this stage the main requirement was to build a fabric cutting machine to cut in two directions at right angles. The first prototype (Fig. 2) was described in BP 1 084 991[4] and the second model is illustrated in Fig. 3a, b, c.[10] It is important to emphasize that this was a pilot plant machine and further development study was required to produce a robust machine capable of continuous high duty processing. This is dealt with in the later part of this chapter. The experimental machine provided a process, preparatory to the recovery of fibre from fibre-containing material, in which this material is cut in two directions into pieces having no dimension greater than the mean fibre length in the material. The optimum size of the pieces depends on various factors, e.g. the speed at which the subsequent "opening" machine is required to work and the nature of the raw material involved.[11] Attention must also be paid to the use for which the recovered fibres are intended. The pieces must not be so small that the average staple length of the recovered fibres is too short, on the other hand the pieces should not be so large that too great a strain is put on the opening machine with consequent damage to the machine. It is felt that by the previous application of the cutting process the action of this machine becomes one of disentanglement of the fibres in the material rather than the severing of the fibres from it.

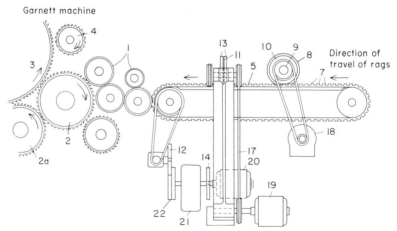

Fig. 2. Diagrammatic side elevations of rectilinear cutter for the CGC process (Mark 1). 1,2,3,4—"Licker in" to garnett machine for cut rags. 8,9,10—Cutting knives for longitudinal cutting of fabric, with drive, 18. 5,7—Lattice with longitudinal and lateral depressions to accommodate cutting knives. 13,11,17—Lateral cutting mechanism from drive, 19. 20—Drive for oscillating mechanism with cam, 14. Also reducing gear box, 21, and, 22, Geneva mechanism to advance lattice movement in conjunction with each completed transverse cut.

TABLE I. Fibre length before garnetting.

| Sample | Mean fibre length cm | Standard deviation | Coefficient of variation % |
|---|---|---|---|
| Fabric | 6.74 | 2.12 | 31.5 |
| 1″ square cut | 2.23 | 0.06 | 2.83 |
| 2″ square cut | 3.40 | 1.45 | 42.7 |
| 2.5″ square cut | 3.64 | 1.91 | 52.5 |
| 3″ square cut | 4.88 | 2.00 | 41.0 |

In order to monitor the optimum size of cut pieces for shredding in the garnetting machine tests were made on a semi-production scale with in-puts of 50–60 lb of material per experiment. 1″, 2″, 2.5″ and 3″ squares were produced on the cutting machine from a tropical weight "worsted" fabric containing 55% polyester and 45% wool (60s quality) and having the following fabric parameters: ends per inch 64; picks 52; warp count 2/64s (R27.5 tex/2); weft count 2/64s (R27.5 tex/2). The squares were fed to the garnetting machine and the mean fibre lengths were estimated. In the square cut samples the measurements were made with a WIRA fibre length machine. Diagonally cut specimens were also prepared and an alternative method was needed for measurements of these on account of the great variation in fibre lengths. Threads were removed from the diagonally cut pieces in groups according to their length and the percentage weight for each group indicated the short fibres present. Tables I, II and III show the mean fibre lengths, the standard deviation and the coefficient of variation before and after garnetting for the different sized pieces.[10]

TABLE II. Fibre length after garnetting.

| Sample | Mean fibre length cm | Standard deviation | Coefficient of variation % |
|---|---|---|---|
| 2″ square cut | 2.46 | 1.41 | 64.0 |
| 2″ diagonal cut | 2.27 | 1.26 | 55.0 |
| 2.5″ square cut | 2.77 | 1.59 | 57.3 |
| 2.5″ diagonal cut | 2.29 | 1.25 | 54.8 |
| 3″ square cut | 2.71 | 1.39 | 51.0 |
| 3″ diagonal cut | 2.62 | 1.34 | 51.0 |

TABLE III. The distribution by length of threads removed from diagonally cut samples before garnetting.

| | Cumulative percentages | | | |
|---|---|---|---|---|
| Class length cm | 1 inch (2.5 cm) % | 2 inches (5.08 cm) % | 2.5 inches (6.35 cm) % | 3 inches (7.62 cm) % |
| 0–1 | 9.14 | 2.91 | 1.02 | 1.1 |
| 0–2 | 35.46 | 9.64 | 4.36 | 3.64 |
| 0–4 | 99.99 | 36.5 | 18.48 | 14.25 |
| 0–7 | — | 99.78 | 60.17 | 42.89 |
| 0–9 | — | — | 100.07 | 72.0 |
| 0–11 | — | — | — | 99.9 |

In square cut samples all threads were of similar length, but in diagonally cut pieces threads varied in length from 0 to 2.81" for a 2" piece, 0 to 3.5" for a 2.5" sample and 0 to 4.25" for a 3" piece. These two geometric examples provided the extremes of dimensional lengths of threads which are likely to occur from rectilinear mechanical cutting.

## A. Fibre distribution in cut and uncut fabric before and after garnetting

In Fig. 4 the mean fibre length of a 2.5" square cut sample is shown as a cumulative frequency curve comparing the percentage fibre distribution in the original fabric, after cutting and after garnetting. Figures 5 and 6 compare the square cut and diamond cut samples. It can be seen that the square cut pieces give curves with ordinates generally slightly higher than those shown by diamond cut specimens. Figures 7 and 8 are frequency polygons showing fibre lengths of the parent fabric and rectilinear cut samples before and after garnetting. Consideration of Fig. 7 shows that the spread of mean fibre length is greater in the uncut fabric than in the square cut specimens of various sizes. The 2.5" squares gives a longer mean staple than the 3" specimens but the difference is only marginal. Figure 8 indicates that the various size squares provide very similar curves after the garnetting process has been applied. For the type of fabric investigated it was considered that cut squares of 2–2.5" were satisfactory as no trouble was encountered due to

FIG. 3. Fabric cutter (Mark 2). (a) Fabric cutter showing cutters to act parallel to direction of flow of rags. The safety guard has been opened to show spacing of cutter discs. (b) Side view of cutter showing driving mechanism. (c) Front of cutter showing shearing blades for cutting perpendicular to flow of rags.

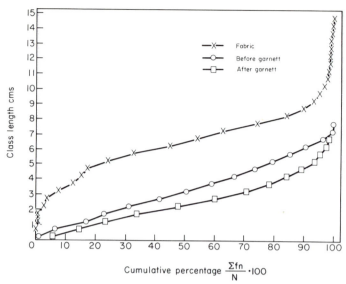

FIG. 4. Fibre length of 2.5″ sample square cut. Points on graph mark centre points of class length. Class length in intervals 0.5 cm. Cumulative % $(\Sigma fn)/(N) \times 100$ where *fn* is number per class length and $N$ = total number of fibres measured.

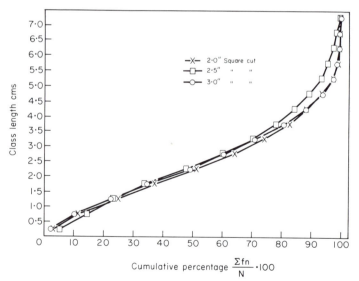

FIG. 5. Fibre length of square cut samples after garnetting. Units as in Fig. 4.

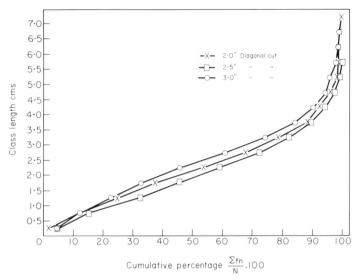

FIG. 6. Fibre length of diagonally cut samples after garnetting. Units as in Fig. 4.

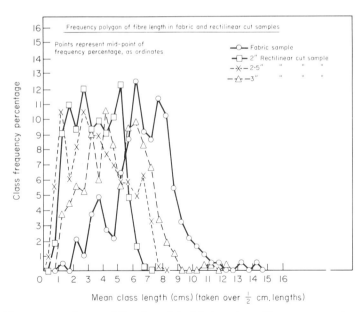

FIG. 7. Frequency polygon of fibre length in fabric and rectilinear cut samples. Points represent mid-point of frequency percentage, as ordinates.

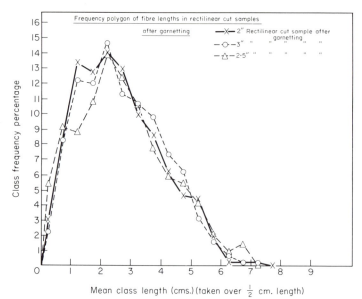

FIG. 8. Frequency polygon of fibre length in rectilinear cut samples after garnetting.

strain on the garnett machine, and the second pilot scale cutting machine was designed accordingly. This did not preclude the adjustment of the blades to produce larger pieces if a more robust garnett became available. Further experiments on comparison of pulled and cut waste again showed a decrease in "fly" in the latter case. Much heat and some smoke were generated during the "pulling" of this fabric and the resulting product showed black specks due to the fusing of some fibres in the severe treatment required to break open the parent fabric. This was not encountered in the products obtained by the CGC process. All the yarns produced by this method confirmed the early findings that the mean fibre length was ~20% longer than that produced by the traditional pulling technique.

The optimum size of the cut pieces also depends on the staple length needed in the recovered fibres and the use for which they are intended. Spinners who blend man-made fibres with virgin wool may receive fibres in staple lengths varying from 2″ to 4½″ for fine worsted blends. When such fibre blends are opened in the traditional way it is inevitable that a large proportion of the fibre yield has a length of only 2″ or less. If, however, the same blends are subjected to the CGC process they are opened efficiently and yield fibres of satisfactory staple length for spinning according to normal shoddy and mungo practice. When cutting it is desirable to group together fabrics containing similar fibre lengths. Generally coarser cloths contain yarns spun from longer staple fibres, and the staple length to be expected in most types of cloth is well known in the trade. The more tightly the yarns are twisted or the fabric constructed the greater the advantage of using the cutting technique before submission to the opening machine. Although man-made yarns such as nylons and polyester are hard to break they nevertheless separate readily into their constituent fibres when opened by the CGC method. This is probably due to fibre repulsion due to static charges acquired by the fibres during separation. In the mass of loose fibres resulting when a blend of tough and gentler fibres is opened by the process the fibres are very well mixed, improving the evenness and quality of the yarn into which they are re-spun. The method can be applied to all types of natural and synthetic materials as shown in the industrial applications described later.

The relationship between size of cut pieces and resultant mean fibre length can be used as a basis to produce a sliver of the desired maximum fibre length. An example would be the production by the CGC process of a web of suitable fibre distribution for direct spinning into an open end spinning unit for yarns from recovered waste. Over recent years open end spinning has increased in importance[12] and shorter fibres are generally required for this process. The need for the production of short staple material from waste fibres is shown later in this chapter, when an industrial application is illustrated where a second cutting of the waste is undertaken to meet the needs of this newer form of spinning (see p. 187). This second cutting may be unnecessary as shown earlier in the chapter when it was pointed out that the size of cut pieces can be selected to produce staple lengths as required, including short staple for open end spinning. Thus it may be that square cut samples at 1″ to 2″ produced by the CGC method will provide fibres of optimum length for this process, as the fibre length can be no greater than the square sizes plus a fraction due to the yarn crimp or untwisting of threads in the parent fabric. An additional advantage with such small cuts would be the greater ease of extraction of the fibres. Table I showed the statistics for the fibres in 1″ cut squares but fibre length measurements after garnetting are not available. It is clear, however, that a fibre web for use in open end and DREF spinning could be produced. Fibres produced by the 2½″ rectilinear cutter followed by condenser/mule or ring spinning gave rise to lean semi-worsted yarns and it is reasonable to assume that a sliver of ~1″ mean fibre length would produce a more bulky yarn for similar counts because of the larger number of fine fibre ends.

Mixture fabrics are now more the rule than the exception and the supply of new raw materials is becoming more difficult and expensive. In the early 1960s fabric was comparatively cheap and the use of man-made fibre appeared to be monitored carefully by manufacturers. This specificity of supply has now gone and man-made fibres are available from all parts of the world. This has caused the original "trade names" of fibres to be of less individual importance and in general descriptions and nomenclature such as polyester, polyacrylic and nylon are accepted in many cases. This has eased the definitions which may be used for recycled materials. Comparatively speaking wool supplies are even more widely stretched than formerly and staple man-made fibres have become more and more a world wide product. It is reasonable to assume that wool/synthetic waste must accumulate together with man-made mixtures and the CGC process of fibre recovery is the one to look to for future progress in recycling fibre waste.

In 1975 study of fibre recycling by the CGC process was discontinued in Bradford University because of misunderstandings which led to the lapse of the patent due to administrative problems in patent renewals. The work carried out in the textile department of the university had nevertheless shown that the CGC process was not only viable but had led to the manufacture for display of yarns and fabrics and the carrying out of wearing trials on recycled worsted/man-made fibrous materials.[13, 14] Quantities of the order of 2.5 cwt of wool/polyester worsteds and 1.5 cwt lots of wool worsteds were processed in 2½"squares by the CGC method and the results were compared with the traditional "pulling" process using materials from the same batch of waste with the vindication of all the claims made for the CGC method with smaller batches. It therefore gave me great pleasure to see this method developed recently on a successful industrial basis.

## IV. Industrial Development of Waste-Cutting Method of Fibre Reclamation

This process has supplemented the "classical low woollen" shoddy and mungo manufacture and the earlier rag-grinding is replaced by "high duty cutting" of the parent materials. The routine of the program and production rate of the individual units are as given below:[15]

    rough cutter—up to 3000 kg/hr
    pulling machine—up to 1000 kg/hr
    carding machine—up to 400 kg/hr
    baling press for delivery of opened material—4 tons/hr

The line of the process is shown in Fig. 9 up to the card-doffing of the web. This can be dealt with in various ways and requires a web stripping process to

Fig. 9. Enclosed industrial waste card incorporating fabric cutter (Haigh Chadwick).

supply the carded web in its various forms as required, e.g.: (1) after stripping the web may be collected for delivery in baled form; (2) it may follow the usual systems of sliver production for spinning by ring or tape collection;[11] (3) to produce a cross-folded web, probably for non-woven tissues; (4) to produce a sliver for spinning from cans via open end or DREF machines. It is reported that recent results on direct spinning on DREF systems have been very encouraging;[12] (5) to produce a sliver for semi-worsted spinning;[16] (6) as the final spinning operation may require a fibre distribution in the sliver with fixed maximum fibre length a second cutting operation may be included to ensure that fibre length distribution is limited to a pre-determined value. This is important in the making of slivers for open end spinning in certain conditions.[12]

An interesting example[15] cites the recycling of high tenacity tyre cord waste by feeding the waste from the bale into the feed chute of the rotary cutter whence the cut material is pneumatically discharged to a silo tower. The delivery from this to the "pulling" machine is monitored by ultra-sonic level indicators. The opened material is conveyed to an automatically emptying storage bin where adventitious ferrous debris is removed magnetically. The product is fed then automatically to the carding system and arrives

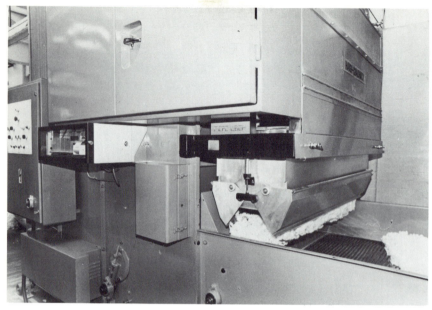

FIG. 10. "Microweigh" unit, a micro-processor controlled weighpan hopper feed complete with control panel and digital display (Haigh Chadwick Trademark).

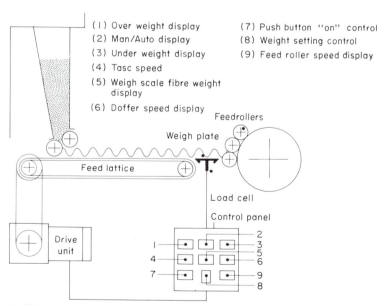

(1) Over weight display
(2) Man/Auto display
(3) Under weight display
(4) Tasc speed
(5) Weigh scale fibre weight display
(6) Doffer speed display

(7) Push button "on" control
(8) Weight setting control
(9) Feed roller speed display

Feedrollers

Weigh plate

Feed lattice

Load cell

Control panel

Drive unit

FIG. 11. "Microfeed" unit system for control of uniform delivery of fibres into card feed rollers (Haigh Chadwick Trademark).

at the end of the carding program as a web ready for removal by the doffing knife. Following this the necessary sequence of operations to produce the required web form or yarn precursor is applied.

It is important to mention that in this process of waste recovery the use of processing oil is not required as in the "older shoddy process". Also the rigid metallic clothing of the card can now become self-cleaning. Any ferrous particles are abstracted from the fibrous layers by magnets fitted over the spiked sheets in the Rota-mix self-emptying bins where storage and mixing occur.

In this field of development electronic controls have been introduced.[17] The first is used to monitor the delivery of specified weights of material at a pre-determined rate into a hopper feed, at approx. 4 drops/min (Fig. 10). From the hopper where the fibre weight has been recorded a micro-feed system (Fig. 11) regulates the weight entering the card feed rollers by monitoring the rate of passage of the fibres over the weighing plate with its

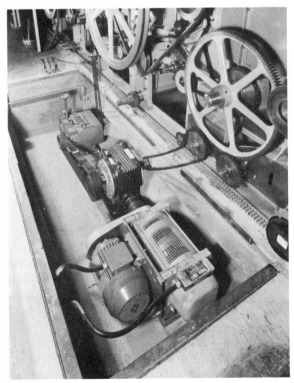

FIG. 12. Variable drive motors for card to adjust input and related speeds of workers and doffers in relation to swifts exposed by removal of walk-way.

two load cells. This system was used earlier in the normal woollen yarn process.[18] To increase the basic adaptability to deal with the energy needs required for the formation of uniform web from whatever raw material the card must be capable of adjustment of running speeds of material input and of speed of workers and doffers in relation to swifts. These are fitted with variable speed drives illustrated in Fig. 12.

## V. Recycling of Man-Made Materials as Fibres

The recycling of fibres from man-made fabric or waste presents several possibilities. Before considering the reconstitution of the fibres as continuous filament it is relevant to mention the formation of unwoven tissues from mixed rags. This requires the breakdown of continuous or staple fibre fabrics into constituent fibrous materials by the cutting, garnetting and carding (CGC) process. Having done this, and knowing the chemical nature of the mixture, several processes for unwoven tissue formation are available. The first need after fibre separation is to card the fibrous complex into continuous sheets. Such methods are described in BP 638 591[19] and BP 640 411.[20] In the former partial acetylation of cellulose, in particular cotton, was cited as a possible method of producing unwoven tissue. Thus cotton and viscose rags could be used to produce unwoven tissues if acetylation were combined with the CGC process probably coming between the cutting and garnetting stages. Considering the quantity of waste cotton and viscose available this could make a major contribution to the re-use of cellulosic materials in sheet form (see Table I, Chapter 1). It should be mentioned that in the fibrous acetylation of cellulose the changes occur to completion first in the less crystalline part of the cellulose and even up to 30% acetylation little change occurs in the crystalline parts of the fibres. Hence the stronger component is unchanged and would form the basis of the unwoven tissue, giving it flexibility and strength.

Alternatively the web may be consolidated by use of a water soluble size which can be dried without the need for solvent recovery. Thirdly the web can be made with a soluble adhesive capable of precipitation as a film holding the fibres together in tissue form.[19] If the first method is preferred a suitable amount of cellulose acetate fibres can be incorporated in the cut mixture of rags to provide the adhesive. This should be done at the primary fabric cutting stage as the intimate mixing of the fibres is then more complete. In certain fibre mixtures where polymers of low melting point are used, e.g. some poly-olefines, the fibrous mass can be carded into thin webs and adhesion achieved by passing these through heated rollers. To finalize this short discussion of the use of man-made fibre waste to produce unwoven tissues it should be noted that this technique may be adapted also to the processing of mixture rags containing a proportion of wool. The normal wool properties of warmth and voluminosity would however be inhibited in the final product, but might be developed by loose adhesion in the fibrous sheet followed by a raising process to free some fibrous ends in the web.

## VI. Recycling of Man-Made Fibres in Filament Form

It has been described earlier how man-made fibres may be recycled in the CGC process using cutting to replace the rag tearing operation. There is a further method of reconstitution of man-made filaments by their re-extrusion as filaments. If this is to be done then it is important that the parent fabrics should be labelled clearly as to the constituents of the material as in normal circumstances only single component systems can be dealt with. At the present time this work is only in the nature of a feasibility study, but even so waste fibres have been used to reconstitute filaments by wet, dry and melt spinning processes. To attempt to carry out large or even medium scale pilot-plant processing would be unwise until such trials and the associated studies of the properties of the reconstituted fibres have been assessed, although some successful work in this field has been carried out.[21,22] In the early work it was impossible to ascertain the fibre forming potential simply by extruding polymers in an uncontrolled manner, and optimization before, during and after extrusion is essential if a satisfactory product is to be obtained. Necessary parameters involved are degree of polymerization, degradation during extrusion, rate of coagulation and jet stretch etc. In addition to all this and the control elements needed for the measurements of conditions during processing a wide range of test facilities are required.[23,24,25] Thus where facilities for feasibility studies of the processes of extrusion recycling of fibres are available it is possible to demonstrate experimentally recycled materials by the wet spinning of acrylic waste, dry spinning of cellulose acetate and tri-acetate waste and melt spinning of nylon and polyester wastes. As in the normal continuous processing of melt spun polymers unpolymerized monomers or oligomers can be extracted from the bulk polymer before extrusion. It is of interest to observe that in the processing of monomer to final yarn any residual low molecular weight material can be extracted before melt spinning and extrusion so that no difficulty should be experienced in the production of regenerated fibre from synthetic fibrous waste. An example of such recovery was shown by the re-spinning of waste commercial acrylic fibres to produce fibres with tensile properties at least as great, and in certain cases greater, than those of the original material. The ultimate strength obtained was 8 gf/tex with a fibre extension at break of approx. 20%.[26,27] This tenacity would be greater than that normally required in a textile yarn, but was thought to be important in the production of high tenacity carbon fibres.

## VII. The Recycling of Glass as Fibres

The raw material is in the form of waste glass such as bottles etc. and is ground into cullet and extruded as continuous filament in the molten state. The production of fibres is achieved by high temperature extrusion of the molten complex followed by stretching in a plastic condition and coagulation into filaments on cooling. There are technical difficulties in adjusting the process as different glasses may have varying chemical content and hence different melting points. Such physical differences may inhibit uniform flow and shrinkage properties have adverse effects during processing, but work is proceeding in this field.[28] If the process can be made capable of replacing even in part the use of standardized glass rods in the manufacture of glass filaments it will be an important step forward in the recycling of glass. An important use of glass fibre is in the production of insulating materials and a development in its handling and manufacture was described in the *Financial Times* of 18.4.77. A final point on the use of glass fibre in this way is the fact that the material is fireproof and cannot emit noxious fumes at high temperatures which gives it a value which cannot be ignored. It is also possible to use staplized fibrous recycled glass in the formation of glass composites where the fibres are held in random array in the adhesive. Once fixed in position if the fibres are not heated to their glass transition temperature little dimensional change is likely to occur in the system.

## VIII. The Reconstitution of Wool Fibres and Other Waste Proteins

In 1946 Lundgren, High and Ward[29] recycled chicken feathers in the form of a continuous filament yarn by a process of wet spinning. This was the first major advance in protein regeneration using waste material. Milk casein had been spun in filament form,[30,31] and the process was developed by Wormell.[32] In 1956 this process was described at the Symposium on Fibrous Proteins at Leeds,[33] but in 1949 Happey and Wormell,[34] had detailed a process applicable to the recycling of wool waste in which the recycled product were demonstrated as hanks of continuous filament $\beta$ keratin. This was extended to the recycling of the protein of hoof and horn or any other source of keratin and provided the resulting fibres are cleared of solvent, there would be no reason not to use such a product as a foodstuff, at any rate for animals. Following this Alexander, Earland and Happey[35a,b] dissolved and re-precipitated wool in the $\alpha$ form. Later Happey and Wormell[36] showed that wool keratin could be recycled in filament form and retain an $\alpha$ structure.[37] In the process of stabilization of these regenerated protein fibres various methods were applied with only limited success. It has not been found possible yet to replace or reconstitute the cystine cross-link which is a major factor in the stability of wool protein. When this has been achieved then the recycling of waste keratin as a textile fibre could become a realistic process. In the meantime certain regenerated protein filaments both from animal and vegetable sources are being used for the production of edible material, in some cases for human consumption.

## IX. The CGC Process in the Wider Field of Textile Waste Recovery

Having traced the history of the development of the CGC process of fibre reclamation which started as an effort to overcome the difficulties of using mixed fibre waste as a new raw material for the shoddy trade it is important to place the method into a wider scheme within the textile industry. In point of fact the method is available, and is likely to become used for, the processing of all forms of textile waste to serve a wide range of different markets and uses. This would even apply to the best of woollen fabrics used in recycling as it is clear that woollen rags give a better product when processed by the CGC process than when "pulled" by traditional methods. Further the process can deal readily with continuous filament fabrics, and using a uni-directional cutter can produce a staple top from a filament tow. The process is almost simplicity itself technically which is a major point in its favour. This was emphasized as long ago as 1963 by the author when lecturing in Dewsbury,[13] and the recent widening of the scope of the industrial developments of the original scheme bear testimony to this.[15]

## Acknowledgements

The author wishes to thank his colleagues in the Department of Textiles of the University of Bradford, especially Dr E. Dyson and Dr J. A. Iredale, for their excellent collaboration in this work. Mr I. Hudson and Mr R. L. Roberts in their dissertations for degree projects, and the latter in an MSc thesis, furnished early experimental work for the CGC process which forms a large part of this chapter, and their work is gratefully acknowledged. The industrial development of the process carried out by Haigh Chadwick of Cleckheaton is a source of pleasure and thanks are due to Mr M. Tetlow, managing director, for his cooperation in supplying several of the figures and illustrations. Finally I should like to acknowledge Mr C. M. Fenton, managing director of the B. B. A. Group Ltd, Cleckheaton, for helpful discussions on the fibrous recycling of glass.

## References

1. *Man-made Fibres for Developing Countries.* SASMIRA. Ed. J. G. Parikh, 1976, Delhi.
2. Lord, P. R. *Contemporary Textile Engineering,* c. 9. Academic Press, London, 1981.
3. *Wool Year Book, 1960,* pp. 76 and 231, Textile Mercury, Manchester.
4. Happey, F. *The Recovery of Fibres from Textile Fabrics and Other Fibre Containing Materials.* BP 1 684 911. 1967.
5. Lord, P. R. *Contemporary Textile Engineering,* c. 4. Academic Press, London. 1981.
6. Speakman, J. B. *et al.* "Conference on Reclaiming of Fibres". *J. T. Inst.* **41**, 187–232. 1950.
7. Dyson, E., Happey, F. and Iredale, J. A. (1975). "The Recycling of Fibres". *Conference on Reclamation,* University of Birmingham.
8. Hudson, I. (1967). "The Effect of Fabric Dimensions on the Reclamation of Processed Wool/Synthetic Blends". *B. Tech. Dissertation,* University of Bradford.
9. Roberts, M. J. (1967). "The Influence of Blends and Fabric Dimensions on Fibre Length Distribution of Processed Rags". *B. Tech. Dissertation,* University of Bradford.
10. Roberts, M. J. (1970). "Studies of New and Improved Methods of Fibre Recovery from Rags and Waste Materials". *MSc Thesis,* University of Bradford.
11. Richards, D. *Contemporary Textile Engineering,* c. 5, Academic Press, London. 1981.
12. Dyson, E. *Contemporary Textile Engineering,* c. 7, Academic Press, London. 1981.
13. Happey, F. (1963). Lecture, Dewsbury Textile Society. "Shoddy from Mixed Rags". *Wool Record,* April 1963.
14. O'Neill, H. (1967). "Textile Waste". *Financial Times,* 15 July 1967.
15. Osman, M. (1981). *Text. Inst. Ind.,* **19**, 154. N.S.
16. Parkin, W. and Iredale, J. A. *Contemporary Textile Engineering,* c. 3. Academic Press, London, 1981.
17. Lyttle, W. J. and Greaves, P. (1981). *Text. Inst. Ind.,* **19**, 175.
18. Schofield, B. *Contemporary Textile Engineering,* c. 13. Academic Press, London, 1981.
19. Happey, F. and Grimes, J. H. *Improvements in the Manufacture of Sheets and Webs of Textile Materials.* BP 638 591, 1950.
20. MacGregor, J. M. and Happey, F. *Improvements in and Relating to the Manufacture of Webs of Textile Materials.* BP 640 411, 1950.
21. Green, D. B. *Applied Fibre Science,* **3**. Academic Press, London, 1979.
22. Happey, F. *Contemporary Textile Engineering,* c. 1. Academic Press, London, 1981.

23. Happey, F. and Dyson, E. *Conference of Production and Processing of Manmade Fibres,* p. 188. Varna, Bulgaria, 1973.
24. Happey, F. *IUPAC Conference on Macromolecules,* p. 752. Madrid, 1974.
25. Happey, F. and Green, D. B. *European Pol. J.,* **13**, 689–694, 1977.
26. Montgomery, D. E. "The Production and Properties of Some High Performance Synthetic Fibres". *PhD Thesis,* University of Bradford, 1970.
27. Blakey, P. R., Montgomery, D. E. and Tregonning, K. J. *Text. Inst.,* **61**, 234, 1970.
28. Fenton, C. M. (1977). *Private Communication.*
29. Lundgren, H., Ward, W. and High, L. (1946). *J. Polymer Res.,* **I**, 22.
30. Ferretti. (1938). BP 483 731 and 483 809.
31. Ferretti. (1939). BP 511 160.
32. Wormell, R. L. (1954). *New Fibres from Proteins.* Butterworth Scientific Publ., London.
33. Wormell, R. L. and Happey, F. (1946). *Proc. Symp. on Fibrous Proteins,* p. 160. Soc. Dyers and Col., Leeds.
34. Happey, F. and Wormell, R. L. (1949). *J. Text. Inst.* **40**.
35a. Alexander, P. and Earland, C. (1950). *Nature,* **166**, 396.
35b. Happey, F. (1950). *Nature.* **166**, 396.
36. Happey, F. and Wormell, R. L. (1953). "Improvements Relating to the Production of Artificial Threads and the Like." BP 673 676, BP 690 566 and BP 692 876, 1953.
37. Happey, F. (1955). *Proc. 1st International Wool Textile Research Conference,* p. 432 (Australia).

# Chapter 7

# Open End Spinning: Development of Present Methods

## E. DYSON

---

Since the late 1960s there has been widespread and very rapid growth in interest in alternatives to ring spinning for the production of textile yarns. Before discussing any of the new techniques it is a valuable exercise to briefly review the position as it was during the mid-1960s. Ring spinning had been in commercial use ever since its invention during the 1820s in the USA, and throughout the intervening years it had steadily taken over an ever-increasing share of the spinning capacity of the world; flyer-spinning had been obsolete for many decades except for a very limited range of specialized applications; cap spinning had only ever appealed to fine-count worsted spinners and was being superseded by ring spinning; and mule spinning had disappeared from the cotton and worsted trades and was used to a decreasing extent in the only area where it had maintained a reasonable presence, i.e. woollen spinning. Thus, in the middle of the 20th century the ring spinning capacity of the world represented an enormous capital investment in thousands of millions of spindles. The technology was, and still is, based upon a vast store of knowledge, originally based upon individual invention and empirical observation and experience and in more modern times supplemented by organized research and development firmly based on applied science, mathematics, economics etc. Ring spinning represented the established order, widely practised and understood in all its details to a considerable extent, but the continuous scrutiny of research and development workers highlighted not only the virtues of ring spinning but also its technical and economic limitations and these aspects of the work stimulated the search for attractive alternatives. This search, and the striking results which it has

197

already achieved, was aided by many features of modern technology, e.g. it was able to build immediately upon the hard won fund of knowledge relating to ring spinning; the advances in every aspect of engineering, electronics, plastics technology and so on, could be instantly drawn into play and the rapid interchange of ideas and information which is such a feature of modern life all played their part.

Why then should there be such interest in new spinning techniques? The answer is almost entirely one of economics, for the technical performance of the ring-spun product represents a very good balance of properties which would be difficult to better unless an improvement in one direction were to be so important that it was acceptable to achieve it by sacrificing some other standard of performance. Economic assessments of the costs of converting fibre into usable yarn reveal that approximately half of these costs are absorbed by the spinning process alone, and in the case of fine-count yarns more than half. In addition, the bobbin or spool which is the output from ring spinning is unacceptable for the majority of the processes which follow and re-winding onto a more acceptable, usually much larger, package can represent a further cost and a process which does nothing to enhance the yarn. Indeed re-winding has been shown to introduce or accentuate undesirable features of the yarn under some circumstances. These high conversion costs are a direct result of the low productivity of each spinning position (spindle), resulting from the very fine strand which is produced at a restricted speed.

In spinning three things have to be done; the roving or sliver has to be drafted to give the desired linear density, the drafted strand has to be given an acceptable level of strength, usually by inserting a considerable amount of twist, and the yarn has to be wound onto a package suitable for handling during further processing. Studies of the technology and economics of ring spinning have demonstrated that the process is operated industrially under conditions close to the optimum and significant economies are unlikely to be achieved. The technical limitations are determined either by yarn tension or the maximum speed at which the traveller can run successfully around the ring. Both these limitations result from the fact that twist-insertion and winding onto the package occur simultaneously. The high rotational speed of the package and, more importantly, of the yarn balloon, are needed to insert twist into the yarn. Because the strand is continuous from supply to take-up it is essential to rotate one or the other if real continuous unidirectional twist is to be inserted. Thus, to achieve a significant increase in speed of production it is necessary to either devise a technique of twist insertion which separates it from winding-on so that the unacceptable yarn tension produced by rotating a long length of yarn (the balloon) about a large radius to encompass the package is avoided or to bind the fibres together in some other way, perhaps using alternating twist levels or even to cease using twist altogether.

By comparison, high-speed drafting presents relatively few problems, for experience has frequently shown that a strand which will draft at low speeds will invariably draft successfully at high speeds.

Economic appraisals of ring spinning have shown that for any given balance of costs there is an optimum system of operation, largely governed by spindle speed and package size, at which the balance of capital investment, power and labour costs gives the lowest production costs and that industrial practice is normally quite close to this optimum. As the balance of costs changes so does the optimum speed and the general trend is for speed to rise gradually over the years. This gradual rise in typical cotton spinning spindle speed in modern times from 8 to 10, 12 or 15 thousand rpm and beyond has not presented any real engineering difficulties. Even at these spindle speeds the production rate of each spindle for typical yarns is only of the order of 25 to 250 gm/hr and this fact emphasizes the high costs of spinning.

No discussion of modern developments in spinning can fail to be dominated by the outstanding success of rotor open-end spinning which, since its commercial introduction at the end of the 1960s, has taken over a very significant and ever increasing share of the short-staple spinning capacity of the world. Table I illustrates this in a quantitative way and it must be emphasized that one rotor has the productive capacity of three to six ring spindles. In addition, the high capital cost of rotor OE machinery means that it is usually operated on a 168-hours per week basis with doffing and so on executed with the machines in operation. No-one should, however, be so overwhelmed by this success story that he is blind to other developments such as alternative open-end techniques, self-twist, twistless and many other proposals which are at various stages of development and exploitation. A two-part paper based upon the author's very considerable experience of

TABLE I. World inventory of cotton ring spindles and OE rotors.

| | Ring spindles × 1000 | | OE rotors | |
|---|---|---|---|---|
| | 1976 | 1977 | 1976 | 1977 |
| Africa | 5498 | 5856 | 14 152 | 14 592 |
| N. America | 21 990 | 21 898 | 200 700 | 212 600 |
| S. America | 8576 | 8665 | 49 350 | 62 200 |
| Asia and Oceania | 67 103 | 68 513 | 336 412 | 350 357 |
| Europe | 47 952 | 47 062 | 1 252 466 | 1 402 299 |
| Total | **151 083** | **151 994** | **1 852 980** | **2 042 028** |

large-scale commercial operation of open-end machinery has given a brief but penetrating introduction to the subject of rotor spinning.[1]

The fundamental concept of open-end spinning has existed, at least in the patent literature, for a century or more. This concept is that there must be a break or open-end in the flow of fibres between supply and take-up in order that twist can be inserted on the down-stream side of this open-end without having to rotate the take-up package and without the twist being dissipated up-stream as false twist. This immediately opens up the possibility of twist insertion by rotating the yarn about a small radius, even about its own axis, and thus the possibility of very high speeds of twist insertion without the development of centrifugal forces which exceed the strength of the yarn. The very first commercially viable open-end machine was the Czechoslovakian-developed BD-200 machine, which in its earliest form had a rotor speed and nominal twist insertion rate of 30 000 rpm thus more than doubling the production rate of the corresponding ring frames.

Very many open-end techniques have been proposed and a review and experimental assessment of many of them was published in the early days.[2] This review was followed by a further Shirley Institute publication[3] dealing with the early years of commercial exploitation and several issues of *Textile Progress*[4-6] have continued to update the literature reviews, supplemented by a wide-ranging work from Germany.[7]

During the decade between Catling's and Hunter's publications there must have been several thousand papers dealing with all aspects of open-end spinning, as is demonstrated by the references in the above reviews, and the flow of information continues unabated.

All successful open-end systems produce the break or open-end by subjecting the incoming strand to a very high draft, so that the fibre flux is reduced to an average thickness of only a few fibres, say ten or less. It should be noted that this implies a very high velocity of fibre flow since the product of fibre flux and velocity must be constant throughout the system. With such a fine flux there will be many regions in which there are no fibres at all since, if we assume that a random lengthwise distribution of the fibres is maintained (i.e., a Poisson distribution of fibres as originally suggested by Martindale[8] in the case of yarns and slivers), at such a low expectation the probability of zero number of fibres is quite high. This tenuous strand is then condensed by allowing many layers to build up on a rotating surface until it achieves a thickness equal to the desired yarn when it is withdrawn along the axis of rotation, so inserting true twist in the final yarn. The many breaks in the flow of incoming fibres ensure that this twist cannot be dissipated as false twist. In rotor spinning the fibres are condensed onto the internal surface of a rapidly rotating rotor which is in the form of a shallow re-entrant frustrum. This surface has a linear speed in excess of that of the incoming fibres and so they are, to a large extent, maintained in a reasonably straight, well-aligned array. If the fibres are condensed onto a relatively slow-moving surface as in the DREF system then the fibres assume more crumpled and less well-aligned configurations and the resultant yarn is more akin to a traditional woollen or condenser yarn.

## I. Rotor Spinning

The majority of commercially successful OE spinners are of the opening roller drafting/rotor type. The first of these to be used on a large scale was the Czechoslovakian BD-200 machine, the principle of which is illustrated in Fig. 1. The incoming sliver of cotton or other short-staple fibres is drawn in by a feed roller operating in conjunction with a plastic condenser and spring-loaded feed plate. The sliver is fed to a rapidly rotating opening roller

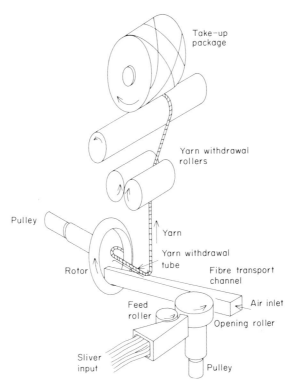

FIG. 1. Principle of operation of a typical rotor spinner (based on the BD-200 machine).

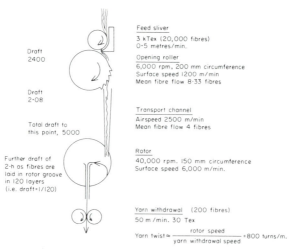

FIG. 2. Speeds, drafts and fibre flows in a typical rotor spinning system. Total draft between sliver + fibres being laid in rotor groove = 12 000, followed by a condensation of 120 (i.e. draft of 1/120). Thus, overall draft = 100.

covered with either rigid card clothing or pins. This opens the sliver and in doing so a very high draft is applied to the fibres as illustrated in Fig. 2. The fibres are carried by the opening roller until they reach the tangential transport channel. This has a rapid air-flow which strips the fibres and subjects them to a further draft of, say, 1.5 to 4 and carries them into the rotor. The air-flow is generated by a ring of pumping holes in the rotor which is driven at maximum speeds of 30 000 to 100 000 rpm, dependent upon the model. A trumpet-shaped separator plate deflects the fibres onto the sloping interior side-wall of the rotor where the high speed imposes the final draft. The fibres slide into the apex of the rotor groove where multiple layers are built up, perhaps 200 or more. To start spinning a seed yarn is fed, either manually or automatically, into the yarn withdrawal tube where it is drawn in by suction until it touches the rotor. This inserts the twist and as the yarn tail is thrown outwards by centrifugal force it contacts the fibres in the ring and starts to twist them, the yarn is immediately withdrawn and fed into the withdrawal rollers and if these actions are correctly co-ordinated continuous spinning commences. A discussion of the action within the rotor has been given by Nield[9] and a very full discussion of theoretical and practical aspects of the BD-200 machine has been published in Czechoslovakian and English by the team responsible for the initial development of the machine.[10]

The early BD-200 machines were very soon followed by similar units from several of the world's machine makers (Figs 3–7), notably Platt-Saco-

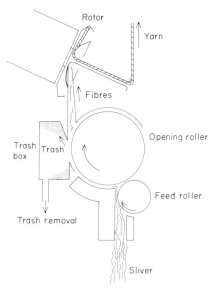

FIG. 3. A rotor spinning machine incorporating a trash removal system (based on the Platt-Saco-Lowell 883 machine).

FIG. 4. Overall view of a modern rotor-spinning machine, the Platt-Saco-Lowell 887.

Fig. 5. A modern version of the Elitex BD-200 rotor spinner.

Lowell, Rieter, Sussen and others. Many of these have been classified as "second generation" machines, their distinguishing features being rotor speeds in the range of 45 to 80 thousand rpm and, perhaps more important, trash removal devices based upon the air-stream cleaning principle, a typical example being illustrated in Fig. 3. The action of the opening roller in producing effective separation of the individual fibres gives ideal conditions for the removal of trash particles which may be in the sliver. With appropriate design and controlled air-flow it can be arranged that these relatively heavy, compact, particles are thrown clear of the fibre trajectory into a trash box which can be cleared by suction to a central collecting system. As is discussed below, trash particles and dust represent one of the potential problems in rotor spinning and their removal can be a very real advantage.

FIG. 6. Sectioned BD-200 spinning head. The knurled feed roller is driven from a horizontal worm shaft, about which the whole unit can pivot, via the plastic worm wheel and magnetic clutch within the cylindrical housing. The helical toothed-wire covering of the opening roller and the sectioned rotor are also clearly visible. Although this is the RC-type, which incorporates a trash removal system, this is not evident in this view.

## A. Economic aspects of rotor spinning

From the earliest days of the technique, rotor spinning has been a successful method of producing short-staple yarns. Initial success was with cotton but minor modifications to the machine and the settings very soon enabled virtually all man-made fibres, either alone or in blends, to be spun successfully. Whatever fibre is being spun, economic appraisals of the relative attractiveness of ring and rotor spinning continue to demonstrate that rotor

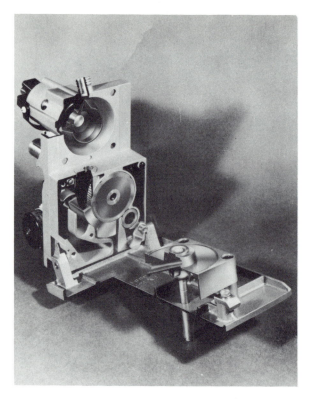

FIG. 7. Sectioned model of the Platt 887 spinning head. The feed roller is in the lower right hand corner of the head, the pinned opening roller throws the trash clear of the transport channel which is in the cover and vertical when the unit is closed and via which the fibres are fed into the rotor.

spinning is economically attractive for medium to coarse count yarns. Rotor machines are several times as expensive as ring frames in terms of capital outlay per spinning position but have between 2½ and 5 times the productive capacity and this cost can be offset against lower power and labour requirements per kg of yarn produced. In addition, rotor spinning eliminates at least one prior operation since it is a sliver-to-yarn system using drafts as high as 250. Thus the roving operation is done away with. Likewise, rewinding of the yarn can frequently be dispensed with, for the cheese or cone which rotor spinning yields may have a weight of 2 kg or more and be perfectly acceptable as the supply package in further processing. The high degree of regularity of rotor yarns and the unusually low incidence of gross faults also contribute to the acceptability of the yarn direct from spinning. These features of rotor yarns have also enabled some users to replace folded ring yarns with singles rotor yarns, thus effecting even greater economies in their processing sequences.

The range of yarn counts for which rotor spinning is most appropriate is probably from 20 tex to 500 tex or coarser. Any purely economic comparison must assume stated values for capital costs, interest rates and payback times, labour, power, space and maintenance and the relative spindle speeds of ring and rotor, the twist levels of the two types of yarns and especially the counts which are to be spun. In general, such appraisals have demonstrated that the coarser the yarn the greater are the attractions of rotor spinning. As time goes by and the relative costs of capital, power and labour change, the trend is for rotor spinning to become economically attractive for a wider range of counts, extending to progressively finer counts, and it has been reported that yarns of 15 tex ($40N_E$) have been rotor spun under commercial conditions in the Far East. It should be noted that it is generally accepted that the finest rotor yarns which can be spun have more fibres in their cross-section than the finest similar ring yarns but it is most unusual for commercial yarns to be spun to counts close to the limit and, therefore, the question of which method can give the finest yarns is to some extent academic.

## B. A comparison of ring and rotor yarns

A short-staple yarn (or, indeed, any type of yarn) which is rotor spun is not an exact replica of the ring yarn produced from the same sliver (Fig. 8). In

FIG. 8. Typical rotor (a) and ring (b) spun yarns of similar fibre type and yarn count.

the early days of rotor spinning these differences led to considerable criticism of the techniques but experience has demonstrated that these differences are not always disadvantageous to rotor yarns and at the same time refinements in technique have enabled some aspects of rotor yarns to be improved. These differences arise primarily because of the differences in fibre arrangement in the two types of yarn and these are in turn the result of very different conditions which obtain at the point of formation of the twisted strand, as is discussed later. The principal differences can be listed as follows:

(1) Rotor yarns are exceptionally regular in linear density and also have very few gross faults. These properties result from the very large number of doublings as the incoming strand is condensed on the rotor wall and the very effective opening action of the opening roller. It should be noted that as with any spinning system long-term evenness of the yarn is determined by the incoming sliver rather than the properties of the system. If the Uster standards of regularity are used for comparison purposes it is found that rotor yarns are almost always equal in regularity to the best 5% of carded yarns. For example, in the case of cotton yarns of 30 tex ($20N_E$) count rotor yarns will typically achieve an Uster value (CV) of 11% whereas a value of 13% for a similar ring yarn would be unusually good.

(2) Rotor yarns have a lower average tenacity and higher extensibility than ring yarns but a lower variability of tenacity because of their more even nature. Early rotor yarns were 20% or more lower in mean tenacity and this frequently resulted in severe criticism of rotor spinning, despite the fact that the yarns were so even that they had fewer weak places and as a direct result gave very few end-breaks during succeeding processes. Refinements in rotor design, etc., now mean that typical rotor yarns are often only 10 or 15% weaker than their ring spun equivalents. These, and other related properties such as the increased bulk of rotor yarns, are the result of the different fibre arrangement which occurs because of the very low tension at the point of twist insertion. In ring spinning it has been well-known for many years that the higher the spinning tension (i.e., balloon tension), the higher the yarn tenacity. This is believed to result from the higher tension drawing the fibres into straighter and more parallel configurations as they emerge in a flat ribbon from the front drafting rollers to be twisted into the yarn. These straighter fibres contribute a greater proportion of their ultimate strength to the strength of the yarn. By contrast, analysis of the forces acting upon the fibre strand in the rotor emphasizes that the strand tension at the rotor wall is very low, even virtually zero, and as a result the yarn which is produced tends to have its fibres in less-well-aligned and more convoluted forms. In addition, rotor yarns have a different twisted form to ring yarns. The latter have

all their fibres with effectively the same twist and during twist-testing such a yarn can be untwisted until the fibres are straight and parallel. If attempts are made to untwist a rotor yarn it is immediately evident that the core of the yarn becomes twist free before the surface fibres and in addition there are many "wrapper fibres" very tightly twisted around the outside of the yarn. Many of these fibres are also looped back upon themselves with reversals of twist. They are almost always regarded as being detrimental to the properties and appearance of the yarn but it has been suggested that in some low twist rotor yarns these wrapper fibres bind the core together and make an important indirect contribution to yarn strength.

(3) Because of the less-well-ordered fibre configurations open-end yarns are 10% or more bulkier than ring yarns and have a more open structure. As a result, it has been found that OE yarns absorb sizes and dyestuffs more readily than ring yarns. Also, to produce a fabric having similar handle either a slightly lower end and pick density must be used or a slightly finer count of yarn. These effects lead to direct economies during fabric production when using rotor yarns. However, the increased bulk and different fibre configurations lead to rotor yarns being stiffer and somewhat harsher in feel. Critics of rotor spinning have claimed that this is a serious disadvantage, leading to a 'boardy' feel in the fabric, although proponents of rotor spinning attempt to use this as an advantage, saying that these fabrics have a 'crisp' handle.

(4) Because twist insertion and package formation are completely separated an open-end spinning machine can, in principle, take up the yarn onto any type or size of package. The majority of rotor spinners yield a narrow parallel cheese weighing between one and two kilos. An alternative which many machines now offer is a narrow-angle cone of similar size, and coarse count machines may yield a much heavier package. These packages are acceptable to further processes without the need for rewinding, an operation which can be expensive and is essentially non-productive. Rotor-spun packages have been used directly in two-for-one twisting or in warping creels and knitting, all of which are operations which would demand rewinding of the relatively small ring bobbins.

## C. The effects of rotor spinning on other processes

The most obvious and immediate effects of the introduction of rotor spinning upon the processing sequence of short-staple yarn manufacture were the elimination of roving and re-winding immediately before and after spinning, respectively, but it also had more far-reaching effects upon the preliminary processes. Rotor spinning quickly demonstrated that it could

**H**

operate with unusually low end-breakage rates and that many of the ends down which did occur were related to the build-up of deposits within the rotor, especially accumulations of trash when spinning cotton. These trash particles, which would go unnoticed in ring spinning, may be trapped in the rotor groove and be held firmly in place by the high centrifugal force. They continue to attract further small particles and as the aggregation continues it can disturb the smooth stripping action of the yarn tail and give rise to a periodicity in the yarn having a wavelength equal to the groove circumference. This may continue for some time and be a more serious problem than an end breakage. A larger trash particle frequently causes an immediate end breakage and several authors have published recommendations as to the maximum acceptable trash content (based upon the Shirley analyser) of slivers for rotor spinning. This, in turn, has led to considerable attention being paid to the cleaning efficiency of blow-room and card alike and the use of additional cleaning points in the blow-room along with trash removal via modifications to the card such as tandem carding, "Crosrol" web clearing units and so on.

A more subtle but equally important effect is that of micro-dust in the rotors and this affects cotton and man-made fibres alike. In all rotor spinning sooner or later the inner surfaces of the rotors will acquire a coating of fine white powder. As this develops the yarn quality will gradually deteriorate, especially its tenacity which may fall by 20% or so. Cleaning the rotors will at once restore the yarn properties. This micro-dust has been extensively studied to ascertain its nature and origin and strenuous efforts have been made to reduce the amount to the smallest possible level. This has been done by ensuring that throughout the earlier processes there are sufficient really effective cleaning points at which airflow is adequate to remove the dust as it is produced by the mechanical action upon the fibres removing particles from them.

Modern rotor spinning frequently operates with such a low end breakage rate that it is desirable to clean the spinning heads sequentially during spinning, say, every few hours, and the travelling automatic cleaning and piecening devices which are available can be programmed to do this in between dealing with any end breakages.

## D. Some detailed considerations of the actions in rotor spinning

### 1. Sliver opening

The action of the opening roller is to reduce the fibre flux to an exceptionally tenuous stream by means of the very high draft which occurs between the feed system and the opening roller. Despite this high draft, ultra-high-speed photographic studies have revealed that the fibres are not straightened to the extent which might have been expected, and this may account in part for the structure of the final yarn.[11] Early rotor spinners used, and many still use, rollers covered with metallic card clothing with steeply inclined teeth as shown in Fig. 9. Such wire gives excellent results with cotton but produces a totally unacceptable level of fibre breakage with man-made fibres, which demand wire having a radial or negative front angle (Fig. 9). The speed of the roller also has an effect upon both the opening action and the degree of fibre damage and typical speeds are in the range 3000 to 8000 rpm for rollers of around 80 mm diameter.

FIG. 9. Opening rollers. Behind are two wire covered rollers, on the left a typical tooth form for synthetic fibres, on the right a cotton type wire, and in front a pinned roller. The wire covered rollers are from an early version of the BD-200 machine, the pinned roller is from the Platt 883 machine.

An alternative type of roller is set with pins and the Platt machines, for example, have always used these and they are tending to be used more widely, frequently as replacements for wire-covered rollers. Their primary advantage is their much longer life, especially when processing abrasive fibres such as heavily delustred polyester. They also appear to have a more gentle opening action, cause less fibre damage and can be run at lower speeds. The reasons why pins last many times as long as metallic card clothing are not understood. Several suppliers of replacement rollers of the latter type offer them with special surface treatments intended to improve the fibre-to-metal frictional effects or increase the surface hardness but even these rollers can have the leading edges of the teeth deeply grooved by the fibres until the tips of some teeth are cut completely off. This can occur after only a few months operation when spinning heavily delustred fibres.

## 2. Action at the rotor groove

All conventional rotor spinners feed the fibres into the rotor more or less tangentially at one point, building up multiple layers in the rotor groove prior to the yarn tail stripping them as they are twisted into the yarn. The yarn tail must rotate at a different speed to the rotor in order to generate this stripping action and under all normal conditions it rotates slightly faster than the rotor ("normal spinning") but it is possible for the yarn tail to rotate more slowly ("reverse spinning"). The latter situation is uncommon but is known to yield poor quality yarn, unstable spinning and frequent breakages. Despite statements to the contrary, work at Bradford has demonstrated that under some conditions reverse spinning can be maintained and can be caused to revert to the normal mode.[12]

The stripping action of the yarn tail has been analysed by various authors (e.g., Nield).[9] Essentially the yarn tail strips a ring of fibres which is equal in thickness to the yarn at the stripping point and the ring then tapers linearly around the rotor groove to zero thickness immediately behind the stripping point.

In the early days of rotor spinning it was predicted that the dominant force on the yarn would be centrifugal and that this would mean that the yarn tail would at all times be radial with a very sharp bend at the rotor wall and that the fibre band would be pressed so tightly into the groove that there could be no possibility of twist passing the peeling point. However, high-speed photography has shown that neither assumption is correct; the yarn tail adopts a curved path and blends smoothly into the fibre band and, furthermore, twist is propagated for some distance along the rotor groove. This peripheral twist extent ranges from a few millimetres to a few centimetres and its presence is believed to be essential if spinning is to be continuous. High-speed photography in the laboratories of the present author also indicates that within the peripheral twist extent the fibre band is so tightly twisted that it forms a compact, well-rounded structure which is not compacted into the corner of the groove (Fig. 10).[12]

Every time the yarn tail and the region of peripheral twist pass the fibre entry point there is a finite probability that any of several related effects will occur and these have unfortunate results upon the yarn.

Firstly, fibres may be laid in the groove on top of the already-twisted structure in the region of peripheral twist, secondly they may be laid partially in front and partly behind the stripping point so bridging the stripping point and, thirdly, they may be caught by and twisted directly around the yarn tail. All such fibres will be twisted around an already-twisted structure and will not be fully integrated. As a result they form loosely twisted surface fibres or very tightly twisted surface fibres which in some cases are twisted back upon themselves and which can have regions in which the twist is so tight that they are perpendicular to the fibre axis. These wrapper fibres are entirely on the surface of the yarn and exhibit little or no fibre migration through the core of the yarn; in addition, because of the low tension in the twist-insertion region within the rotor, the core fibres also have a lower degree of migration than have the fibres in ring yarns and this reduced migration may be a factor contributing to the lower tenacity of rotor yarns.

## 3. The importance of the withdrawal tube

Photographic and stroboscopic observations of the rotating yarn tail within the rotor show that the pressure of the yarn against the entrance to the withdrawal tube results in the twist building up in the yarn tail to a level considerably in excess of the final twist level. (This effect is similar to the twist-blocking action of the traveller in ring spinning.)

FIG. 10. High speed photograph of the yarn tail and fibre ring within a transparent rotor. The peripheral twist extent is clearly visible as the region in which the fibre band is compact and not lying in the apex of the rotor groove (Seow, 1980).

Early withdrawal tubes were always highly polished and perhaps chromium plated but it was then found that a grooved or roughened surface on the withdrawal tube entry (the "navel") resulted in a further build-up of twist in the yarn tail and this means that the tail can more easily twist-in the fibre ring with an increased peripheral twist extent. This in turn means that with such a modified navel spinning can be maintained under adverse conditions despite the fact that the yarn is almost certain to be weaker and much rougher and more hairy in appearance than when a smooth navel is used. Hence, one machine manufacturer offers both smooth and grooved navels with the advice "spin with the smooth tube if possible, but if an excessive end-breakage rate is experienced then change to the grooved tube provided the more hairy yarn is acceptable". This effect can be so pronounced that under experimental conditions it is sometimes possible to spin yarns which, when wound onto the package, are so weak that they can only be unwound with great difficulty.

## E. Further developments in rotor machinery

Rotor spinning machinery has developed to a very sophisticated level within a decade of its first introduction. The obvious developments, once the basic machinery was firmly established in the spinning industry of the world, were increases in speed, increasing levels of automation and technological refinements which meant that the yarns were more closely akin to their ring-spun counterparts.

The production rates have increased as a direct result of both the first and last of these developments. The first BD-200 machine had rotor speeds of 30 000 rpm and the yarns needed 20% to 30% more twist than the equivalent ring yarns. Speeds have risen to 40 000; 60 000 and even 80 000 rpm under production conditions and at the same time twist levels have dropped by 10 to 15%.

This has meant that yarns of medium to coarse counts, say, 30 to 100 tex, are frequently produced at speeds well in excess of 100 m/min and under many conditions these speeds are so high that manual piecing after an end breakage is very difficult or well-nigh impossible. Thus, automated or semi-automated piecing becomes a near-necessity. Allied to the fact that the high production rate means that relatively few spindles are required for a given production this means that the high cost of the mechanically and electrically complex devices needed for automatic piecing can be justified also on economic grounds.

A further reduction in manual supervision can be achieved by automatic doffing of full bobbins of yarn and the majority of rotor spinners are now offered with both automatic piecing and doffing either integrated into the machine or offered as optional additions (Figs 11 and 12). Apart from maintenance and general supervision the most sophisticated machines only require manual intervention for the replacement of empty cans of sliver, all the other requirements of normal operation being fully automated, the repairing of end breakages, rotor cleaning and doffing of full packages all being included in the automation facilities. The full packages are transported to the end of the machine and stacked on pallets or in large containers as well as being doffed from the spinning positions.

The automated cleaning/piecening devices can also be programmed to undertake rotor cleaning duties in a regular cycle. The Sussen "Cleancat" and "Spincat" are a pair of associated units which were the first to be capable of this type of duty. Because of the desirability of regular cleaning of the rotor to avoid a deterioration in yarn properties and the very low end-breakage rates which can be achieved in rotor spinning it may be desirable to break each end down at intervals of a few hours in order that this cleaning can be executed. The Sussen units are travelling ones which can service a

Fig. 11. The Schlafhorst "Autocoro" rotor spinner. On the left is the travelling automatic doffer which places the full packages on to a central conveyor for transport to the end of the machine, and replaces them with spools carrying a short pre-wound length of yarn ready for the automatic piecener/rotor cleaner on the right to restart spinning.

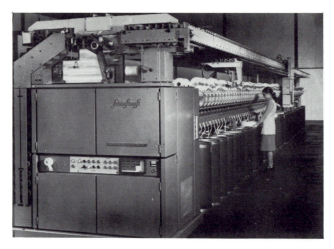

F<small>IG.</small> 12. General view of the "Autocoro" machine. The equipment above the headstock prepares the pre-wound spools by winding a short length of yarn on to each one prior to it being loaded into the automatic doffer.

number of open-end frames, provided that they are built using the Sussen spinning heads, and which service one head at a time. The units can be programmed to patrol the complete installation to deal with any broken ends, they then return to the first frame and open every head in turn, cleaning the rotor and its housing and then piecening the end. The complete installation is then patrolled again for the repair of any broken ends before the cleaning routine is applied to the second frame and so on until the routine is completed.

## II. The Commercial Position of Open-End Spinning Techniques other than Short-Staple Opening Roller/Rotor Machines

Of the many alternative methods which have been described or patented most of which were reviewed in the Shirley Institute publication,[2] few have come anywhere near commercial viability with the exception of those reviewed below.

The first of these is rotor spinning utilizing roller drafting, although there is no fundamental difference between this and opening-roller drafting. The first full-scale rotor machine to be publicly exhibited, the Czechoslovakian KS-200, was of this type and although it never went into series production similar machines have been constructed in Japan by Howa and in France by

SACM. Conventional double-apron drafting is used with the front rollers rotating at a high enough speed to reduce the strand to the necessary tenuous form. Applied in the first instance to machines suitable for the spinning of coarse-count short-staple yarns, sometimes spinning directly from a card sliver to produce yarns which could compete with condenser-spun types, the technique has also been used for open-end spinning of long-staple materials.

A rather more unusual machine of this type is the SACM ITG 300 (Fig. 13). Intended for the rotor spinning of long staple materials this machine is unique amongst production units not only in its design but also in the claim of its manufacturers that it will spin wool or wool rich blends successfully to medium counts. The unique feature of the design is the top front roller of the drafting system. This, designated the "disintegrator roller" by the manufacturers, is built up from a series of segmented discs so that the nip is intermittent. It is claimed, apparently with considerable justification, that at the high draft which is used this construction gives good drafting and at the same time enables the fibres to be drawn into the feed tube which conveys them into the rotor without the disturbing air currents which are generated by solid drafting rollers. The first machines of this type had a layout similar to that of a ring frame, the drafting system feeding the fibres into a short vertical tube which conveys them into the rotor which also has its axis vertical. The yarn is withdrawn vertically downwards through a stationary withdrawal tube, which enters the rotor via its hollow spindle, to be packaged onto a parallel spool. In the second version of this machine a complete redesign is evident with a layout virtually identical to a conventional rotor spinner, the drafting system and rotor being housed in a compact unit in which the rotor axis is

FIG. 13. ITG 300 long-staple rotor spinner of the SACM Company.

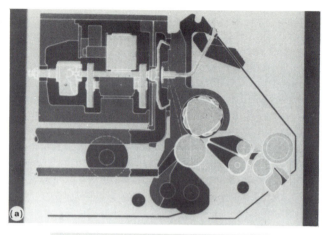

FIG. 14. (a) General arrangement of later type of ITG 300 machine. The drafting system is lower right, the rotor and its indirect bearing system is upper left and yarn withdrawal tube is to the top right. (b) The disintegrator roller.

horizontal with both fibre feed and yarn withdrawal at the front of the rotor, spinning taking place in an upwards direction (Fig. 14).

## A. Air vortex spinning

Air vortex devices were amongst the earliest open-end devices to be studied and even today remain one of the simplest of devices which are easy to set up as demonstrations of open-end spinning. Nevertheless, and despite intensive study, the original type of device based upon a helical vortex which is considerably longer than its diameter, has never produced satisfactory yarns and has not progressed beyond the stage of an interesting laboratory curiosity. And this is despite the fundamental attraction of the technique that there are no mechanical components rotating at high speed, the rotation of the fibres and the open end of the yarn being a direct result of their being carried around by a rapidly rotating current of air. In the 1970's work in Poland resulted in the production of an air vortex spinner of a somewhat different type, the Polmatex PF1 machine. This machine has come into industrial use in the socialist countries to a limited extent for the production of coarse count fairly softly twisted short-staple yarns. These yarns are claimed to be competitive with both condenser-spun and some carded yarns whereas the best that could be achieved with the earlier experimental vortex devices was yarns which were so soft, hairy and irregular as to render them commercially completely unacceptable.

The principle of the early vortex spinners was that the yarn was formed in a relatively long vortex tube. The Shirley Institute tried many elaborations of this principle in attempts to increase the twisting efficiency, reduce the loss of fibres which failed to be twisted into the yarn tail and to try to ensure that the majority of the fibres were caught by the rotating end of the yarn instead of being trapped along the yarn closer to the yarn withdrawal point. This last feature means that the yarns have a very high proportion of loosely-attached surface fibres which are not integrated into the structure.

By contrast, the PF1 machine (Fig. 15) utilizes a vortex in which the axial direction of the airflow reverses. The airstream enters the vortex tube, carrying the fibres, and forms an almost planar vortex near to the closed end of the tube. The air then reverses its direction of axial motion as it is withdrawn along the centre of the tube, leaving the fibres as a rotating ring in

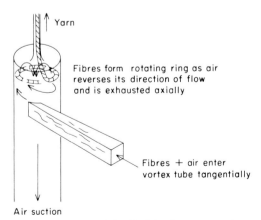

FIG. 15. Principle of the Polish air vortex spinner.

the initial vortex. The yarn tail can thus gather and twist-in the fibres in a manner very similar to that of a rotor spinner. Indeed, this machine has been described as a "rotor spinner without a rotor".

## B. DREF spinning

Throughout the 1970's there has been considerable interest in this method of spinning, invented and commercially exploited by the Austrian company, Dr Ernst Fehrer. The method has the potential merit of being seemingly independent of fibre length, so that a very wide range of types of fibres can be handled. However, the process has what for many purposes is a serious disadvantage, although it might equally be used advantageously in other end-uses, and this is that the fibres in the yarn are always in a highly disoriented array, giving the yarn many of the characteristics of a condenser type of material (Fig. 16). Hence the yarns, in common with woollen or cotton condenser yarns, tend to be weak, bulky, hairy and of low strength. This characteristic arises because of the nature of the yarn-forming mechanism. The tenuous and necessarily rapidly-moving stream of fibres which emerges from the drafting system is condensed onto the outer surfaces of two closely-spaced perforated rotating drums as the air-stream which carries the fibres is drawn into the interior of the drums. Because the drums revolve in the same direction the fibres are rolled into a twisted yarn in the narrow gap between the drums and can be withdrawn axially. As the fibre stream is condensed in this way it must necessarily suffer a high deceleration, just as do the fibres as they pass from cylinder to doffer in a card, and therefore the fibres are deposited in non-aligned and largely random configurations. Furthermore, just as in woollen spinning, there is then little or no draft applied to the strand to straighten and align the fibres.

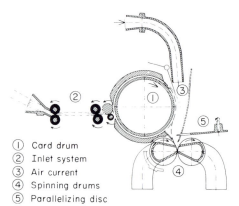

①   Card drum
②   Inlet system
③   Air current
④   Spinning drums
⑤   Parallelizing disc

FIG. 16. Principles of the operations of the DREF spinning system.

FIG. 17. 48-Spindle DREF spinning machine.

Although large-scale commercial utilization of DREF machinery has been slow to occur, considerable success has been achieved in preliminary trials with the spinning of fibres as diverse as wool on the one hand and asbestos on the other (Fig. 17). More recent versions of the process have combined the staple fibres with a continuous filament core so as to obtain the benefits of the strength of the core allied with the softness of the staple-fibre surface.

## III. The Prospects for Yarn Production by Novel Techniques other than Open-End Spinning

In many respects, and despite its explosive growth during the period 1970 to 1980, open-end spinning can be regarded as a conservative development in that the product is very similar to its ring-spun counterpart. In particular, the yarn contains continuous, unidirectional, twist, and open-end spinning however novel the technology, is but an alternative method of producing such yarns, in which the differences from the traditional yarns are relatively minor "second order" differences. Although both traditional yarns and open-end ones too derive their coherence and tensile properties very largely from the inter-fibre forces generated by twist, in many respects twist has become an unfortunate necessity which textile technologists have had to defeat or learn to accept by using great ingenuity. Alternative methods of binding the fibres together are to use variable or even alternating zones of twist, either along the length of the yarn or in the different radial zones, or to bind a twistless strand together by means of a temporary or permanent adhesive. It can be argued that modern advances in the technology of synthesizing organic polymers have led to developments in man-made fibre production which could render the age-old technology of staple fibre processing obsolete by enabling textile production to concern itself only with continuous-filament products. However, the advantages of staple fibre products are such that really they have no serious competitors, and for very many end-uses such fibres are preferred for a multiplicity of reasons. This is borne out by the very large production of man-made fibre in staple form, all of which is produced by cutting or breaking what is initially continuous filament. Textured continuous filament yarns have gone part way to bridging the differences between "flat" filaments and staple, but their widespread acceptance is subject to fashion and customer preference as well as economic and technical comparisons.

During the 1970s there have been a number of proposals aimed at utilizing the best features of both continuous filament and staple by combining both types of fibre into one yarn in what are effectively variations of the older core-yarn concept and which, in some cases, reverse the configuration so that a staple fibre core is wrapped with a continuous filament binder. However they are produced, and whatever their configurations, all these yarns are designed to retain those features of staple yarns which give them "customer appeal" such as warmth and softness of handle, bulk and covering power, whilst at the same time deriving strength and coherence from the filament component. With such a combination it may be possible to produce a yarn of acceptable strength at an unusually low level of twist, leading to a high production rate which may offset the economic disadvantage of all these yarns which is that the fine-count continuous filament component represents an inherently expensive material.

It is just as difficult and problematical to attempt to predict the future in any area of technology as it is in any other area of human endeavour, for whilst it is reasonably easy and safe to predict the creeping progress of already established technology, there is always the very real possibility that a completely new invention or discovery will come along with no prior warning and be so successful that it renders obsolete the previous state of the art. The commercial success of rotor spinning is almost in this category, but if we look for higher speeds we should remember that the self-twist technique has been effectively five or more times faster than rotor spinning ever since it was first introduced, and adhesive bonding, either temporary as in the TWILO process or permanent as in the Bobtex technique, is potentially even faster.

Whilst these new processes may give concern to spinners with heavy investment in more conventional machinery, they make an absorbing study for students of technology who may see the next two decades occupied by technical developments equally as fascinating as those during the last 20 years or so. This prospect of even more changes in yarn production in the coming years means that any country which wishes to remain in the forefront of technology must ensure that it develops and maintains the Universities, Colleges and Research Institutes which will provide a steady flow of those technologists and applied scientists who will be needed to exploit the commercial possibilities of these ever-more-sophisticated processes and to further the research needed to back up this important industry. Examples of the type of work which is required to educate these technological leaders of

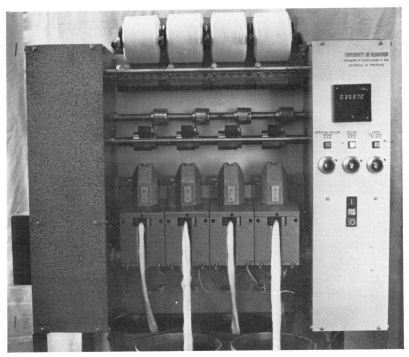

FIG. 18. Small scale rotor spinning machine utilizing four heads of the BD-200 machine, specifically developed for research, rapid trials and student demonstrations.

the very near future have come from the laboratories which are under the present author's supervision, where the most sophisticated state-of-the-art instrumentation has been used in a variety of research studies, for example, Behzadan (the use of ultra-high-speed photography via laser techniques), Afshari (applications of high frequency tension measurements) and Seow (further studies also using computerized data analysis).[11-13] Such work has also led to the development of a small-scale rotor spinning machine, eminently suitable for research, demonstration and teaching, various versions of which have been supplied to institutions in many countries (Fig. 18).

## References

1. Worrall, G. (1978). *Textile Month*, May 29 and June 69.
2. Catling, H. (1968). *Break Spinning*. Shirley Institute, Manchester.
3. Bancroft, F. and Lawrence, C. A. (1975). "Progress in OE Spinning, World Literature Survey 1960–74". *Shirley Inst. Publication S16*, Shirley Institute, Manchester.
4. Smith, P. A. (1972). *Textile Progress*, **1**, No. 2. Textile Institute, Manchester.
5. Dyson, E., Iredale, J. A. and Parkin, W. (1974). *Textile Progress*, **6**, No. 1. Textile Institute, Manchester.
6. Hunter, L. (1978). *Textile Progress*, **10**, No. 1/2. Textile Institute, Manchester.
7. Artz, P. and Egbers, C. (1979). *Technologie des Rotorspinners*, Meliand Textilberichte, Heidelberg.
8. Martindale, J. (1945). *Journal Text. Inst.*, Manchester.
9. Nield, R. (1975). *Open-end Spinning*. Textile Institute, Manchester.
10. Rohlena, V. *et al.* (1975). *Open-end Spinning*. Elsevier, New York.
11. Dyson, E. and Behzaden, H. (1976). *In The Yarn Revolution*. Ed. Harrison. Textile Institute, Manchester.
12. Seow, Y. S. (1980). *Ph.D. Thesis*. University of Bradford.
13. Afshari, G. and Dyson, E. (1979). *ASME Paper*. 79-Tex-7, American Soc. Mech. Engineers, New York.

# Chapter 8

# Quality Control and
# Data Logging Techniques

K. DOUGLAS

## I. Introduction

In the last 25 years, the textile industry has changed from a wage intensive to an extremely capital intensive industry. At the present time, the investment associated with a modern textile spinning or weaving unit is comparable with that of a high quality machine manufacturer. This results in the fact that although the textile industry has tended to expand more rapidly in the developing countries of the world, its existence is still assured in the more developed countries, particularly where highly trained staff are necessary to run machinery which is becoming extremely complex. Figure 1 compares weaving and spinning operator work-load over the last 250 years in terms of amount of yarn or cloth produced per hour worked. It would seem that with both spinning and weaving, the average production increase per year during this period was approximately 3%, i.e. production was doubled every 20–25 years. The last 25 years, with the introduction of projectile and air jet/water jet weaving, has brought about quite a revolution in woven cloth production and a reduction in the hours worked to produce a length of cloth.

Production in the future will depend on the development already in progress with respect to open-end spinning and shuttleless weaving, and the application of these products world-wide. In 1978/79, 65% of all spinning machinery shipments referred to OE-rotor spinning machines, and 40% of all weaving machinery shipments referred to shuttleless weaving machines. It is quite evident that as a consequence of the progress being made in these

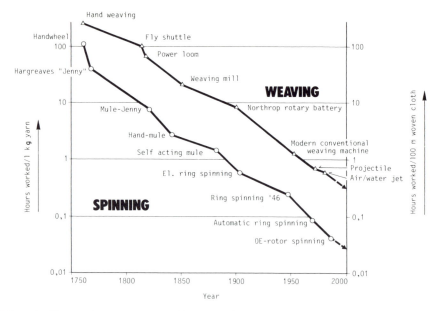

Fig. 1. Change in the amount of human effort required in spinning and weaving during the last 250 years.

fields, quality control procedures will also have to be intensified and in some cases updated, in order to keep pace with production requirements.

The high production shuttleless weaving machine, for instance, demands a much higher yarn quality than its conventional predecessor, and the OE-rotor spinning machine a much more intensive preparation of the raw material.

Inspection of woven cloth covers every metre of yarn in both warp and weft, whereas in spinning, sampling techniques are normally applied. Sampling techniques will always be successfully applied as long as the parameter measured is a frequent event, e.g. within a relatively short sample length of yarn, roving or sliver, enough "events" are available to statistically represent the "population".

To an increasing amount, however, continuous data logging and automatic material correction procedures are being introduced into staple spinning. As shown in the diagram (Fig. 2), sampling techniques are applied in the quality control of yarn strength and elongation, count, evenness and the incidence of yarn faults (slubs, fly, piecings, etc.). Certain spinning preparation processes, on the other hand, can be equipped with autolevelling devices which ensure an improved evenness or count constancy of the

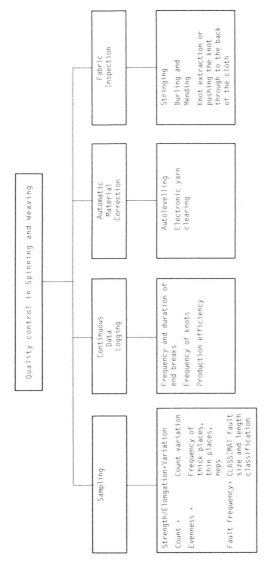

FIG. 2. Quality control in spinning and weaving by means of various sampling, monitoring, autolevelling and inspection techniques.

material, and therefore better count and strength properties in the yarn. With the increasing reliability of such devices, sample sizes and sampling intervals can be reduced.

Continuous data logging is necessary where the parameter required to be controlled is very infrequent and where manual sampling methods are inapplicable. An example of this would be the occurrence of end breaks in terms of their frequency and duration. In this respect, it has been shown that there is a direct correlation between the end break frequency and the quality of the material being processed, and this with respect to both spinning and weaving. The same parameters are also used in assessing productivity and the efficiency of any particular process, article or operative.

Finally, data logging techniques provide a means of reducing processing costs. Table I shows an analysis of the costs resulting from machine stops due to yarn breaks in spinning, winding and weaving. The table indicates that both in cotton and worsted spinning, quite considerable costs are involved as a result of yarn breaks in terms of waste material, downtime, operator work-load, downgrade quality, etc.

In this chapter, data logging techniques for determining end break frequencies in spinning and weaving will be considered. For each process the incidence of random and systematically occurring end breaks and the reasons for such end breaks will be analysed. Reference will be made to data logging techniques in general and, based on these, to methods for reducing end break frequency, improving quality and reducing costs.

## II. Frequency of End Breaks in Spinning

### A. Introduction

Ring spinning is the most wage-intensive part of staple yarn production. End breaks at the ring spinning machine with respect to an individual spindle should be considered as "seldom-occurring events". However, translated into the total number of spindles in the mill, these end breaks can become a major factor from a personnel point of view and consequently also with respect to costs. A much more serious situation is available when the end breaks produce roller lapping!

The manual detection and recording of end breaks necessitates a considerable amount of work. Furthermore, the data is often unreliable because the analysis is carried out over too short a period of time. Results which are not based on several 10 000 spindle hours are statistically unreliable. Furthermore, it is possible that short-term deviations may occur. Repeating such end break trials would, in most cases, be advisable. However, due to the excessive costs and amount of work involved, it is not usually carried out.

TABLE I. Costs resulting from machinery stops due to yarn breaks.

|  | Cotton SFr. |
|---|---|
| *Ring spinning* |  |
| —Cost of 1 yarn break | 0.10 |
| —40 000 spindles produce approx. 8000 end breaks/shift |  |
| —Costs/shift | approx. 800.00 |
| *Automatic winding* |  |
| —Cost of 1 knot | approx. 0.005 |
| —40 000 spinning spindles, feeding approx. 700 winding positions produce approx. 70 000 knots/shift |  |
| —Costs/shift | approx. 350.00 |
| *Weaving* |  |
| —Cost of 1 end break | 0.50 |
| —40 000 spinning spindles feeding approx. 150 weaving machines produce approx. 1200 end breaks/shift |  |
| —Costs/shift | approx. 600.00 |
|  | *Worsted* SFr. |
| *Ring spinning* |  |
| —Cost of 1 knot | 0.15 |
| —20 000 spindles produce approx. 12 000 end breaks/shift |  |
| —Costs/shift | approx. 1800.00 |
| *Automatic winding* |  |
| —Cost of 1 knot | approx. 0.005 |
| —20 000 spinning spindles feeding approx. 500 winding positions produce approx. 100 000 knots/shift |  |
| —Costs/shift | approx. 500.00 |
| *Weaving* |  |
| —Cost of 1 end break | 0.50 |
| —20 000 spinning spindles feeding approx. 100 weaving machines produce approx. 1800 yarn breaks/shift |  |
| —Costs/shift | approx. 900.00 |

The number of end breaks at a ring spinning machine is also a measure of the quality of the yarn. Unevenness introduced during the processes prior to spinning inevitably affects the quality of the yarn being spun and consequently the number of end breaks. In the example shown (Fig. 3), two spinning mills are compared with respect to ends down frequency, and this

230 K. DOUGLAS

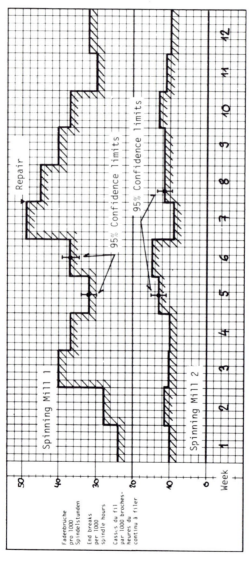

Fig. 3. Comparison between two spinning mills (weekly average of ends down frequency). Spinning mill 1: Carded cotton Nec 30, Nm 50, 20 tex. Spinning mill 2: Combed cotton Nec 80, Nm 135, 7.5 tex.

over a period of 12 weeks. Spinning mill 1 refers to a badly-run mill and spinning mill 2 to a well-run mill. The reason for the high ends down rate in spinning mill 1 is shown in Fig. 4, i.e. poor quality sliver produced at the drawframe affecting ends down frequency of a complete machine and, to some extent, the complete mill. Figure 3 shows: up to week 6, normal end break conditions; after week 6, increase in the number of end breaks by approximately 30%; evenness testing of roving and drawframe sliver (Fig. 4); reason, worn-out bearing at the front roller of the second passage of drawing; measures undertaken, repair of the drawframe (week 7); after week 10, effective improvement in end break frequency at ring frame.

Special attention should be paid to those spinning positions where the number of end breaks exceeds the tolerance limit. These spindles necessitate not only an increased amount of work on the part of spinning personnel, but the quality of the yarn falls far below the average. Special tests carried out on such yarns showed that a higher number of faults is available, causing considerably more downtime in subsequent processing.

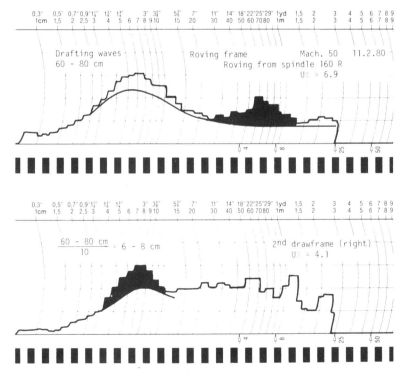

Fig. 4. Spinning mill 1 (see Fig. 3). Effect of drawframe sliver unevenness on the end break frequency of the ring spinning machine.

The few spinning positions with end break frequency values above the average also influence the overall mean value of the number of end breaks, and this to quite a high degree. Investigations have shown that 3–10% of such spindles can increase the overall end break frequency value by as much as 30%

Spinning mills which monitor and systematically record the number of end breaks are in a much better position to improve quality and retain their position in a competitive market. Better quality yarn justifies higher selling prices or at least helps to secure follow-up orders. Furthermore, this yarn will present fewer problems during the subsequent manufacturing processes.

It is to be expected that considerable advancements will be possible with the help of data logging systems since manufacturers of ring spinning machines, drawframes and roving frames, etc. as well as textile research institutions, will be capable of studying the origin of end breaks.

## B. Theoretical aspects

It is generally assumed that a small percentage of the spindles in the ring spinning mill is responsible for the majority of end breaks. Figures are quoted of 1–3% of spindles being responsible for 10–30% of the end breaks.

Unfortunately, such figures are normally based on extremely short and statistically inadequate sampling sizes. This results in the fact that the 1–3% of spindles having a high number of end breaks in one shift will not necessarily be the same spindles having 10–30% of the total number of end breaks in the next shift. Furthermore, this assumption does not normally take into consideration the fact that end breaks in ring spinning can also be due to the material and therefore occur according to a statistical random distribution. In the case of end breaks, this random distribution is a Poisson distribution which refers to the "distribution of a number of events occurring over a long period of time when these events occur infrequently".

The basic parameter, in this respect, is the mean number of end breaks/spindle such that if the mean number of end breaks/spindle is high, then it can be concluded that only a certain percentage of these end breaks is due to random events, and that a further percentage of the end breaks is due to some systematic reason. Figure 5 shows the relationship between the number of spindles and the number of end breaks/spindle with reference to a mill example. The diagram shows that with, for instance, seven end breaks/spindle or more, a total of approximately 4.2% of the spindles have more than this number of end breaks/spindle, but that approximately 2.5% of the total spindles having seven end breaks/spindle or more is due to random (Poisson) conditions. Consequently, the example illustrates that even with a

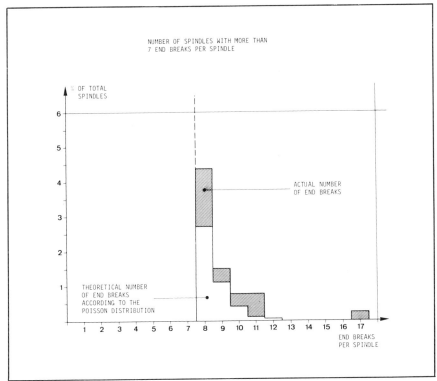

FIG. 5. Diagram showing the number of spindles in a cotton spinning mill having more than seven end breaks/spindle (mean value) taking into consideration a statistically representative sampling size.

high number of end breaks/spindle, only a small percentage of these end breaks is due to systematic reasons, which can be determined and corrected.

In order to better understand the conditions pertaining to the random distribution of end breaks at ring spinning and the systematic number of end breaks which can be due to "outsider" or "rogue" spindles, a calculation can be set out. This calculation is shown in Fig. 6 and refers to an example arbitrarily chosen from a "Letter to the Editor" on the subject of jute yarn spinning as published in the *Journal of the Textile Institute.*[6] The calculation shows the following. Taking the probability of more than two end breaks occurring, in the ideal case, 4.7% of the spindles will be responsible for 29% of the end breaks. According to observed results as shown in the example, in actual fact 6.25% of the spindles will be responsible for 29% of the end breaks. Consequently, it can be concluded that $6.25 - 4.73 = 1.52\%$ of the

ENDS DOWN FREQUENCY AT THE RING SPINNING MACHINE

Frequency distribution of end breaks per doffing:

| Number of end breaks | 0 | 1 | 2 | 3 | 4 | 5 | 6 | 7 | 8 | 9 | 10 and more |
|---|---|---|---|---|---|---|---|---|---|---|---|
| Observed frequency | 337 | 31 | 17 | 5 | 3 | 2 | 2 | 1 | 1 | 1 | 0 |

Example:    138 end breaks at 400 spindles. Mean number of end breaks/spindle = $\frac{138}{400}$ = 0.345

Probability:

$$p\ 0 = \frac{0.345 \cdot e^{-0.345}}{0!} = 0.7082$$

|  | No. of spindles (ideal) | | Observed | End breaks |
|---|---|---|---|---|
| p 0 = 0.7082 | = 283 | Sp = 70.8% | 337 | 0 |
| p 1 = 0.2443 | = 97.73 | Sp = 24.4% | 31 | 1 × 97.73 → 70.8% |
| p 2 = 0.0421 | = 16.86 | Sp = 4.21% | 17 | 2 × 16.84 → 24.4% |
| p 3 = 0.0048 | = 1.94 | Sp = 0.48% | 5 | 3 × 1.94 → 4.22% |
| p 4 = 0.0004 | = 0.17 | Sp = 0.04% | 3 | 4 × 0.17 → 0.49% |
|  | approx. 400 | approx. 100% | approx. 400 | 100% |

29.1%   4.73%   0.48%   0.04%

14%   6.25%

29%

Ref: J.Text.Inst.,1976,No.1,p30.

Fig. 6. Calculation of the probability of 0, 1, 2, 3 and 4 end breaks taking place on a certain number of spindles (ideal Poisson distribution conditions) and a comparison of the calculated with the observed results with reference to a jute spinning mill.[6]

spindles producing two or more end breaks throughout the testing time covered should be considered as "rogue" spindles.

Taking the example of the spindles with one or more end breaks, it will be seen that in the ideal case, 29.1% of the spindles are responsible for 100% of the end breaks, or in the observed case, 14% of the spindles were responsible for 100% of the end breaks. This arbitrarily chosen example refers to a typical mill trial based on *manual observations on ends down*. With a mean number of end breaks/spindle of 0.345, it has not been possible for every spindle to have recorded at least one end break, so that the statistical significance of the results is questionable.

In Fig. 7, a graphical arrangement based on the Poisson distribution has been set out to illustrate that in order to take all spindles into consideration, a minimum of three end breaks/spindle (average value) must be available, e.g. a sample size of 300 end breaks when checking 100 spindles, in order that "statistically representative" conclusions can be drawn.

If we now draw in these conditions on the graphical arrangement shown in Fig. 7, and taking this practical example of only 0.345 end breaks/spindle, i.e.

$$\frac{138 \text{ end breaks}}{400 \text{ spindles}} = 0.345$$

under ideal conditions, 14% of the spindles would be responsible for approximately 65% of the end breaks. As, in the observed case, 14% of the spindles

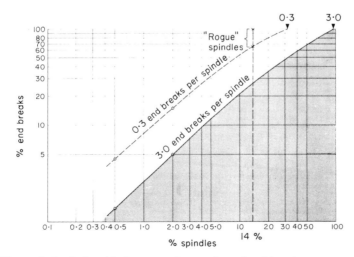

FIG. 7. Theoretical relationship between the number of end breaks as a percentage and the number of spindles as a percentage.

are responsible for *all* the end breaks, then it can be concluded that this extra 35% of the end breaks are due to the "rogue" spindles.

The diagram (Fig. 7) also shows the ideal Poisson distribution line taking the case of three end breaks/spindle, i.e. an average of end break frequency until every spindle has produced at least three end breaks. In order to achieve this "representative" testing time with, for instance, an end break frequency of 30 end breaks/1000 spindle hours, three end breaks/spindle would only be reached in 100 hours which, with a three-shift system plus doffing time, would refer to a continuous testing time of five days!

For this reason it is recommended that in mills where the end break frequency is low, a suitable enough testing time must be taken into consideration in order to ensure that enough "events" will be recorded. In mills where the end break frequency is high, a representative sample will be available in a much shorter period of time. Nevertheless, the diagram shown in Fig. 7 illustrates the importance of "round-the-clock" data logging and of obtaining an end break/spindle value which is representative.

## C. Yarn quality and end breaks

The reduction of the end break frequency value does not only result in a production increase, it also brings about an improvement in yarn quality. An evaluation of the trials carried out with USTER testing instruments as described in the following confirms this statement.

### 1. Evenness and strength

It is of importance to determine whether there is a difference in yarn quality between yarns from cops spun at spinning positions with a high number of end breaks, and those from cops spun at spindles where there were no end breaks. First of all, trials were carried out to check the two parameters of evenness and strength. In this respect, the procedure was such that cops were selected and tested from spindles with a significantly high number of end breaks. The corresponding measured results are shown in Table II. In the last column of this table, the most probable reason for the high number of end breaks is referred to. At the bottom of the table, the monthly averages are given as determined over a period of a few months. The measured results which significantly deviate from these average values are in italic type.

### 2. Yarn faults (CLASSIMAT)

The influence of end break frequency on subsequent processing and on the appearance of the yarn in the finished product was checked with the CLAS-

TABLE II. Quality analysis of yarn from spindles exhibiting a higher than average number of end breaks.

| | USTER® RINGDATA | | | USTER® AUTOSORTER | | USTER® DYNAMAT | | | | USTER® TESTER II | | | |
| Prod. date | Spindle number | End breaks per day | Mean per spindle and day | Ne | CV (%) | Rkm g/tex | CV p (%) | E (%) | CV (%) | Thin places per 1000 m (−50%) | Thick places per 1000 m (3) | Neps per 1000 m (3) | Reason |
|---|---|---|---|---|---|---|---|---|---|---|---|---|---|
| 30. 9.78 | 48 | 7 | 1.57 | 31.6 | 0.86 | 13.83 | 7.28 | 6.31 | 18.63 | 38 | 268 | 360 | Roving quality |
| 4.10.78 | 138 | 9 | 1.08 | 43.4 | 1.06 | 18.78 | 9.69 | 6.22 | 18.48 | 50 | 245 | 375 | Combed roving instead of carded roving |
| 5.10.78 | 53 | 9 | 0.97 | 34.8 | 2.09 | 9.90 | 14.82 | 5.03 | 24.11 | 206 | 426 | 492 | 2nd passage drawing |
| 6.10.78 | 124 | 11 | 1.20 | 30.9 | 1.47 | 11.88 | 19.56 | 6.04 | 19.30 | 59 | 225 | 373 | 2nd passage drawing |
| 18.10.78 | 138 | 22 | 0.80 | 31.1 | 1.76 | 12.51 | 7.77 | 5.76 | 19.02 | 48 | 294 | 405 | Yarn satisfactory |
| Monthly average | | | 0.92 | 30.0 | — | 12.20 | 9.80 | 6.10 | 18.50 | 43 | 250 | 370 | |

SIMAT. In this respect, the fault frequency was determined in the various fault classes (A1–D4):

Yarn lot no. 1: 70 cops containing two or more end breaks
Yarn lot no. 2: 70 cops without end breaks.

In Fig. 8, a schematic representation of the USTER CLASSIMAT Grades shows the fault classes according to the lengths A–D and cross-sectional sizes 1–4. With each class (e.g., A4) reference is made to the fault frequency (number of faults/100 000 m) as determined by means of the USTER CLASSIMAT test. The lower figure (e.g., "4" in class A4) shows the fault frequency of the yarn lot with two or more end breaks, and the upper figure (e.g., "1" in class A4) shows the fault frequency in the yarn lot without end breaks.

In all fault classes, quite considerable differences were available between the two yarn lots. According to experience, this difference in fault frequency can even be much higher, i.e. ten times more faults in yarns with end breaks than in yarns without end breaks. The fact that, even with the smaller size and length faults, large differences can be available would seem to indicate that these are not only due to random occurrences such as the incidence of fly

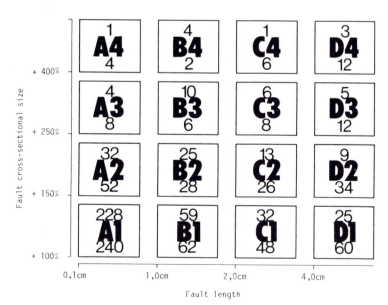

FIG. 8. Yarn faults/100 000 m (worsted yarn Nm 40, 25 tex) in spinning cops taken from spindles with two or more end breaks (lower figures), and in cops from spindles without end breaks (upper figures).

in the ring room, but that these cops with a high end break frequency are in general of a poorer quality. The detection of such spinning positions, therefore, is extremely important.

### 3. Other factors

*(a) Ring traveller.* It is worthwhile, when determining the best type of ring traveller for a particular purpose, to consider the end break frequency under normal operating conditions. In this respect, it is advisable to determine average conditions over a number of time intervals. Figure 9 shows a graphical representation of such a trial. The continuous increase in end break frequency during the operating time is quite clearly indicated. Towards the end of the time of application (cycle time) of the ring traveller, the end break frequency has increased by approximately 20%.

Such trials provide the basis for determining the optimum cycle time of the ring traveller. They are also of value if, for instance, different types of ring travellers have to be compared. Table III refers to such an example. Two different types of ring traveller, the one type costing much more than the other, were compared. In the table, the weekly end break figures are shown for both types of traveller. Throughout the complete time of testing there is practically no difference, i.e. the two types of ring traveller are of equal value. The price difference is therefore not justified. Considering the quite

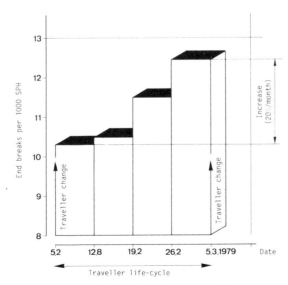

FIG. 9. Weekly average values of end break frequency with respect to the cycle time (life-time) of the ring traveller (Yarn Nec 80, 7.4 tex, combed cotton).

TABLE III. End break frequency with respect to two types of
ring traveller, the one more expensive than the other (Nec 58,
10 tex combed cotton. 100% cotton).

|  | End break frequency (End breaks/1000 spindle hours) | |
| --- | --- | --- |
| Week | Ring traveller A | Ring traveller B |
| 1 | 25.7 | 25.2 |
| 2 | 26.8 | 25.7 |
| 3 | 28.6 | 29.8 |
| Average value | 27.0 | 26.9 |

long testing time of more than 300 000 spindle hours/type of ring traveller, this conclusion can be considered as based on significant results.

*(b) Spindle speed.* Besides the determination of the life-cycle for ring travellers, trials with respect to end break frequency as a function of the spindle speed can also be considered to be of equal importance. This would present no particular problem except for the fact that, in this case, a much higher accuracy is required. As the end break frequency varies from day to day, reliable results can only be obtained when the end break frequency is checked for a period of at least a week.

Figure 10 shows an example of increasing end break frequency with an increase in the spindle speed. At first, a spindle speed of 10 600 rpm was applied. In this case, an average value of 27.8 end breaks/spindle hours was obtained. The spindle speed was then increased to 14 100 rpm. Measurements showed that with this speed an average value of 41.6 end breaks/1000 spindle hours was obtained. From the variation in the daily end break frequency, one can calculate the 95% confidence range for the average value of 27.8 to be ±3.3 and for the average value of 41.6 to be ±4.4. The difference in end break frequency of 13.8 end breaks/1000 spindle hours, i.e. an increase of 49% as a result of the higher spindle speed, refers to a significant difference (in terms of a 95% confidence range) of ±5.5 end breaks/1000 spindle hours, i.e. 20%.

Although this example refers to rather a severe speed increase and correspondingly a production increase of more than 30%, under normal conditions such a large increase in end break frequency would not be acceptable.

*(c) Further reasons.* An increased level of end break frequency can be due to

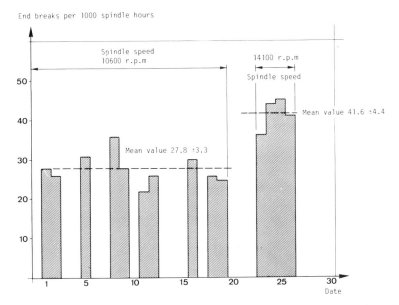

FIG. 10. Increasing number of end breaks with an increasing spindle speed. (Nm 100, 10 tex combed cotton 100% cotton yarn).

other reasons. Such possible reasons could be: the evenness (U%/CV%) of the roving, resp., drawframe sliver; the settings at the drafting elements of the ring spinning machine; fly or dust build-ups due to defective suctioning or blowing arrangements; insufficient maintenance work; unsuitable aprons and rubber covering material; the occurrence of draughts due to open windows, doors, etc.; local heat concentrations; unsuitable blowing intervals of the travelling blower; properties of the raw material, such as the maturity, fineness, strength, etc., where these do not correspond with the nominal values.

## D. Data logging systems for spinning

### 1. Hardware

PILOT installation (Fig. 11), for determining: optimum machine settings such as speed, twist, draft etc.; the most suitable machine units such as the

J

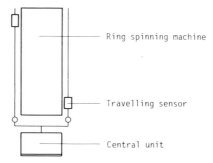

FIG. 11. PILOT installation.

drafting elements, rings, ring travellers, rubber covered rollers, etc.; best types of yarns, fibres blends, etc.; optimum climatic and environmental conditions at reference machines on which the most important or most critical yarns are being spun.

MOBILE installation (Fig. 12) represents: an extension of the PILOT installation for checking the more important yarns and machines; an arrangement whereby the stationed central unit constitutes a reference installation for eliminating environmental influences.

The OVERALL installation (Fig. 13) is for determining: productivity (at all connected machines); quality (only at those machines equipped with travelling sensor arrangement).

These provide information on: optimum conditions of personnel allocation; production data for the whole mill; an assessment of wage levels; a means of dispositioning.

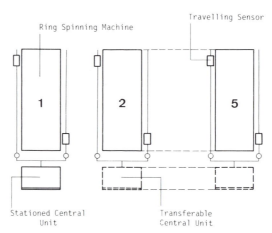

FIG. 12. MOBILE installation.

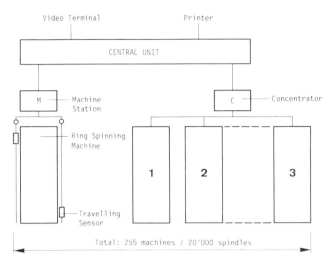

FIG. 13. OVERALL installation.

## E. Economic aspects of data logging in spinning

| Savings in end break costs | Spinning system | |
| --- | --- | --- |
| | *Cotton* | *Worsted* |
| End breaks per 1000 SPH | 40 | 80 |
| Annual production hours | 6000 | 6000 |
| Efficiency | 90% | 90% |
| Effective annual production hours | 5400 | 5400 |
| End breaks per spindle year | 216 | 432 |
| End break costs per spindle year | SFr. 21.60 | 64.80 |
| (Ex. 1 end break = SFr. 0.10 cotton)  1 end break = SFr. 0.15 worsted) | | |
| Reduction of the end breaks by means of  data logging | 25% | 25% |
| Savings in end break costs per spindle year | SFr.  5.40 | 16.20 |
| Savings in end break costs per machine year | SFr. 2160.00 | 7290.00 |

*Savings in production costs*

Taking into consideration the same spindle allocation and work-load for the spinning personnel, after achieving a reduction in the number of end breaks the spindle speed can be increased, which results in an increase in production.

| | | 150 kg | 125 kg |
|---|---|---|---|
| Production per spindle year | | 150 kg | 125 kg |
| (Cotton: Nm 34, 15 m/min) | | | |
| (Worsted: Nm 40, 15 m/min) | | | |
| Production costs per spindle year | SFr. | 250.00 | 450.00 |

Cotton: SFr. 1.50/kg
Worsted: SFr. 3.60/kg

| Increase in the delivery speed | | 5% | 5% |
|---|---|---|---|
| Reduction of production costs | | | |
| (3% cost reduction with | | | |
| 5% production increase): | | | |
| per spindle year | SFr. | 7.50 | 13.50 |
| per machine year | SFr. | 3000.00 | 6075.00 |

*Total savings:*

| Total per spindle year | SFr. | 12.90 | 29.70 |
|---|---|---|---|
| Total per machine year | SFr. | 5160.00 | 13365.00 |

Of these savings per machine and year the sum of SFr. 2160 (cotton) and SFr. 7290 (worsted) is due to increasing production by 5% and the sum of SFr. 3000 (cotton) and SFr. 6075 (worsted) is due to improving quality (25% reduction in end breaks). Total savings per machine and year are SFr. 5160 (cotton) and SFr. 13 365 (worsted).

Of comparable importance is the continuous quality, machine and personnel supervision on a "round-the-clock" basis. The advantages offered in this respect are difficult to assess numerically for inclusion in a "return on investment" calculation.

## F. Conclusions

A data logging system for the ring spinning mill offers the following possibilities:

*Reduction of piecing-up costs.* In a cotton spinning mill with, for instance, 40 000 spindles, the piecing-up costs per year will be approximately SFr. 650 000 (e.g., 30 end breaks/1000 spindle hours, 5400 hours worked per year and a cost per end break of SFr. 0.10). A 10% reduction in the number of end breaks will result in a reduction of SFr. 65 000 in the piecing-up costs per year. With increasing difficulty in obtaining personnel, particularly night shift personnel, the amount of piecing-up will have to be reduced.

*Reduction of costs by increasing production.* A 5% increase in production in ring spinning will result in an approximately 2% increase in production costs, i.e. a resulting 3% cost reduction. Experience has shown that with a data logging installation, a 5–10% increase in production is possible (Fig. 14). For example, production costs in a spinning mill with 40 000 ring

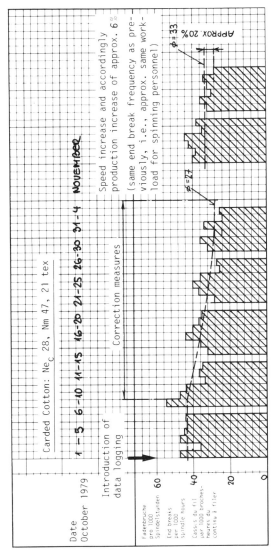

FIG. 14. Production increase of approximately 6% as a result of the introduction of data logging at ring spinning.

spindles are approximately 10 million SFr./year. A production increase of 5% results in a 3% costs reduction, i.e. SFr. 300 000/year. A continuous supervision of the end break frequency indicates how much production can be increased.

*Quality control.* The break frequency value provides an indication of the suitability of operation of the ring spinning machine and all the preparation processes. Furthermore, it has been shown that yarns spun with less end breaks result in better running conditions in subsequent processing.

Data logging at ring spinning also provides a reference to the "spinnability" of different raw materials, a reference to the running conditions of the ring spinning machine as a whole, and to the various spinning and drafting elements, and a means of setting productivity levels for operating personnel.

## III. Frequency of End Breaks in Weaving

### A. Introduction

End breaks, whether in the warp or the weft, can result in wage costs and a reduction in efficiency of weaving, and can negatively affect the quality of the woven fabric. Consequently, some reference with respect to their frequency of occurrence is necessary and, based on this, preventive or corrective measures can be undertaken.

Such preventive or corrective measures necessitate, however, some reference to the source of the fault, and as these faults are relatively infrequent, such information must be gathered over considerably long periods of time. Consequently, only by means of data logging techniques is it possible to undertake end break checks in weaving and to differentiate between the various factors which influence end break frequency.

In this chapter, an attempt will be made to quantify and qualify the stops which take place in the weaving process based on results obtained operating with a micro-computer controlled data collection installation. In particular, the stops due to breaks in the warp and weft will be analysed as these are directly correlated to the quality of the yarn.

### B. Warp, weft and machine stops

First of all, consideration should be given to the importance of differentiating between the three basic reasons for weaving machine stoppages, i.e. breakdown of the warp yarn, the weft yarn or the machine. Depending of course on the type of material in the warp and weft, the method of weaving (weft insertion), the warp preparation, the suitability of sizing, ambient conditions affecting the yarn, the weaving machine settings and the weaver ability, there will either be more downtime due to the warp than the weft, or vice versa.

Figure 15 refers to an example of the conditions available in a cotton weaving plant (conventional shuttle weaving) with ring-spun cotton yarn in both the warp and weft. The figure shows that short stops were responsible for 12.9% loss in efficiency and that 66% of this efficiency loss was the result of warp stops. Weft stops, on the other hand, were only responsible for 1% loss in efficiency or 7% of the total loss in efficiency due to the short stops.

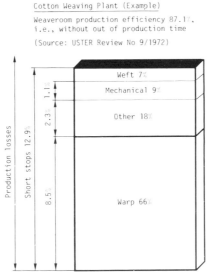

FIG. 15. Cotton weaving plant. Warp, weft, mechanical and other stops as a percentage of the total short stops.

This ratio between the warp, weft and machine stops (machine and other stops are usually grouped together) can vary quite considerably according to the reasons given above and also as a function of the number of short stops per machine. Figure 16 shows an analysis of more than 1000 short stops undertaken in a cotton weaving mill covering a period of non-stop production. The figure relates the actual number of weaving machines and the number of short stops per machine. As an example, 16 machines stopped five times, i.e. total stops $16 \times 5 = 80$, of which:

>10 were due to warp stops (12.5%),
>40 were due to weft stops (50%) and
>30 were due to machine or other stops (37.5%).

This figure illustrates that only those machines with a low number of short

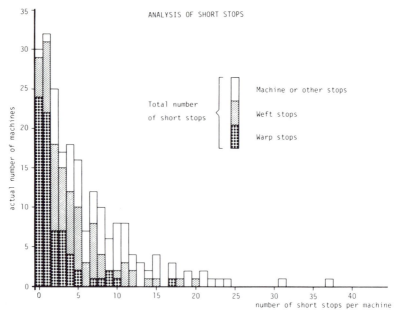

Fig. 16. Analysis of short stops on weaving machines. The figure shows the actual number of machines with respect to the number of short stops/machine for warp, weft and other stops.

stops per machine have the normally expected ratio between warp, weft and machine stops (see Fig. 15). As the number of stops per machine increases, systematic reasons for weft stops associated with the weft insertion affect this ratio, as do systematic reasons associated with the weaving machine itself, until all short stops are the result of systematic machine faults.

If a large percentage of the short stops is due to systematic machine faults, there must be a certain small percentage of the weaving machines in the mill which are responsible for the majority of short stops. Figure 17 shows a further analysis of the results referred to in Fig. 16 such that the cumulated values of the number of short stops is compared to the cumulated number of machines, and each is represented on a percentage basis.

In this particular example (Fig. 17):

1.4% of all the machines have no warp stops; 50% of all machines are responsible for less than 25% of all warp stops;

7% of all the machines have no weft stops; 50% of all machines are responsible for less than 18% of all weft stops; and

33% of all the machines have no machine stops; 50% of all machines are responsible for less than 7% of all machine stops.

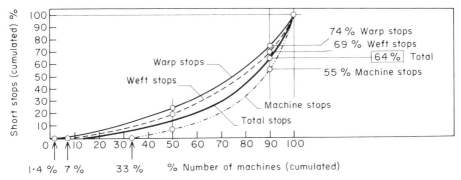

FIG. 17. Analysis of short stops. The figure shows short stops cumulated in percentage against number of machines cumulated in percentage for warp, weft, machine and total stops.

In terms of a system of "management by exception" and taking into consideration the worst 10% of the machines, 90% of all the machines are responsible for 64% of all short stops and consequently *10% of all the machines are responsible for 36% of all short stops*. By means of data logging techniques, these 10% of all the machines can be singled out and, based on an analysis of the reasons for the short stops, corrective measures instituted at these machines.

## C. Reasons for short stops

Numerous observers have recorded quite a range of possible causes for short stops, particularly where this refers to warp stops. Figure 18 shows a more

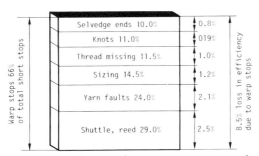

FIG. 18. Cotton weaving plant. Reasons for warp stops expressed as a percentage of the total short stops.

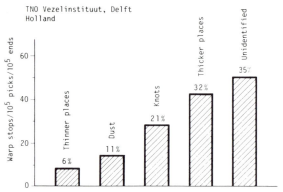

Fig. 19. Reasons for warp breaks in cotton weaving.

detailed analysis of Fig. 15 with respect to the warp stops in the same cotton weaving mill. Figure 19 shows a similar survey carried out some years ago by the TNO Vezelinstitut in Holland and Table IV a similar survey carried out by Zellweger Uster. Three basic factors can be concluded from these various surveys covering the reasons for warp breaks in weaving: (1) "unidentified" stops can be interpreted as being due to weak places in the yarn; (2) excessive stress is induced on the yarn by the weaving process; and (3) the yarn has inadequate "stress capacity" to resist this stress.

TABLE IV. Reasons for warp breaks (in %) in cotton weaving.

| Type of fault | Cotton weaving mill No. 1 % | Cotton weaving mill No. 2 % |
|---|---|---|
| Knots | 65 | 48 |
| Thicker places | 13 | 34 |
| Unidentified | 22 | 18 |

It can be that the weaving machine is abusing the yarn when the shuttle or the reed is quoted as being responsible for warp breaks. It can be that the basic properties of the yarn, e.g. strength and elongation, number of thin and thick places, knot content, etc., limit the "stress capacity" of the yarn. Or it can be that the preparation of the warp in terms of crossed ends, double threads, size application, etc. have negatively affected the "weavability" of the warp. The following refers to a list of reasons for end breaks in weaving:

*Quality and arrangement of the yarn material*
Irregular yarn with many thick and thin places.
Badly cleared yarn.
Yarn with a too high number of knots as introduced during winding or beaming.
Pirns or cones with poor drawing-off properties due to incorrect build-up.

*Warp preparation*
Incorrect conical setting at the section warper drum.
Incorrect setting of the yarn tensioners at the warp creel.
Too softly-wound warp beams.
Crossed ends due to incorrect setting-up of the warps before drawing-in or tying. Incorrect setting or handling of the tying machine.

*Sizing*
Unsized warp yarn.
Too strongly or too weakly sized warps.
Use of an unsuitable size material or size recipe.
Warps which have been sized to be too damp or too dry.
Faulty setting of the sizing machine.

*Setting of the weaving machine*
Too high warp tension.
Incorrect shedding geometry.
Incorrect beat-up setting with shuttle type weaving machines.
Damaged and dirty healds, drop-wires or reeds.
Unsuitable types of healds and drop-wires.
Too few shafts in relation to the warp density.

*Ambient conditions*
Temperature.
Room humidity.
Dust.
Fly.
Draughts.

*Operation of the weaving machines*
Knotting of broken ends with unsuitable knots.
Yarn rests left in the shed.
Drawing-in of a broken end (crossed ends).
Allowing fly and dirt to collect on the warp yarns.

## D. Yarn properties influencing warp breaks

The quality of the yarn is in direct correlation to the number of end breaks during weaving. Quality, in this respect, can be considered in terms of yarn strength and elongation, yarn fault and knot content and, to some extent, yarn irregularity (short-term unevenness) and count variation (long-term unevenness).

### 1. Yarn strength and elongation

The probability of end breaks in the warp can be calculated (see References). This calculation is based on the tension placed on the warp yarns by the shed opening motion of the weaving machine, and the strength variation in the yarn. The overlapping of these two characteristics determines the probability of end breaks occurring. Figure 20 and the corresponding Table V show the conditions existing with changes in yarn strength (examples 1 and 2) and with changes in loom tension (examples 3 and 4). In both cases, there is an immediate reaction in the number of estimated warp end breaks per 1000 ends and 100 000 picks. In example 1, the average yarn strength is 250 cN with a standard deviation of 30 cN. If the average loom tension at weft insertion is approximately 80 cN and varies with a standard deviation of ±10 cN, an end break frequency of 4.2/100 000 picks and 1000 ends is to be expected. Weaving with a stronger yarn would result in the end break frequency being reduced by a factor of five (example 2 with average yarn strength of 260 cN). A reduction in end break frequency can also be achieved if, as with example 3, the variability of yarn strength is reduced from 30 cN to 28 cN with a constant mean strength value. According to the calculation (see References), an end break frequency of only 0.7/100 000 picks and 1000 ends is possible. Example 4 indicates that a reduction in machine tension variation from 10 to 7 cN does not affect the end break frequency to the same

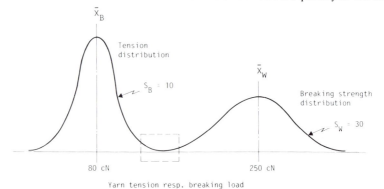

Fig. 20. Relationship between loom tension variation and yarn breaking strength variation.

TABLE V. Four examples illustrating the effect on yarn break frequency of changes in loom tension and/or yarn strength.

| Example | Loom tension warp | | Yarn strength count 20 tex | | Yarn break frequency |
| | Peak tension $\bar{X}_B$ (cN) | Standard deviation $S_B$ (cN) | Breaking load $\bar{X}_W$ (cN) | Standard deviation $S_W$ (cN) | Breaks/100 000 picks and 1000 ends |
|---|---|---|---|---|---|
| 1 | 80 | 10 | 250 | 30 | 4.2 |
| 2 | 80 | 10 | 260 | 30 | 0.5 |
| 3 | 80 | 10 | 250 | 28 | 0.7 |
| 4 | 80 | 7 | 250 | 30 | 1.8 |

extent but results in an improvement of from 4.2 to 1.8/100 000 picks and 1000 ends. The table shows, quite clearly, however, that small changes in the yarn strength properties can result in large changes in the end break frequency.

The same applies with yarn elongation. This is particularly evident when considering the yarn elongation properties of a wool worsted yarn. Due to the fact that the wool fibre has an inherent length and fineness variation within any one quality grading, the yarn elongation at break variation value ($CV_E\%$) is correspondingly high. The high variation value for elongation results in the fact that the minimum values (below the $3\sigma$ limit) refer to extremely low elongation values. The variation of elongation value ($CV_E\%$) for wool worsted yarns is, on average, twice that of the variation of strength ($CV_P\%$), which can be extremely critical.

The sketch of a frequency distribution diagram of wool worsted yarn (Fig. 21) compares the $CV_P\%$ values of 30 randomly chosen samples in the count

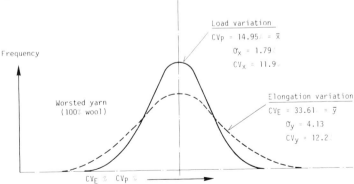

FIG. 21. Comparison between the $CV_P\%$ and $CV_E\%$ values of 30 randomly chosen wool worsted yarns in the count range Nm 24 to Nm 67 (41 to 15 tex approximately).

254                                    K. DOUGLAS

range from Nm 24 to Nm 67 (41 to 15 tex approximately) with the $CV_E\%$
values. It will be noticed that although the variation between samples is
reasonably constant throughout a quite wide count range ($CV_X$ and $CV_Y$),
the elongation variation refers to a much higher value than the load variation.

The second sketch (Fig. 22) shows a frequency distribution diagram of
elongation variation for one of the samples referred to in Fig. 21. Taking the
$3\sigma$ limits of lengths of yarn with a too high or too low elongation, it would seem
that because of the high $CV_E\%$ figure, there is a certain number of yarn
lengths which can even exhibit an elongation value of less than 0.61%! For
this reason, when considering the elongation value of, in particular, wool
worsted yarns, the problem of the high $CV_E\%$ value should always be taken
into account particularly as this can influence the end break frequency value.

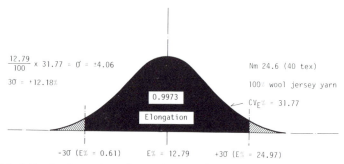

FIG. 22. Variation in elongation of a wool worsted yarn can result in a number of yarn
lengths (below the $3\sigma$ limit) having extremely poor elongation properties.

## 2. Yarn faults

Yarn faults and the knots which they replace can be responsible for warp
breaks during weaving (see Fig. 18). "Optimum yarn clearing" is a compromise between the extraction of those yarn faults which would cause
breaks during weaving (and which would also be considered "disturbing"
with respect to the appearance of the woven fabric) and those yarn faults
which can be left in the yarn and need not be replaced by a knot (approximately 1% of knots result in warp breaks). To illustrate this, an example of a
poplin cotton "shirting" material is shown in Fig. 23. Knots due to bobbin
changes (value "$c$") amount to 10/kg considering a ring bobbin containing
approximately 100 g of yarn. As faults are extracted and replaced by knots
(value "$b$"), the losses ($G\%$) in winding (loss in efficiency) and weaving
(knot causing end breaks) increase.

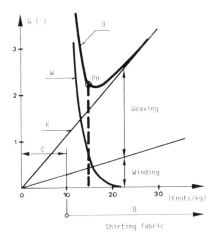

FIG. 23. "Optimum" yarn clearing as a function of the number of knots per kg or per 100 000 m.

The curve "W" refers to a fabric appearance rather than the effect of subsequent processing. If, for instance, no faults are extracted at winding, then a high percent loss ($G\%$) will be the result because the fabric will be downgraded. As more faults are extracted (and more knots as a consequence introduced), the losses due to fabric quality will be reduced.

If now a cumulative curve ("O") is drawn out which combines the cumulative values at any knots/kg position of the winding/weaving losses ("K") and the fabric quality losses ("W"), then the lowest point on this curve will establish the *optimum clearing characteristics* and the *optimum number of knots* which should be placed in the yarn.

With this "shirting" example, the optimum position would be at approximately 15 knots/kg which for Nm 50 (tex 20) yarn would refer to approximately 30 faults/100 000 m of yarn. For weaving yarn, it would seem that there is an extremely limited range of, in this case, between 14 and 16 knots/kg within which the optimum value lies and the corresponding electronic clearer settings will have to be chosen.

Consequently, data logging at the winding process can play an extremely valuable rôle, and particularly from a quality point of view in differentiating between knots placed in the yarn for different reasons.

## 3. Yarn evenness and count variation

Yarn irregularity (short-term weight/unit length variation) and count variation (long-term weight/unit length variation) can be considered together.

Their influence on end break frequency whether in spinning, winding or weaving is shown in Table VI. This table illustrates that with a better yarn count variation (as obtained by autolevelling at carding), the ends down frequency in spinning, winding and weaving can also be improved. Although with this particular example the difference in the results is by no means significant, a trend is evident at all three processing stages. The possible explanation for the different end break conditions is illustrated by Fig. 24. This sketch shows that if short-term irregularity is superimposed on long-term irregularity, the thin places will be thinner according to the amplitude of the long-term variation. The more accentuated the short-term irregularity, the more accentuated will be the thinness of the thinnest places in the yarn. Although thin yarn does not necessarily represent weak yarn, it is most probable that the yarn on both sides of the thin place will be weaker than the rest of the yarn, and yarn breaks will result.

TABLE VI. Comparison of end breaks in spinning, winding and weaving with respect to two cotton-spun yarns, one produced from autolevelled and the other from non-autolevelled card sliver.

| Process | Characteristic | Yarn (combed cotton) | |
| --- | --- | --- | --- |
| | | Without card autolevelling | With card autolevelling |
| Yarn Count Checks | Nominal count ($N_E$) | 60 (tex 9.8) | 60 (tex 9.8) |
| | Count deviation 3% ($N_E$) | 58.25–61.86 | 58.25–61.86 |
| | Actual count ($N_E$) (6200 bobbins) | 62.85 | 60.63 |
| | Count variation $CV_{T(100\ m)}$% (200 bobbins a 120 yds, 1 test/bobbin) | 3.25 | 2.37 |
| Spinning | End breaks/1000 SP hours ring spinning machine (10 000 SP hours) | 10.35 | 10.00 |
| Winding | End breaks/100 000 m in yarn due to thin places, automatic winder ($1.5 \times 10^6$ m yarn) | 1.41 | 1.11 |
| | Faults/100 000 m in yarn due to thick places, automatic winder ($1.5 \times 10^6$ m yarn) | 14.19 | 14.05 |
| Weaving | Weft breaks/mach. hour weaving machine (more than 200 mach. hours) | 0.24 | 0.22 |

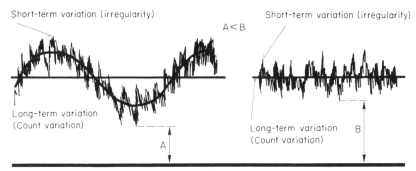

FIG. 24. Short-term variation superimposed on long-term variation results in thinner thin places than those which would be present in the yarn if no long-term variation is available (i.e., $A < B$).

## E. "Repeater" end breaks in weaving

As in spinning, the reasons for end breaks in weaving are difficult to determine. If a higher than average number of warp breaks results at all weaving machines using a particular yarn, the fault can be determined by checking the yarn. If, on the other hand, it can be shown that warp breaks take place repeatedly at one particular weaving machine, the warp yarn at this weaving machine can be examined and the reason for the fault determined. In a study carried out by the Institute of Textile Technology in Charlottesville, it was found that only 6.1% of the total ends in a warp caused all the warp end breakages. Even more important, only 22 ends (less than 1% of the 3600 ends in the warp) resulted in 37.7% of the total end breaks. As shown in Table VII 5 ends broke 10 or more times, 5 ends broke

TABLE VII. "Repeater" breaking ends have contributed significantly to the total number of warp stops which occurred.

| Number of times an end broke | Number of ends involved Number | % | Number of breaks Number | % |
|---|---|---|---|---|
| 10 or more | 5 ends | 0.14 | 66 | 16.7 ⎫ |
| 5–7 | 5 ends | 0.14 | 34 | 8.6 ⎬ 37.7 |
| 3–4 | 12 ends | 0.33 | 49 | 12.4 ⎭ |
| 1–2 | 197 ends | 5.47 | 247 | 62.3 |
|  |  |  |  | ———— |
|  |  |  |  | 100.0 |
| Zero breaks | 3381 ends 93.92% |  |  |  |

5–7 times and 12 ends broke 3–4 times. There were 197 ends which broke 1–2 times whereas 3381 ends never broke at all.

The conclusion which can be drawn from the theory of "repeater broken ends" is that: (a) repeating conditions represent a systematic fault rather than random conditions, so that if those weaving machines can be singled out which have "repeater" conditions, corrective measures can be instituted; and (b) a relationship (correlation) between the various yarn physical properties and the yarn "weavability" can be established.

## F. Data logging systems for weaving

### 1. Hardware

Data are collected at each weaving machine by means of a single universal sensor attached to the main shaft or other rotating element which is in direct relationship to the pick entry frequency. A separate indication of warp and weft stops is available from the signals provided by the respective stop motions. The signals from a certain number of weaving machines are transmitted by cable to one concentrator. This concentrator provides a means of preparation and storage of the signals. The concentrators are connected to a central unit by means of a circular bus system. The concentrators are periodically "called-up" by the central unit which arranges a signal output in the form of reports via a monitor and printer terminal. The general arrangement of the system is shown in Fig. 25.

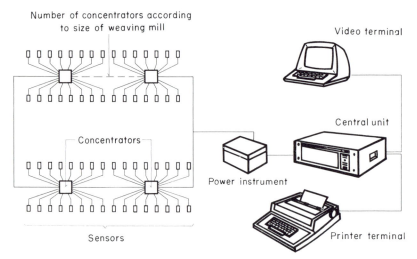

FIG. 25. General arrangement of a data logging installation for the weaving mill.

## 2. Software

The more important data are provided in standard reports which, in order to facilitate evaluation, are always arranged in the same way. In the title lines of each report, the following are normally available: type of report; date; time; shift number; monitored time. Finally, the following data is printed out or displayed:

MACH    Machine number

$$\text{ACT\%} \quad \text{Actual efficiency} = \frac{\text{running time}}{\text{monitored time}}$$

$$\text{PRD\%} \quad \text{Production efficiency} = \frac{\text{running time}}{\text{monitored time} - \text{out-of-prod. time}}$$

\*        Number of completed shifts in which at least one characteristic has been continuously out of tolerance (indicated with an \*)

WST      Number of warp stops
FST      Number of weft (filling) stops
TST      Total number of short stops
ST/H     Stop frequency/running hour
M/ST     Mean down-time/stop (in minutes)
CD       Stops code
OPH      Out-of-production time (hours)
P/M      Speed (picks/minute)
P:1000   Number of picks (in thousands)
METER    Length of material (in metres)
STYLE    Article or style number
WE       Weaver group
F        Fixer or overlooker group
G        Experimental or freely-chosen group

If the efficiency or speed of a weaving machine is below the pre-entered limiting value, or if the stop frequency is too high, the corresponding value is indicated with an asterisk (\*). In this way, "repeater" or "rogue" units can be quickly identified.

At the end of each report are the total figures which contain the mean, resp. the sum values, of the data in the respective columns. In order to be able to carry out purpose-orientated trials, selection possibilities are available according to the following: machine; article (style); weaver; fixer (overlooker); group. For further data concentration, the following selection possibilities can be arranged: only the sum values per selected criteria; only machines with an asterisk (*); only machines out-of-production.

All these reports can be called-up both for the running shift and for the last shift. With an article (style) report, a report covering the running week is available.

A number of other call-up possibilities covers further functions: out-of-production times (for the complete weaving mill listed according to code); warp change prediction (for 15 days with graphical representation); warp store; article (style) data; control data (e.g., shift change-over times); service reports. The reports which are automatically printed out at the end of the shift can be arranged according to requirements.

In spite of the multiplicity of call-up possibilities, the operation of such an installation is simple because the calling up of the reports and the entry of data is undertaken in dialogue form. This means that the calculator always tells the user what he should enter next. It institutes a question and offers, according to requirements, a so-called menu (list of entry possibilities).

## 3. On-line reports and information

The most important prerequisite of a computer-controlled production-supervisory installation is the immediate application of the data obtained to correcting or improving the production process. Although continuous data collection provides valuable information for long-term evaluations, the primary purpose of the installation is the detection of incorrectly running machines in order to save production losses, to reduce the work-load of the weaver and overlooker, and to correct machine settings before too much low-grade material has been produced. The information for this form of control is provided by the "on-line" reports.

The most important information available is certainly that which can be obtained from the reports printed out at any time during the shift. These reports can be obtained in terms of the weaver, overlooker, style, or a freely chosen group, or can be printed out when a special event occurs. Such reports provide the overlooker with directives as to the reason for unsuitable production on certain machines. Machines that do not reach the required efficiency can be selected and the separate stops and downtimes analysed to determine the reason.

From the numerous "automatically-printed-out" or "on-call" reports available, two have been chosen for a detailed description because of their importance and application to a system of "management by exception". These are a below standard machine report for the last shift (Fig. 26) and a warp-out prediction report (Fig. 27).

*(a) Below standard machine report.* A report of this type is shown in Fig. 26.

```
MACHINE       / MACH WITH * / LAST S

USTER LOOMDATA      18-08-80    07:00    SHIFT 3   MMIN 540

MACH ACT% PRD%  * WST  FST   TST ST/H M/ST CD OPH P/M   P:1000   STYLE WE F G
  17*87.9 88.6  9  20    6   29* 3.6  2.1     .0 260   123.9    32162  1 1 1
  39*35.3 99.9  9             .0   .0 10 5.8 261    49.9    10345  2 1 2
  56*87.0 87.7  9  22    5   29  3.7  2.2  3 .0 262   123.3    10500  2 2
  71*86.1 99.9                .0   .0 66 1.2 265   123.6    10400  2 2 3
  74*70.8 71.9  9  57    8   72*11.2  2.0  3 .1 271   103.6    10400  3 2 3
  75*71.6 72.6  9  16    3   21* 3.2  6.9  3 .1 259   100.6    32162  4 2 1
 110*72.4 73.4  9  36    6   46* 7.0  3.0     .1 265   103.9    10345  4 3 2
 122*74.3 75.2  9  50    5   63* 9.4  2.0     .1 268   107.7    11181  4 3

T132 77.2 84.5   237   39  308  4.4  2.4     7.7 263  1098.2
```

FIG. 26. Below standard machine report for the last shift.

The machines 17, 39, 56, 71, 74, 75, 110 and 122 ran below standard during the night shift, in each case for one reason or another. In the case of machine 17, the total stops (TST) were below standard (indicated with an asterisk). This was also the reason for the below standard condition of the machines 74, 75, 110 and 122. Machine 39 was out-of-production (OPH) for most of the shift, as was machine 71. Machine 56 was below standard because of the low actual efficiency value (ACT%).

The report provides the total or mean values at the bottom (T) and reference on the right to the STYLE being woven, the weaver group (WE), the overlooker group (F) and a freely-chosen group reference (G).

*(b) Warp-out prediction.* The warp-out prediction report (Fig. 27) shows which machine will require a new warp at a certain time up to 15 days ahead to an accuracy of 1 shift within the first 4 days and an accuracy of 1 day within the following 11 days. The importance of this report is significant when a concentration of warp change-overs occurs on one particular day. With this

```
WARP CHANGE

USTER LOOMDATA     15-08-80    22:01    SHIFT 3    MMIN   1

      STYLE MACH METER RUN-H  1234 1234 1234 1234  5 6 7 8 9 0 1 2 3 4 5
      10345   38  1978  365 .    .    .    .    . .
              42     6    1 .  X .    .    .    . .
              43  1740  321 .    .    .    .    . .                      X
             112  1635  302 .    .    .    .    . .                    X
      10400   76  1702  272 .    .    .    .    . .                  X
              77  1603  256 .    .    .    .    . .                X
      10500   52   320   38 .    .   .X    .    . .
      11181  120  3635  178 .    .    .    .    . .        X
      32162   10   567   78 .    .    .    .    . .X
              16   684   94 .    .    .    .    . .X
```

FIG. 27. Warp-out prediction 15 days ahead. A concentration of warps running out in one day will be known well ahead so that warp-preparation requirements can be organized in good time—also for the night shift!

knowledge, the necessary steps can be undertaken in the warp-preparation department to ensure that this concentration of work can be dealt with.

## 4. Economic aspects of data logging in weaving

The economic aspects of a data logging installation for weaving can be determined from a number of possible savings according to the various application requirements. In Table VIII, only the evaluation of on-line data and the savings which can be directly achieved from on-line information have been considered. This does not necessarily suggest that no further savings can result from off-line processing of the obtained information.

The most important *savings* are obtained from: (a) reduction of the frequency of the short stops (warp breaks, weft breaks and other stops); (b) reduction of the repair and warp change times; (c) optimizing weaver allocations; (d) reduction of warps on stock; (e) savings in personnel carrying out manual data collection. These savings can best be explained with the help of Table VIII.

TABLE VIII. Example of worsted weaving mill with 100 Sulzer weaving machines.

| Production: | 5 million m/year |
| Efficiency: | 82% |
| Costs + profits/machine/year | SFr. 45 000 |
| Total: | SFr. 4 500 000 |
| 1% increase in efficiency = 1.22% increase in production | |
| | = SFr. 54 900/year |

| Production losses in % | | Personnel | No. | Shifts | Costs person/year SFr. |
|---|---|---|---|---|---|
| Short stops: | 10.4 | Weaver | 10 | 3 | 27 000 |
| Setting + repairs: | 1.0 | Assistant weaver | 2 | 3 | 24 000 |
| Warp + article changes: | 5.2 | Overlooker (fixer) | 2 | 3 | 42 000 |
| Waiting times, etc. | 1.4 | Warp change pers. | 3 | 3 | 27 000 |
| | | Time study | 3 | | 30 000 |
| **Total** | **18.0** | Wages office | 1 | | |

*Losses due to second quality* (weaving faults)
| Amount of second quality: | 3.5% |
| Losses/m: | SFr. 2.50 |
| Losses/year: | SFr. 437 500 |
| Percent of these losses due to stops: | 60% |

*Costs as a result of the storing of warps:*
1.5 warps/machine = 500 kg/machine
at SFr. 20/kg     = SFr. 6 666.60/warp
150 warps        = SFr. 1 million at 5% interest rate
                 = *SFr. 50 000/year*

*Savings resulting from less short stops:*
Less short stops result in less machine downtime and therefore more efficiency.
20% less short stops = 0.2 × 10.4%
$$= 2.08\% \text{ efficiency increase}$$
$$= 2.08\% \times \text{SFr. } 54\,900$$
$$= \text{SFr. } 114\,200/year$$

*Savings resulting from a better weaver allocation:*
Less short stops result in less work for the weaver and therefore higher machine allocations.
Allocation to the weaver of 12 machines instead of 10 machines
$$= 2 \text{ weavers less/shift.}$$
$$= \text{SFr. } 27\,000 \times 6 = \text{SFr. } 162\,000/year$$

*Savings resulting from less weaving faults:*
In each case according to the type of material, 40–60% of all weaving faults are a result of machine stops. A reduction in the stop frequency will result therefore in a reduction in the fault frequency.
8% less weaving faults = 0.08 × SFr. 437 500
$$= \text{SFr. } 35\,000/year$$

*Savings resulting from a reduction of the repair and warp change times:*
By means of a more effective delegation of the work carried out by the overlooker, and by means of a reduction in the warp change times, the machine efficiency can be increased.
0.2% efficiency increase = 0.2 × SFr. 54 900
$$= \text{SFr. } 10\,980/year$$

*Savings resulting from a reduction in the warp stock and the elimination of machine waiting times due to shortage of warps:*
A reduction in the number of warp beams on stock, and simultaneously the assurance that the actually required warps are available, will result in a saving in interest rates when the number of warp beams on stock can be reduced.
20% less warp beams on stock = 0.20 × SFr. 50 000
$$= \text{SFr. } 10\,000/year$$

*Savings in office personnel undertaking mill statistics work and time studies*
A data logging installation cannot replace time study personnel but can eliminate such routine tasks as reading the pick counters or counting the number of stops.
One person less in the statistics office = SFr. 30 000/year

**Total savings = SFr. 363 180/year**

Of these savings, the sum of:
SFr. 125 180 is due to increasing efficiency by 2.28%
SFr. 193 000 is due to reducing personnel by seven persons
SFr.   35 000 is due to improving quality (8% less faults)
SFr.   10 000 is due to saving on interest rates

**SFr. 363 180 total savings/year**

## G. Conclusions

The running conditions of a mill are made clear by the use of a data logging installation. Faults and sources of losses can be quickly determined and correspondingly quickly corrected. Management has the mill under control, and decisions can be made that are based on concrete and accurate values.

The processing of the on-line data, as provided by a data logging system, in an EDP installation produces off-line data. These off-line data are used by an additional category of management to that using the on-line data. Finance and organization personnel are mainly interested in the off-line data for purposes of: material disposition; calculation of production costs; determination of trends; supervision of personnel costs; wage calculations and product-costing.

Data logging installations can be applied to all branches of textile processing for the provision of the most important characteristic data. Such installations are, in every case, an economic proposition where there is high production but small batches or where there is limited production but large batches.

Quality control, the attainment of production levels, and a knowledge of machine and worker performance are the fundamentals of present-day manufacture in the textile industry. Small on-line data logging installations are the modern means of realizing these.

## Envoi

*I often say that when you can measure what you are speaking about and express it in numbers, you know something about it; but when you cannot measure it, when you cannot express it in numbers, your knowledge is of a meagre and unsatisfactory kind; it may be the beginning of knowledge, but you have scarcely, in your thoughts, advanced to the stage of science, whatever the matter may be.*

Attributed to Lord Kelvin, 1883.

## References

1. Krause, H. W. (1979). "Textilmaschinenbau im Zeichen der Automation". *Neue Zürcher Zeitung*, July 26, 1979.
2. Douglas, K. (1979). "Ends down in ring spinning". *UMIST Symposium on "Yarns and Yarn Manufacture" Book of Papers*.
3. Felix, E. and Harzenmoser, I. (1978). "Fadenbrüche in der Ringspinnerei", *Melliand Textilberichte*, **10**.

4. Douglas, K. (1979). "The detection of end breaks in ring spinning". *USTER News Bulletin*, **27**, August, 1979.
5. Beaulieu, C. "What can I learn from a computer in my spinning room". *Book of Papers, 13 Int. Canadian Textile Seminar*, August, 1972.
6. Sen Gupta, P. and Dutta, P. "The distribution of end breaks in small scale jute-spinning trials". *J. Text. Inst.* **1**, 1976.
7. van Harten, K. "The prediction of weaving performance". *1er Symposium International de la Recherche Textile Cotonnière*. April, 1969. (Institut Textile de France, 35, rue des Abondances, 92 Boulogne-s-Seine).
8. Krause, H. W. "Ueber die Wahrscheinlichkeit von Fadenbrüchen ("Frequency of thread breakages"), *Melliand Textilberichte*, July, 1979.
9. Locher, H. "Machine control and monitoring by computer". *6th Shirley International Seminar (Book of Papers)*, Sept., 1973.
10. Barcley, I. "Quantification of weaving shed problems". *Textile Institute and Industry*, **18**, No. 6, June, 1980.
11. Melling, K. G. "Warp preparation in Europe—yesterday, today and tomorrow". *USTER Review*, **9**, 1972. Zellweger Uster Ltd, Uster, Switzerland.
12. Cahill, C. "Impact of warp preparation defects at weaving—developing a predictive process control system for slashing". *Textile World*, October, 1979, p. 101.
13. Felix, E., Douglas, K. and Harzenmoser, I. "Garnfehler in Maschenware", *Melliand Textilberichte*, **55**, No. 12.
14. Locher, H. and Douglas, K. "Computerized control in the textile industry". *Textile Institute Annual Conference Papers, "New ways to produce textiles"*, 1972.

## Addendum

Since writing this chapter approximately two years ago, many changes have taken place in both electronic and textile technology which necessitate certain corrections to the statements and calculations made. For instance, a great deal of reference is made to replacing a yarn fault with a knot at winding, and the negative effect of too many knots in weaving and knitting. In the last two years, winding machine manufacturers have been able to offer a splicing instead of a knotting device. Consequently, many of the arguments referring to the negative aspects of knots in this chapter have now been somewhat weakened by the commercial and technological advantages offered by the increasing introduction of small-size and strong splices.

The cost calculations were based on conditions of approximately 3–4 years ago, so that here again, the figures given should be considered as relative and only intended to signify amounts in general.

Data collection at ring spinning has also made considerable progress in the last 2–3 years in that more information, e.g. production, twist, end breaks, production speed and efficiency, can be provided, and this to cover longer periods of time and more extensive groupings. The figure shows a machine report covering one 8 h-shift (SHIFT 2) and eight machines with reference, at the end, to total, resp., mean values.

```
REPORT · MACHINE
USTER RINGDATA V 2.3 08-07-82 TH 14:01   SHIFT 2   MHRS    8:00

 MA STYLE   NM     T/M    KG GR/H  AEF%  CD  STM   DFM   D   EBR EBR/H EM  RPM  GA GB
  1 14SC   50.0   714    65 19.3  91.0  10  20     6   1   125 34.6  27  11.0  1
  2 14SC   50.0   708    52 17.6  93.3  10  20     6   1    95 31.7  25  10.5  1
  3 14SC   64.0  1110    22  9.6  70.4  22 125     8   1   138 60.4  26  11.9  2
  4 225TRI 58.0   900    43 12.3  96.1      10             130 36.9  28  11.0  2
  5 14SC   64.0  1118    36 10.1  98.5                     110 31.0  28  12.3  1
  6 225TRI 58.0         53 15.2  97.8                                          2
  7 14SC   50.0         61 15.8  94.0  10  15                                  1
  8 14SC   64.0         78 10.2  91.0              20    2                     2

* 8        57.0        370 12.9  91.7     200    50    5   598 37.3  27


REPORT · MACHINE
USTER RINGDATA V 2.3 08.07.82 TH 14:01   SHIFT 2   MHRS    8:00

TOT STYLE   NM        KG GR/H  AEF%  CD  STM   DFM   D   EBR EBR/H EM  RPM  GA GB
  8        57.0      370 12.9  91.7     200    50    5   598 37.3  27
```

Fig. A.1. Typical machine report covering eight ring spinning machines. *Key:* MHRS—Monitored hours; MA—Machine number; STYLE—Article number; NM—Count (metric); T/M—Twist (turns/meter); KG—Production since beginning of shift (kg); GR/H—Mean production in grams per spindle hour; AEF%—Actual efficiency; CD—Code; STM—Stop minutes; DFM—Doffing minutes; D—Number of doffings; EBR—Number of end breaks; EBR/H—End break frequency (end breaks per 1000 spindle hours); EM—Mean end break duration in minutes; RPM—Mean spindle rotational speed; GA, GB—Group number A or B.

The chapter also concentrates on ring spinning whereas in the last 3–4 years considerable advances have been made in applying production data collection also to rotor spinning. In this respect, the importance also of providing a length-measured spool has been realized, and this combined with the data collection function. Length accuracies of below 0.5% have ensured that re-winding costs can be eliminated and waste is reduced to a minimum.

The chapter also makes no mention of production data collection at the cone winding process where the importance of data with respect to number of knottings, efficiency of winding and the number of knots required to replace disturbing yarn faults via electronic clearing is of prime importance.

And finally, in the last few years the possibility of storing data at reasonable cost on diskettes or magnetic tapes has made available a much better off-line evaluation. Via an interface and such memorying capabilities as referred to, therefore, it is now possible to process on-line production data on EDP units in order to provide, for instance, long-term statistics and wage premium systems.

Progress in electronics and its influence on data logging techniques moves so quickly that what was up-to-date four, three or even two years ago cannot today be considered as "contemporary textile engineering", and therefore the necessity of this addendum.

# Chapter 9

# Textured Yarns and Fabrics

P. R. LORD

---

## I. Texture, Bulk and Cover

It is useful to define terms carefully before proceeding because in so doing one exposes some of the essential requirements relating to whatever is defined. For the present purpose, "texture" is defined as a term describing those attributes of an object which can be recognized by the human sight or touch. It will be seen that it encompasses not only bulk and cover, but also lustre, appearance, hand and other tactile sensations, such as warmth.

Bulk is important in the present context because one of the important properties of a textile material is its ability to cover a given area. Bulk is here defined as the specific volume of a yarn, but it could equally well be defined as the specific volume of a fabric. One might regard the specific volume of the yarn to be the potential specific volume of the fabric. The potential is never fully realized because the yarns squash when they are assembled into fabric. Cover may be regarded as the percentage of area covered by one or more yarns in a fabric.[4] Fabric structure determines how often two or more yarns are piled on top of one another and the resultant pile of yarn covers no more area than one yarn. The structure also determines the amount of squashing of the yarns. Hence fabric structure is very important in determining cover, and bulk alone cannot define it. Nevertheless a bulky yarn provides better cover in a given fabric structure than a non-bulky one.

## II. Optical Properties

The appearance of a surface depends on the way light is reflected from it. A

267

matt surface reflects light randomly whereas light reflected from a lustrous surface is distributed but has an organized character. A shiny surface tends to have hard highlights because of the concentration of the organized reflected light beams. An example of a lustrous surface is provided by an array of parallel and closely packed filaments. Each filament has a highlight which is parallel to the filament axis and thus there are many tiny parallel lines of light distributed over the surface covered by the fibre array. In an array of fibres, the geometric disposition of the fibres helps to determine the degree of lustre achieved. A random array of fibres, such as in paper, gives a matt surface. A roughly parallel array such as that produced by weaving twistless yarns in a satin weave gives a lustrous silky appearance. A similar philosophy can be applied at the molecular and yarn levels. Reflective particulate matter embedded in the surface of a fibre acts as a delustrant. Titanium dioxide is frequently used for this purpose. Variations in a fibre cross-section also tend to scatter the reflected light but in so doing they can produce local light concentrations that appear as sparkles on the surface. Needless to say variations in the fabric structure can produce strong optical effects. A good example of this is given by the figuring in a Damask weave.

## III. Hand

The tactile sensations of handling a piece of fabric are difficult to define. The sensations from a light touch are controlled by different fabric parameters from those relating to a heavy grasp. The latter involves the bending and shear properties of the fabric which are heavily dependent on the fabric structure. A light touch involves the deformation of the surface fibres. This sensation is, in turn, dependent on the yarn structure and the fibre properties. A hairy yarn made of fine flexible fibres makes a soft surface, whereas a tightly twisted yarn tends to produce a harder smooth surface. A bulky yarn feels warm because of the entrapped air, but a hairy yarn made of heavy stiff fibres produces a prickly surface.

A cylindrical filament or fibre subjected to an axial load will buckle when

$$F > \left(\frac{A}{l}\right)^2 KE$$

where $l$ is the length of the fibre, $A$ is the cross-sectional area, $E$ is the modulus, $K$ is a constant and $F$ is the applied load.

For a given material, the length and the fineness of the outstanding hairs are very important in determining the surface softness to the light touch. If the surface is composed of loops then

$$F = \frac{KEA}{(\Delta R)l}$$

where $K$ is a constant dependent on the cross-sectional shape and other geometric aspects of the fibre, $\Delta R$ is the change in radius of curvature of the loop and the other symbols have their previous meaning. Thus, there is a difference in behaviour of the surface according to whether it is composed of hairs or loops. The single hair collapses under load and the force of touch ($F$) diminishes sharply after buckling occurs. A loop does not collapse in the same way and ($F$) tends to decrease more slowly with loop deflection. Furthermore the force is a function of loop radius. Hence the response to a light touch is greatly affected by the geometry of the surface hairs and loops.

## IV. Stability, Cover and Comfort

Fabric behaviour is strongly related to the characteristics of the yarn from which it is made. Figure 1 shows unit cells of the simplest forms of woven and knitted fabrics. Apart from the obvious relationship between fabric structure, yarn diameter and cover, it should be observed that a fabric tension in

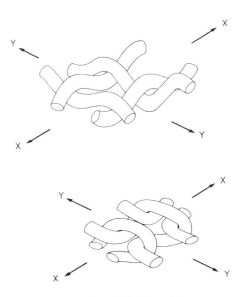

FIG. 1. Simple fabric cells.

direction "X" produces a contraction in direction "Y" and vice versa. The contraction is resisted by adjacent yarns pressing against each other. This resistance is augmented by the squashing of the yarns at the intersections which tends to push the yarns apart. Thus it becomes evident that the lateral compressibility of the yarn plays a part in determining fabric characteristics.

It is possible that a yarn will shrink in treatment or in subsequent use. Once again the fabric shrinkage is, in part, determined by the lateral compressibility of the yarns. It is also partly determined by the ability of the fabric structure to change and relieve the stress. In the case of woven fabric this usually involves change in yarn crimp, and therefore the bending stiffness of the yarn is of some importance. In the case of knitted fabrics there is usually a change in loop shape.[5] This not only involves the bending stiffness but also the torsional characteristics of the yarn. A twisted yarn causes the loop to incline relative to the "X" direction and to rotate out of plane. This gives rise to spirality in single jersey fabric and shrinkage in all types of knitted fabrics. The amount of shrinkage is, in part, controlled by the jamming of the structure in at least one direction, and thus the lateral compressibility of the yarns is once more an important factor.

The rate at which a fabric adjusts itself is affected by the frictional characteristics of the component yarns after finishing. Thus one cannot rely on a fibre finish to provide lubricity in the fabric. (One often uses wax or other lubricant to facilitate the manufacture of the fabric.) Furthermore, the frictional characteristics are affected by the yarn structure. For example, an air-textured yarn tends to have the filaments locked together by friction, whereas a false-twisted "stretch" yarn tends to allow considerable filament movement during extension. In the one case there is much more recovery after extension than with the other. Thus, insofar as the yarn can control the fabric behaviour, the tensile and lateral compressive characteristics of the yarn are important; in many cases so are the bending, torsional and frictional properties. The complexities of the case can be reduced by using energy rather than force as the working parameter. It can then be seen that a fabric will always try to reach the minimum energy state. This minimum energy concept is very useful and will be referred to again with reference to yarn behaviour.

In the matter of cover, the greater the specific volume of the yarn the better. However such bulk can only be achieved by promoting filament disorder. As the yarn structure becomes more bulky so the yarn usually, but not always, becomes more extensible under tension and more compressible under lateral pressure. It also entraps more air and thus there is an improvement in thermal insulation. On the other hand, it becomes less easy for moisture to be transported by capillary action between adjacent fibres. Hence not only does a change in bulk affect the mechanical properties but it also alters the parameters that affect comfort and other tactile properties.

## V. Colouration and Quality Control

The colour shade of a fabric is dependent on the optical properties of the surface and the dye affinity of the fibres. As was explained in Chapter 4, the thermal and mechanical histories of the fibre affect the dye affinities of the polymer from which the fibre is made. Not only do the extrusion and drawing processes involved in producing the feed yarns affect the dye affinity,[8] but so do the texturing and finishing processes which occur in the primary sector of the industry. Hence great care has to be taken to ensure even temperature and tension at every stage of production. This evenness has to be maintained from machine to machine, spindle to spindle at a constant level. The problem is that the results of any unevenness cannot be seen until after the dyeing process is completed.

Modern extrusion technology has led to the development of POY yarns.[8] The feed yarns are extruded at very high speeds (up to 5000 m/min) and they are supplied to the texturing machine in a partly oriented state. It is cheaper to complete the molecular orientation in the texturing machine. However such POY yarns age fairly rapidly and, amongst other things, there is a tendency for the dye affinity of the finished product to vary with age of the feed stock. Hence it becomes necessary to control stocks of feed yarn so as to ensure that over-aged yarns are not used and to reduce waste by careful control of the flow of stocks.

To control quality it is necessary to be able to measure the parameters involved. In the case of textured yarns there is little problem with unevenness of linear density. The main difficulty is with colouration, and a secondary problem is associated with changes in bulk and yarn structure. Two methods of checking colour are now available. The first, and simpler, routine is to knit a sample tube from the subject yarn and to dye it under standard conditions. Various forms of colour comparison are then used depending on the resources available. The second method is to use a special machine to dye the yarn under standard conditions and then to measure any variations along the yarn length. This is an expensive system but it is capable of high throughput and analytical accuracy. Both methods are capable of showing the colour patterning (barré) that has caused so much trouble in the past. The fabric sleeve may also show variations in the yarn structure, although it takes a certain skill to be able to differentiate between the various faults. This is in part due to the fact that differences in structure can change the apparent colour of the surface. Variations in the shapes and direction of the fibres on the surface of the yarns as well as variations in fibre cross-sectional shape can produce significant effects. Variations in bulk obviously affect cover and this, too, affects the reflection of light. Yarn strength is sometimes used as a proxy for molecular disorientation. A yarn of low strength usually dyes differently from one of high strength.

## VI. Fabric Durability

Fabric durability can be measured in a number of ways, but no single way is in itself sufficient to describe fully the durability of the material. The simplest and most reliable is the simple tensile test but this does not uniquely relate to the tear strength or any of the characteristics of the surface. Also it cannot be used for knitted samples and the ball-burst test is usually substituted. The tear test is dependent on the yarn strength, yarn modulus and fabric structure (possibly also the fabric weight). The surface characteristics include resistance to pilling, snagging and abrasion. Usually laboratory test results relating to these are highly variable and do not correlate well with in-use testing. This is because in-use tests involve proportions of several mechanisms of failure which do not coincide with the proportions emphasized by the various laboratory testing systems.

In a woven fabric the tensile strength is usually little different from that calculated from the yarn strength and the number of yarns involved. Where twistless staple yarns are used, failure can be by fibre slippage. With continuous filament yarns there can be no continuous slippage and the tensile strengths (or ball-burst strength) are related to the yarn strength. For most modern filament yarns this strength is more than adequate. Nevertheless it is noticeable that the tensile strength of a woven fabric declines as more bulky yarns are used even for the same fibre. This is because of the fibre disorder and is similar to, but at a different level from, the loss in strength due to molecular disorientation. In a knitted fabric the ball-burst test strength is also related to the yarn strength provided there is no undue fibre slippage.

The strength of a fibre can sometimes be a disadvantage. In pilling and snagging, the pill or snag cannot break off in service because of the fibre strength. Indeed the snag would not have formed had the filament been weaker. Breakdown of the surface of a fabric made from strong textured yarns usually shows up in the form of pills or snags. Fabric structure plays a part because projecting crests of yarn are easily caught and are subject to extraordinary pressures. Yarn structure is also significant. Tightly twisted staple yarns have few projecting hairs and practically no projecting loops. Many textured yarns have surfaces replete with loops made of strong filaments. The disorder of the structure often makes it possible for initially small loops to be pulled into larger loops which are then even more vulnerable. A textile article may often be discarded due to deterioration of the surface rather than to tensile or tear failure. Thus such surface failures are of considerable importance. Changes in the surface texture of a fabric can also result in apparent colour shade changes. Sometimes these changes are acceptable (e.g. jeans), and sometimes they are not.

## VII. Types of Texture

Staple yarn texture can be produced when the filaments or fibres are set into the desired shape or are held in shape by friction. These two classes can be further subdivided.

I. *Frictionally bound fibres*
   A. Twisted yarn (usually staple) in which the fibres migrate to give an interlocked structure. This has been the historical norm by which other structures have been judged.
   B. Twistless yarns in which the characteristics are determined largely by the fabric structure.
   C. Air-jet textured yarns (usually filament) in which the fibres are intermeshed in such a way that they lock into place.

II. *Fibres set into zig-zags, helices, snarls or combinations thereof*
   A. Stuffer-box and other systems which usually involve the heat setting of a zig-zag fibre shape.
   B. False-twist systems which involve heat setting of fibres into helices or snarls.
   C. Bicomponent systems which produce fibre helices by differential shrinkage caused by mechanical or thermal disorientation of one layer (e.g. edge-crimp) or by the use of two different polymers.

It is possible to combine elements of the two classes in various ways. Thus one can air-jet texture a false-twist yarn so as to modify its characteristics.

Class I has the property that fibres follow such a complex path through the yarn structure that they accumulate enough capstan tension to prevent them from being pulled from the structure. They will usually break before they slip. The same frictional mechanism applies as with a good knot. This means that yarns in this class tend to have good snag resistance. When the fibre strength is low and the fibres are well bound into the structure, the pilling performance is usually good because the pills break off. (Sometimes with strong fibres and poor binding forces, defects can be produced akin to pilling.) To secure adequate fibre locking it is necessary to have sufficient lateral forces (e.g. such as are caused by high twist), and this usually means

K

that yarns in this class are relatively lean and hard. Thus cover and hand tend to suffer.

Class II is typified by high-bulk yarns made from strong thermoplastic filaments. Thus cover is good but there is a tendency for the fabrics to snag and pill. The fact that the surface is mainly comprised of loops or coils rather than hairs means that there are differences in hand which are not universally acceptable. Furthermore the variable ability of the fibres and the structure to hold and transmit moisture brings the question of comfort into play. Nevertheless the production rates of these systems are so high that there are now economic reasons for their use. Since there has been adequate discussion of twisted staple yarns in previous chapters the remainder of this chapter will be devoted to yarns in Class II with some extra discussion of air-jet texturing.

## VIII. Heat Setting of False Twist

The essential steps in making a heat-set yarn[6] are: (i) heat the thermoplastic fibres above $T_g$ (but below $T_m$); (ii) deform the fibres to the desired shape; (iii) cool the fibres to below $T_g$ while they are held in the desired shape; (iv) rearrange the cooled fibres so that they occupy a greater volume. Phases (i) and (ii) can be reversed or may be coincident.

The original way of doing this was to use separate operations of twisting, heating and untwisting. A breakthrough was achieved when it was realized that this could be done continuously. In a false-twist system the twist is normally locked in the yarn between feed roll and the twisting element (Fig. 2). The yarn leaving the twisting element should contain no twist because the yarn carries twist from the upstream side, which should exactly cancel the twist projected downstream by the false twister. As far as the moving yarn is concerned, it first experiences twist on arrival at the feed roll. At the twisting element, the yarn experiences an untwisting. Thus a heater placed between the feed roll and the false-twister can cause the yarn to be heat set in the twisted state. If the yarn is cooled prior to reaching the twisting element and is suitably overfed, the emerging parallel filaments tend to coil or snarl separately and thus produce a bulk in the yarn which is greater than that of the feed yarn.

The design problems in such machines fall into several classes: (a) gripping the yarn at the feed- and take-up rolls; (b) heating the yarn sufficiently and evenly; (c) cooling the yarn sufficiently and evenly; (d) twisting the yarns at ultra-high speed without damage, and without generating excessive tension; (e) controlling the overfeed rates and tensions; (f) winding.

Of these, (b) and (d) are the most difficult, although (c) can be a problem

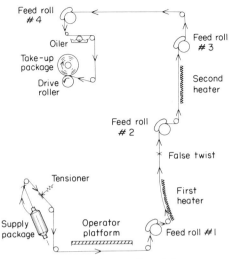

FIG. 2. False twist in a running yarn. (Reproduced from *Economics, Science and Technology of Yarn Production* by Peter R. Lord, with permission.)

in ultra-high speed machines. The following discussion will be limited to these items.

A running yarn is not easy to heat. If the yarn runs in contact with a hot surface it can drag, especially if its surface melts. If the yarn runs remotely from the surface, so that the heat transfer mechanism is that of thermal radiation, then distance from the surface can be an important parameter. If a large radiant surface is used to minimize the sensitivity to distance, then the system tends to be uneconomic as far as energy consumption is concerned. The average temperature at a cross-section rises exponentially with distance travelled (see Fig. 3a), and at normal operational speeds the heaters have to be very long before the desired temperature is reached. Heaters 2 m in length are quite common (see Fig. 4). This poses a problem in controlling the yarn path, and thus the distance between yarn and heater. The rate of temperature rise along the length of the yarn is controlled by, amongst other things, the heater temperature. Since there is a thermal lag in heating the core of the yarn in respect to the surface, it is quite possible to raise the surface temperature above $T_m$ without raising that of the core to $T_g$ (see Fig. 3b). Thus the fibre surface could melt and the core would still not be heat set, a situation rather like burning the surface of a rare steak. Hence there must be a limit to the surface temperature of the heater. A melted yarn surface would cause many operational problems and an unacceptable product. Even if surface melts do not occur and all the fibres are raised to some temperature

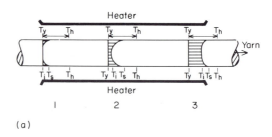

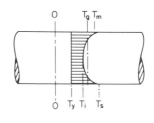

Fig. 3. (a) Variation in temperature profiles in running yarn. Note that $T_y$ = temperature of ingoing yarn. (b) Effect of too high a temperature gradient between heater and yarn. (Reproduced with permission from *Economics, Science and Technology of Yarn Production* by Peter R. Lord.)

above $T_g$, there can still be problems. Although most of the fibre migration occurs at twisting some takes place at untwisting. If the core fibres (which have been heated less and perhaps drawn less) come to the surface, then there are differences in colour between these and the non-migrated surface fibres. Naturally it is the thermal history of the surface fibres in the final product that determines the colour shade. It is clear that considerable care must be exercised to ensure that all positions used to produce a given product are exactly set to give the same temperature and migration patterns at all times.

With low speed machines it is possible to rely on natural cooling to reduce the yarn temperature below $T_g$, but with high speed machines other means of cooling become necessary. One possibility is to use a cooling block which absorbs heat from the yarn and re-radiates the energy to its surroundings by virtue of its large surface area. The use of cooling air or gas is also possible. The main objective is to get the temperature down quickly so that the yarn path length is not unduly extended.

The first false-twist machines used pin-twisters and these were capable of running up to $0.5 \times 10^6$ rpm. Although the spindles were small they absorbed appreciable amounts of energy and created unacceptable noise

FIG. 4. False-twist draw-texturing machines (Barmag).

levels, especially after prolonged use at very high speeds. The pressure of government regulations has tended to force these into early obsolescence. The rate of change-over was hastened by the fact that it was easy to replace them by stacked-disc or other forms of friction twisters. A series of discs (Figs 5a, b) or other elements works directly on the surface of the yarn. At first sight, the "gearing ratio" looks extremely high but it must be appreciated that there is considerable slippage between the disc and the surface of

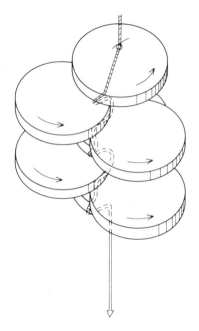

FIG. 5. (Top) Stacked-disc friction twister. (Bottom) Components of a stacked disc assembly (Barmag).

the yarn. It is better to consider these devices as torque generators rather than twist generators. One of the design problems was to create a system that would give an accumulation of torque through the stack of discs without a corresponding rise in the yarn tension. To discuss this further, consider an element of yarn sliding over the surface of a rotating disc as shown in Fig. 6a. If the yarn moves in direction $B$, then there is a drag force ($F_y$) acting on the yarn in the opposite direction. The disc surface moving in direction $A$ produces a drag force ($F_d$). As shown in Fig. 6b the resultant of these two forces is ($F_R$) and this may be resolved into two components, one parallel to and one perpendicular to the yarn element. The component $OQ$ represents the increment of yarn tension and the component $OP$ produces a torque in the yarn. As the angle $\alpha$ is changed so the ratio between tension and torque alters. When it reaches a specific value $\alpha_2$, the increment of tension is zero as shown in Fig. 6c. If the angle $\alpha$ is further altered to $\alpha_3$, it is possible for the resultant tension component to act in the same direction as the yarn movement (Fig. 6a). In other words, the discs "pump" the yarn through the system. Both the forces $F_y$ and $F_d$ depend on the normal force between the yarn and the surface of the disc. Thus although the magnitudes of the torque and tension components change with normal pressure, the ratio does not alter appreciably. It follows that the relative positions of the various discs have to be carefully set to give the correct angles so as to give the best torque/tension relationship through the disc stack. Since the amount of twist set into the yarn greatly affects the texture, this is a very important parameter. Some designs use smooth surfaced intermediate discs which merely serve the purpose of guiding the yarn onto the torque-producing discs. Other designs use active discs throughout. Many designs use hinged assemblies that permit easy yarn manipulation

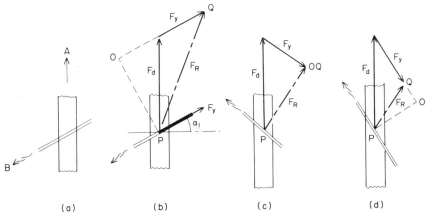

FIG. 6. Force vectors acting on a yarn at the surface of a disc.

A more recent development has been the use of crossed belts to insert false twist at very high speeds. These belts are rather like aprons in staple spinning except that the two belts are placed at an angle to one another. At the intersection, the belts move at different angles in roughly the same plane with the yarn passing between them. They move in such a way as to not only induce twist in the yarn by frictional contact with the surface of the yarn, but also to "pump" the yarn through the system. This system improves the twist control as compared to the stacked disc, but the belts tend to have a rather short life.

## IX. The Yarn as a Viscoelastic Material

A viscoelastic material is characterized by showing two responses to application of stress. The first is a function of the elongation of elements of the material and the second is a function of the rate at which the elements move over one another. Thus it is possible to have a viscoelastic fluid that has recovery properties (e.g. the "bulge" in the polymer emerging from a spinnerette), and it is also possible to have a viscoelastic solid in which relative movement of elements during the deformation causes a dissipation of energy. Since we are concerned with a solid, the system can be represented by a spring and dashpot. The load-extension curve carried through a complete loading cycle produces a hysteresis curve, the enclosed area of which represents the energy dissipated in the dashpot during one load cycle. Similarly a torque-twist curve shows a hysteresis curve and the area enclosed indicates the energy used in overcoming friction (or its equivalent in molecular terms).

Referring to Fig. 7a the torque/twist curve indicates that the state changes from $A$ to $B$ when the bundle of filaments is twisted. When the yarn is heat set, the torque is relieved and each filament has a finite twist but zero torque. If the set fibres are now untwisted (as they are as they pass through the false-twister), the condition or state changes from $C$ to $D$ (Fig. 7b). They now have zero twist and a reversed torque.[6] If the fibres are separated and relaxed, they tend to go to one of two possible minimum energy states. In one case, each filament tries to go into a coil whose helix angle reverses periodically along its length so that the overall twist is zero. In the other case, the filament attempts to go into a series of snarls whose direction of twist alternates from one snarl to the next. If the reversed torque at state $D$ is high and the yarn is fully relaxed, the filaments tend to go into a series of snarls. If the relaxation is limited and the yarn is heat-set for the second time at $E$, the filaments tend to adopt the reversing coil configuration which tends to give greater bulk. The bulk depends to a large extent on the radius of the coils

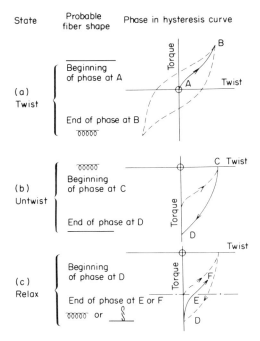

| State | Probable fiber shape | Phase in hysteresis curve |
|---|---|---|

FIG. 7. Torque-twist characteristics of a filament bundle. (Reproduced with permission from *Economics, Science and Technology of Yarn Production* by Peter R. Lord.)

produced, since a series of larger coils laid side by side occupies more volume than a series of smaller coils. The residual "twist" in segments of unidirectional coiling is related to the number of coils/cm. Each coil when extended to the point of being a straight fibre has one turn of twist by virtue of the coiling. Thus the "twist" density at that point is a function of the coil radius. It may be said that the potential bulk is dependent on the local twist densities in the set yarn. The second heat treatment reduces torque and the set "twist" density (point $E$, Fig. 7c); it also increases the potential coil radius and thereby increases the potential bulk of the yarn.

## X. Stretch and Bulk Yarns

A yarn which consists mainly of coils tends to be bulkier than other yarns and has moderate extensibility. This type of yarn is known as "bulked" yarn and it is generally made on a two heater false-twist texturing machine as shown in

Fig. 4. A yarn which is comprised mainly of filament snarls has very much less bulk but normally has great extensibility and is known as a "stretch" yarn. The extraordinary stretch characteristics arise because of the amount of filament stored in the snarls. Only one heater in the false-twist zone is needed in the manufacture of stretch yarns but the standard two heater machine can be used with the second heater turned off. It will be realized that a two heater machine makes for a complicated yarn path, especially if there is limited head room. If both heaters are 2 m long, then the machine could easily be 5 m high and accommodation in normal factories would be difficult. To overcome this, many machines have the heaters inclined so as to reduce the head room needed. However the more complex the arrangement the more difficult it is to "string up" the machine and to operate it.

## XI. POY (Partially Oriented Yarns)

As was mentioned earlier, it is cheaper to draw the polymer during texturing than at a previous and separate stage. A draw-texturing system thus has to be arranged. Two possibilities exist, first the machine may have a drawing section followed by a texturizing section (sequential), and secondly the drawing and texturing may be carried out in the same machine section (simultaneous). The latter is the more common, and machines have been developed capable of texturing at speeds of up to and beyond 1000 m/min.

Simultaneous draw-texturing involves the drawing of a twisted bundle of filaments as they pass through a heated zone.[2,3,9] There is likely to be a difference in draw ratio between the outer and the core filaments. The combination of torque and tension applies a greater elongational strain on the outer filaments of the bundle and these are also likely to be heated to a higher temperature. However the work done on the filament bundle appears as heat inside the bundle and this helps to overcome the heat transfer problem. Nevertheless, it is quite possible for sheath and core filaments to have quite different mechanical and thermal stress histories by the time that the process is completed. This affects the consistency of dye uptake.

A further peculiarity is that by drawing the twisted bundle in the softened state, the individual filaments develop flat surfaces[7] such that the filaments are roughly pentagonal or hexagonal in cross-section. This produces a sparkle in the finished product and also affects the handle. One can imagine that a bundle of hexagonal rods do not have the same lateral mobility as a bundle of cylindrical rods. Certainly the area of a hysteresis curve is increased and this shows up in the increased friction in the system. Furthermore, since bulk depends on the position round the hysteresis curve, variations in the production of flats not only affect sparkle and handle but also affect bulk. As a matter of interest, one can inspect yarn cross-sections and see variations in the extent of the flattening that has taken place across the section. This can give some idea of the variations in temperature and filament tension across sections. It is also possible to see some evidence of fibre migration.

As mentioned earlier, polymer aging is an important phenomenon to take into account. The drawing characteristics of the polymer alter quite markedly as the polymer approaches the end of its shelf life and working with over-age filaments can cause considerable operational difficulty. Thus great care has to be taken to manage the stocks of feed yarn properly.

## XII. Air Texturing

Originally air texturing was used to produce a filament yarn similar in characteristics to a staple yarn.[1] It was also employed to produce texture in non-thermoplastic yarns. It will be recalled that this form of texturing falls into Class I in which the fibres are locked into position by interfibre friction. A form of air texturing is shown in Fig. 8. The essential point is that the air separates, twists and tangles the fibres before allowing them to come back together. In trying to relieve the stresses thus created, the fibres tend to twist or otherwise move together into a structure which is self locking. There are other means of inducing the fibre distortion and tangling but the result is somewhat similar even though the structure changes. To get the interlocking

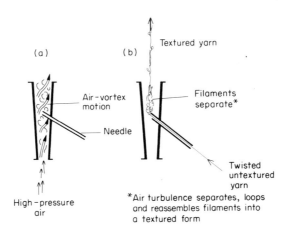

Note: (a) Shows air only; (b) Shows filaments only

FIG. 8. Air-jet texturing device. (Reproduced with permission from *Economics, Science and Technology of Yarn Production* by Peter R. Lord.)

characteristics, it is necessary for the yarns to be rather "lean" in type. The machines are capable of high speed operation.[10] Bearing in mind the remarks made earlier about texture, it will be realized that there is sometimes an advantage in modifying a false-twist yarn even though this sometimes involves a sacrifice in the amount of bulk produced. In the recent past, a number of proposals have emerged to run the false-twist yarns through an air-jet texturing nozzle before the yarn is wound up (see Fig. 9). The products of such a combination machine more nearly approach a staple yarn in characteristics and may prove attractive in the marketplace. Also, some staple yarns are air textured to produce a hairy yarn.

FIG. 9. Air-jet/false-twist texturing (Heberlein).

## XIII. Stuffer Box

The stuffer box is used for many different purposes. In the manufacture of tow and stretch-broken sliver, it is used to produce crimp, which is a sort of texturing operation. In the manufacture of carpet and heavy denier yarns, it is used for texturing filament yarns.[10] It is occasionally used for other purposes. The idea is illustrated in Fig. 10. Fibres are overfed into a restricted space and collapse, giving a zigzag crimp. If thermoplastic fibres are

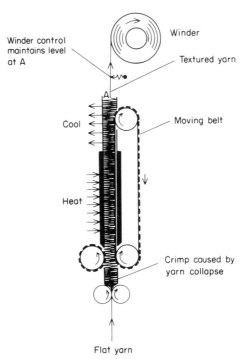

Fig. 10. Stuffer box. (Reproduced with permission from *Economics, Science and Technology of Yarn Production* by Peter R. Lord.)

used they can be heated above $T_g$ and then be cooled before the compression is removed; they are then heat-set into the zigzag shape.

The amplitude of crimp is determined by the size of the stuffer box and the Euler critical length, $l$, where

$$l = 2\pi\sqrt{\left(\frac{EI}{F}\right)}$$

and $E$ = fibre modulus, $I$ = moment of inertia of fibre cross-section, and $F$ = load applied.

Since the moment of inertia of the fibre cross-section depends on the fibre dimensions and shape, the crimp amplitude is likely to vary with denier and compressive force distribution. Thus the crimp amplitude can be somewhat variable. If the stuffer box is cylindrical, the zigzag crimp could be three dimensional, but if the box is flat, the crimp would be roughly planar. This affects the behaviour of the fibre aggregate. Furthermore, the friction between the fibres and the stuffer box sides affects the distribution of forces and thus the crimp amplitude, especially for large bundles. This is overcome to some extent by using a feed system such as those shown in Fig. 10. This also helps to reduce the problem of intermittent fibre chokes. Quality can be improved by using live steam inside the stuffer box, presumably because this improves the heat transfer and allows the set to take place under conditions of high humidity. The systems described are capable of operating at up to 1200 m/min so that high productivities are possible.

## XIV. Yarn Faults and Practical Problems

Generally the stacked disc type of friction-twisting machine can operate between 15 and 150 denier at threadline speeds approaching 1000 m/min with good tension and twist control.[3] As speeds exceed 1000 m/min, the control plateau becomes smaller and it becomes increasingly difficult to maintain stable conditions. The limit probably lies around 1200 m/min. The torque-speed curve of such a stacked disc twister is not linear, as shown in Fig. 11. If the entry and exit yarn angles relative to the flanks of the disc vary (average value = $\theta$), then so does the maximum torque that can be produced from a given tension. If the disc speeds vary (surface velocity $V$), this also affects the torque. Thus it can be seen that for a given speed, there are correct values of the "run on" and "run off" angles to achieve the highest torque. There is only a given range of variation that will allow the machine to run. This is referred to as the control plateau. The run on and run off angles are determined by the spacing and penetration of the three sets of discs. These are usually adjusted to give a tension ratio across the discs of about 1:5. The slippage is usually around 50% and there can be considerable wear on the discs. Various disc profiles and materials can be used. Certain "soft" surfaces, e.g. polyurethane, can give very good friction properties but wear fairly rapidly and can easily be damaged by inexpert handling. Other "hard" surfaces, such as metal coated with aluminium oxide ($Al_2O_3$), are much

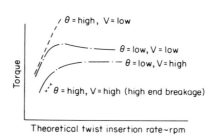

FIG. 11. Torque-speed curve for stacked-disc twister. (Reproduced with permission from *Economics, Science and Technology of Yarn Production* by Peter R. Lord.)

more durable but give more slippage or need more discs. The chance of a change in surface characteristics is much less than with the soft surfaces but the possibility of fibre damage is increased. A further important point is that the material chosen for the disc should tend to wear itself clean and not become "poisoned" in use.

Atmospheric conditions play a part and too humid an atmosphere can cause the yarn to drag through the system and produce erratic tensions and torques and thus affect texture. Too low an RH value causes buildup in static electricity which can be a considerable nuisance. Normally the work space is operated at $24°C \pm 3°$ and $65\%$ RH $\pm 2\%$. The fibre finish can also affect the atmosphere. The fibre finishes are subjected to high temperatures as they pass through the heater and it is necessary to deal with the fumes produced. Also fibre finish and particles of polymer, monomer and oligomer can be deposited as a white powder. Accumulations of this material on sensitive working surfaces (e.g. godets) can cause mechanical errors which show up in the yarn, or production of fumes if they fall into the heater. When the finishes or fibres contain titanium oxide ($TiO_2$) as delustrants, the increased wear on the operating parts can be expected. The wear also shows up as mechanical faults in the yarn and produces strict periodicities in the final product which give rise to moiré patterns.

The level of twist in a false twist yarn affects the yarn characteristics and the fabrics made therefrom. In friction twisting, the level of twist is affected by the various parameters already discussed. Varying the tensions can produce similar effects. If the torque gets too high, or the tension too low, tight spots are produced. Twist slips through the discs and a concentration of real twist appears in the final product as shown in Fig. 12. The fact that it is cancelled by a longer length at a low and opposite twist does not alter the fact that the tight spot is present. At the point of high local twist the bulk is unable to develop and the surface appearance changes. Usually tight spots are a warning of impending surging and such instability usually causes severe filamentation or end breakage. Under these conditions, a single external stimulus, such as the passage of a knot, can set up the instability. The twist also affects the texture. A high level twist tends to produce a fine soft

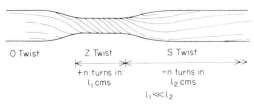

FIG. 12. Tight spot in a false-twist textured yarn.

texture, whereas a low twist produces a rough pebbly appearance. A high level twist weakens the yarn, especially with nylon, and produces high crimp contraction which is associated with fairly high yarn breaking extensions. The tensions are controlled by the overfeed; normally 2–3% overfeed is used in the false twist zone and some 4–5% in the second heater zone, although these figures must be adjusted according to the fibre in use. If the overfeed in the first zone is too high, the tension falls and tight spots and poor set are produced. If the tension drops too low, there are excessive end breaks. Also if the tension is too high, breaks are produced and it is thus necessary to keep within the control plateau.

In draw texturing, an inferior crimp resilience may be found and the products have a crisper handle because of the flattening of the fibre surfaces. Since draw-texturing is associated with the modern generation of ultra-high speed machines which work with a restricted control plateau, great care has to be taken accurately to set and to maintain the machines at the proper values for the filaments being processed. High productivity has its price in the need for superior quality control.

## References

1. Brehm, G."Air-jet Texturing of Polyester and Other Filament Yarns." *Chemiefasern/Textilindustrie*, English Edition, 1978, 28/80, **6**, 552; **8**, 706.
2. Brehm, G. "Draw-texturing Polyamide". *Int. Tex. Bull.* (Spinning), 1978, **1**, 11.
3. Brookstein D. and Backer, S. "Mechanics of Texturing Thermoplastic Yarns". *T.R.J 1978*, **48**, No. 4, 198.
4. Lord, P. R. and Mohamed, M. *Weaving; Conversion of Yarn to Fabric* (2nd Edition), Merrow, Shildon, 1982.
5. Postle, R. "Dimensional Stability of Plain-knitted Fabrics". *J.T.I*, 592, Feb., 1968, 65.
6. Lord, P. R. *Economics, Science and Technology of Yarn Production*. The Textile Institute, 1981.
7. Lünenschloss, J. and Fischer, K. "Texturing Methods—Their Development and Importance". *Chemiefasern/Textilindustrie*, English Edition, 1978, 28/80, **6**, 536.
8. Mark, H., *et al. Man-Made Fiber; Science and Technology.* 3 Vols., Interscience, NY 1967.
9. Weinsdorfer, H. "Aspects of Friction Twisting". *Chemiefasern/Textilindustrie*, English Edition, 1968, 28/80, **6**, 540.
10. Wilson, D. K. "The Production of Textured Yarns by Methods Other Than the False-twist Techniques". *Textile Progress*, **9**, 3, 1977.

# Chapter 10

# High Speed Automatic Weaving

## WALTER S. SONDHELM

## I. Introduction

Fabrics are manufactured from an assembly of fibres and/or yarns which has sufficient mechanical strength to give the assembly inherent cohesion. Fabrics made from yarns may be woven, knitted, or made by a combination of weaving and knitting, knitting together bands of yarns (e.g. Malimo[1]), lace-making, net-making, or tufting. Some of the resulting structures are shown in Fig. 1.[2]

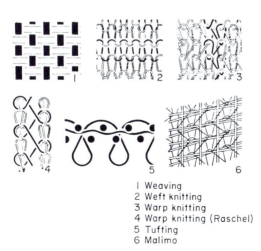

1 Weaving
2 Weft knitting
3 Warp knitting
4 Warp knitting (Raschel)
5 Tufting
6 Malimo

FIG. 1. Fabrics made from yarns.

289

In spite of spectacular gains made by various knitting processes during the nineteen sixties and early seventies weaving remains the most common process. Knitting, because of its shorter production sequence, has an advantage when fashion fabrics have to be produced quickly. It may also need a lower capital investment to produce a specified area of cloth. Woven fabrics, however, are more stable, less prone to snagging and for a given weight of fabric can usually be produced from cheaper yarns.

In weaving a fabric is produced by interlacing two or more systems of parallel threads. The warp yarns run along the length of the fabric and some of them are lifted each pick in accordance with a predetermined pattern to form a shed. A weft yarn is then inserted from selvedge to selvedge and beaten into the fell to form the fabric. The next shed is then formed by lifting a number of different warp yarns and the process is repeated.

## II. Development

Weaving is one of the most ancient crafts. Fine woven fabrics have been found in the tombs of Egypt and weavers are shown at work on pottery from ancient Greece. The Worshipful Company of Weavers is the oldest of the livery companies of the City of London and has been in existence for at least 850 years.

For thousands of years the hand loom developed slowly until in the 16th century inventors began to automate some of its motions. In 1733 John Kay invented the fly shuttle which greatly improved the productivity of the loom and of the weaver. Inventions made during the next few decades made it possible to produce more complicated fabrics at greater speed. These included the dobby,[3] the multiple box for inserting mechanically different weft yarns into one fabric, and the jacquard[3] which is a shedding mechanism giving individual control to up to several hundred warp threads and thus enabling large figured designs to be produced.

In 1785 Cartwright patented the first power loom or, to use modern terminology,[4] non-automatic singlephase weaving machine. Although it was basically a power operated hand loom a weaver could now look after a number of these machines as he had no longer to provide the motive power for shedding and picking. He still had to replenish the shuttle with weft, supervise the functioning of the loom, adjust its various motions, and repair all cloth faults. Over the next century many motions were improved and a number of new motions, including supervisory motion which stopped the loom in case of mechanical malfunctioning or yarn breakage, were developed. This made it possible to increase the range of fabrics which could be produced, improved cloth quality and protected the loom and the yarns.

The next major breakthrough was achieved in 1894 when J. H. Northrop devised a satisfactory mechanical method for replenishing the weft supply in the shuttle. This eliminated the largest element of the work of most power loom weavers and again greatly increased the productivity of the weaver although it had little effect on the productivity of the loom. To enable the weaver to be away from an individual loom for a longer period without increasing the amount of faulty cloth produced if a thread breaks or if the shuttle is not properly replenished it also became necessary to fit to most automatic looms warp and weft stop motions. Mechanization did not change the basic principles of weaving and modern automatic high speed weaving machines with shuttles have been directly developed from the Northrop loom. Improvements in yarn quality, methods of yarn preparation, and in engineering and the replacement of various mechanical motions by electric or electronic equipment, which is simpler and faster in action, have however made it possible to increase loom speeds and reduce further the work of the overlooker and of the weaver. Lancashire's failure during the first half of the 20th century to recognize and fully utilize the effect on labour productivity and fabric quality of the automatic shuttle loom is probably one of the main reasons for the dramatic reduction (shown in Table I) in looms weaving cotton, silk, and man-made fibres in the United Kingdom between 1914, when the industry's peak was reached, and the early 1970s when more productive shuttleless weaving machines started to be installed in considerable numbers. The new generation of weaving machines which is now being developed is likely to produce a change of similar magnitude in the structure of the industry.

TABLE I. Weaving machines installed in the United Kingdom for weaving cotton, silk and man-made fibres.

| Year | Weaving machines in place × 000 | Comparison with base year 1914 = 100 |
|---|---|---|
| 1899 | 649 | 85.5 |
| 1914 | 805 | 100 |
| 1928 | 768 | 95.3 |
| 1937 | 504 | 62.7 |
| 1945 | 302 | 37.5 |
| 1953 | 367 | 45.6 |
| 1963 | 138 | 17.2 |
| 1973 | 55 | 6.8 |
| 1979 | 33 | 4.1 |

The patent literature shows that work on shuttleless looms, that is weaving machines which do not carry a supply of weft for more than one pick in a fly shuttle, has been in progress for more than 150 years. The basic principles of weft insertion by rapiers has been known for more than a century and during the 1920s Brooks described the first air jet weaving machine and Rossmann was already working on the projectile machine. Further developments in engineering techniques and in the methods of warp and weft preparation were, however, required before shuttleless weaving machines could become economically attractive. Sulzer commenced production of their projectile machine based on Rossmann's work in 1953, and continued technical improvements have enabled them to remain in the forefront of weaving machine design for the past 30 years. No loom maker has achieved a similar dominance in any of the other fields of shuttleless loom design. An analysis of the catalogues of the main international textile machinery exhibitions in 1979 and in 1980 (see Table II) shows that only 5 machinery makers were offering projectile machines whilst 20 were still offering single shed machines with shuttles and 35 had a wide range of rapier machines for sale.

Towards the end of the 1970s weaving machines with nozzles, commonly referred to as air jet and water jet looms, were beginning to make an increasingly important impact on weaving. Although jet machines are still at a relatively early stage of their development they have already reached very high weft insertion rates and could well become the dominant weaving machine before the end of the 1980s.

TABLE II. Weaving machines listed in ITMA 1979 and ATME-I 1980 catalogues[a]

| Machinery maker (abbreviated name) | Country | Machines with shuttles | Machines without shuttles | | | |
|---|---|---|---|---|---|---|
| | | | Projectile | Rapier | Nozzle | Multi phase |
| Agache Wilmot (Saint Freres) | France | C | | | | |
| Barber Colman | USA | | | T | | |
| Bentley Weaving | UK | | | | | X |
| Bonas | UK | S | | | | |
| Crabtree, David | UK | S | | | | |
| Cretin | France | S | | X | | |
| Crompton and Knowles | USA | | X | | | |
| Desveus | Spain | | | X | | |
| Dornier | GFR | | | X | | |
| E. F. I. | Portugal | C | | | | |
| Elitex (Investa) and ZVS | Czech. | | X | | A | W |
| Finatex | Italy | | | X | | |

| Machinery maker (abbreviated name) | Country | Machines with shuttles | Machines without shuttles | | | |
|---|---|---|---|---|---|---|
| | | | Projectile | Rapier | Nozzle | Multi phase |
| Galileo | Italy | | | X | A | |
| Gunne | GFR | | | X | A | |
| Gusken | GFR | | | X | | |
| Hardaker | UK | S | | | | |
| Hoeck | Belgium | S | | X | | |
| Iwer | Spain | | | X | | |
| Jager | GFR | S | | X | | |
| Jumberca | Spain | | | X | | |
| Jurgens | GFR | S | | X | | |
| King Kong | Taiwan | | | X | | |
| Lentz | GFR | S | | X | | |
| Macart | UK | | | X | | |
| Mackie | UK | | | X | | |
| Mayer et Cie | GFR | | | | | X |
| Menegatto | Italy | | | X | | |
| Mertens and Frowein | GFR | S C | | X | | |
| ME.TE.OR | Italy | | | | W | |
| Mueller Jacob | Switzerland | | | X | | |
| Neotex | GFR | | X | | | |
| Nissan | Japan | | | | A W | |
| Northrop | UK | S | | X | | |
| Nuovo Pignone Smit | Italy | | | X | | X |
| Picanol | Belgium | S | | X | A | |
| Pymsa | Spain | | | X | | |
| Rockwell/Draper | USA | S | X | X | | |
| Ruti | Switzerland | S | | X | A | |
| SACM | France | | | X | | |
| Saurer and Saurer-Diederichs | Ch & France | S | | X | | |
| Somet | Italy | | | X | | |
| Starlinger | Austria | C | | | | |
| Sulzer | Switzerland | | X | | A | |
| Terhaerst | GFR | | | X | | |
| Texo | Sweden | S | | X | | |
| Toyoda | Japan | S | | | A | |
| Tsudakoma | Japan | S | | X | A W | |
| Vamatex Spa | Italy | | | X | | |
| Van de Wiele | Belgium | S | | X | | |
| Weba Karl-Marx-Stadt | GFR | S | | X | | |
| Wilson and Longbottom | UK | S | | X | | |

[a]A = Air jet machines; C = Circular machines with shuttles; S = Single shed machines with shuttles; T = Triaxial; W = Water jet machines.

Over the past two decades increasing interest has been shown in multi-phase machines in which several picks are inserted simultaneously so that several phases of the working cycle of the weaving machine take place at the same time. Such machines hold out the prospect of even greater weft insertion rates than singlephase jet machines because high weft insertion rates can be realized without the high rates of yarn acceleration or the rapid movement of parts which is necessary with all high speed singlephase machines.

Multiphase machines can either form multiple sheds in the weft direction, when they are generally referred to as wave shed machines, or in the warp direction, when they are called linear shed machines. By 1980 only circular wave shed machines with shuttles had found commercial application. They are used to weave tubular fabrics in fairly open structures. Flat wave shed machines have reached high weft insertion rates under trial conditions but require further development before they become commercially viable. Linear shed machines, which received little attention until recently, are making rapid progress.

Table III shows the improvement in machine and weaver productivity

TABLE III. Productivity of weaver and weaving machines weaving simple staple fibre fabrics.

| Type of weaving machine | Year invented | Productivity (metres of weft inserted per minute) | | Main duties of weaver |
|---|---|---|---|---|
| | | per machine | per weaver | |
| Ancient hand loom | | 12 | 12 | Form shed; Pass weft through warp; Move weft to fell |
| Hand loom with fly shuttle | 1733 | 72 | 72 | Picking; Beating up |
| Power loom | 1785 | 330 | 2000 | Start and stop loom; Reshuttle; Repair faults |
| Automatic shuttle machine | 1894 | 500 | 16000 | Start loom; Patrol; Repair faults |
| Projectile | 1924 | 1100 | 35000 | Start machine; Supervise; Repair faults |
| Jet (air or water) | 1911 1945 | 1500 | 42000 | |
| Wave shed; Linear shed | | 1800+ | ? | |

which has been obtained from progressive stages of weaving machine development. Apart from weavers, operatives are required to service the weaving machines and to carry out the operations preparatory and subsequent to weaving. Many of the modifications to weaving machines have been concerned with rendering these operations unnecessary or with reducing the manpower required for them.

In 1977, nearly two and a half centuries after the introduction of the power loom, there were more handlooms in operation in India alone than there were power looms throughout the world. Most of the cloth was produced on just over 3 000 000 weaving machines of which only 3½% were shuttleless. During the 1970s, however, shuttleless machines were gaining rapidly in importance and by the end of the decade most new machines were of this type (see Table IV). The impact of shuttleless machines is even greater than indicated by their numbers because they tend to be more productive and because new machines are often operated for longer hours.

TABLE IV. Weaving machines delivered in 1974–78 (excluding USSR and China).

| Area | Machines delivered in | | | | | Percentage of shuttleless machines delivered in | | | | |
|------|------|------|------|------|------|------|------|------|------|------|
| | 1974 | 1975 | 1976 | 1977 | 1978 | 1974 | 1975 | 1976 | 1977 | 1978 |
| Europe | 13532 | 13687 | 13758 | 13358 | 13308 | 60 | 54 | 75 | 81 | 84 |
| N. America | 5942 | 6059 | 6114 | 4685 | 4826 | 62 | 65 | 51 | 78 | 90 |
| Latin America | 2137 | 4364 | 3086 | 3733 | 1734 | 25 | 30 | 38 | 46 | 69 |
| Africa | 3457 | 5272 | 5043 | 5046 | 3488 | 8 | 15 | 4 | 13 | 25 |
| Asia | 36216 | 20085 | 19985 | 11908 | 9516 | 8 | 16 | 9 | 21 | 36 |
| Total | 61284 | 49440 | 49854 | 38730 | 32872 | 26 | 33 | 37 | 50 | 64 |

Source: ITMF (IFCATI) Enquiry.

## III. The High Speed Weaving Machine

All automatic weaving machines consist of a number of essential parts which have to be fitted together to make the machine which produces the fabric. Additional elements may be added to improve the performance of the machine or to make more complicated fabrics.

### A. The standard machine

Figure 2 shows the most common arrangement of the different parts of a high speed weaving machine intended to weave fabrics requiring only one warp

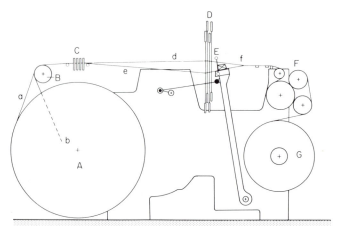

FIG. 2. Section through automatic weaving machine. See text for explanation of lettering.

beam. The main parts of the weaving machine, commencing at the warp entry side, are given below.

## 1. The beam

The warp yarn is generally supplied to the weaving machine on one or more beams[5] ("A" in Fig. 2) which are mounted in brackets within the machine frame. The flanges of the beam normally protrude outside the frame line and when determining the dimensions of the machine this overhang has to be taken into account. If large diameter beam flanges are used, the warp entry end of the weaving machine may have to be raised and the machine slanted. It makes the work of the weaver easier if the cloth delivery side of the machine is kept at the normal working level.

## 2. Let off

To obtain an evenly woven cloth it is essential to maintain a steady tension in the warp throughout the life of the beam and to release the yarn at the exact rate required to form the cloth. The tension in the warp sheet is measured as the yarn passes over a roller (the back rest or whip roller "B") and the let off then releases the amount of yarn required. Friction let off motions where the beam is tensioned by means of a rope or chain are unsuitable for high speed weaving machines, and automatic motions which control the unwinding either mechanically or electrically are used. As the beam empties the yarn line changes from "a" to "b" and the beam will have to be turned at a faster rate because of the reducing circumference of the remaining yarn. For heavy and close fabrics a second whip roller may have to be fitted to ease the strain on the let off. As the width of the machine is increased the diameter of the whip roller is increased, and like many other parts of the machine, may have to be strengthened to withstand greater stresses.

## 3. Warp stop motion and warp protection

A weaver tending a large set (number) of machines can visit each loom only infrequently and it is therefore essential for the machine to stop automatically if an end breaks. A stop motion "C" is fitted to the machine and each end (one warp thread) is threaded through a drop wire[6] (for schematic drawing see Fig. 3). If the end breaks it no longer supports the drop wire which drops on to the bars[7] of the stop motion and stops the machine. In high speed machines stop motions are generally electrically operated whilst older automatic machines use mechanical ones. For very strong yarns, which are unlikely to break, no stop motion is required. For filament yarns a photo-electric stop motion can be used. It requires no drop wires and relies on the broken end to drop into the path of the beam of light.

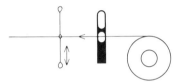

FIG. 3. Schematic drawing of drop wire and heald (Zellweger, Uster).

To prevent the breakage of large numbers of ends if the shuttle fails to reach the shuttle box the machine has to be stopped before beat-up or the reed has to be swung out of the way quickly. In fast reed looms, that is in machines where the reed cannot move out of the way before beat-up, the swell in the shuttle box was formerly used to actuate the bunter lever which brought the machine to an immediate stop. This action placed great strain on the frame of the machine and imposed limitations on the picking speed. In most modern shuttle machines this has been replaced by electronic shuttle flight monitors which, with the help of an electromagnetic brake, can stop the machine faster, without imposing localized strains of the same magnitude. Similar protective mechanisms are required for rapier machines and machines with projectiles and weft carriers.

## 4. Shedding motion

The shedding motion "D" is required to lift and lower the ends of the warp so that sheds can be formed for interlacing with weft. When the heals are level the warp line in most weaving machines is nearly horizontal (c). For weft insertion a shed has to be formed which requires that some ends move to a higher level (d) and the remaining ends to a lower level (e) so that the weft can be inserted between the two sheets formed by the warp on the cloth delivery side of the reed.

For most designs each end is drawn through a heald (Fig. 3) and the healds with ends which have to be lifted together are collected in heald frames.[8] For fairly simple designs lifting and lowering of the heald frames is controlled by a cam box and for more complicated designs by a dobby. For designs which cannot be woven on between 20 and 24 heald frames a jacquard machine has to be used. A new tappet motion which combines the simplicity of the cam box with the versatility of the dobby has recently been developed (Fig. 4). In this motion the movement of the jacks is electronically actuated from a mini-computer which stores repeats of up to 4000 picks and can be reprogrammed quickly.

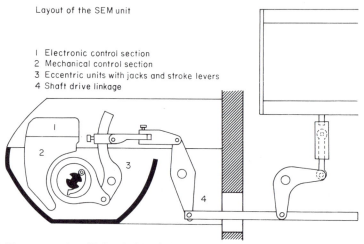

Layout of the SEM unit

1 Electronic control section
2 Mechanical control section
3 Eccentric units with jacks and stroke levers
4 Shaft drive linkage

Fig. 4. Tappet motion (Sulzer's SEM) for the electronic control of shafts and colour change.

## 5. The reed and reed drive

The reed "E" serves two functions: it keeps the warp sheet straight and at the specified width. After a pick has been inserted it also beats the weft into the fell "f" of the cloth. In the case of the Ruti "te Strake" air jet it also acts as an air tunnel through which the weft is propelled.

In shuttle weaving machines the sley moves the raceboard and the reed whilst in many shuttleless machines pick insertion and beating up have been separated. Modern shuttleless machines use torsionally rigid light metal

Fig. 5. Reed drive of Saurer 400.

sleys which carry the reed (Fig. 5) and are designed to work with little vibration and to beat up the weft at high speed.

## 6. Weft insertion system

After an open shed has been formed the weft can be carried or drawn through the warp by a shuttle, projectile, or rapier or can be blown across by air or water. Yarn may be inserted unilaterally or bilaterally and single or super-posed sheds may be used. Looms are generally classified according to their weft insertion system. Some of the available alternatives are listed in Table V and a few will be considered in greater detail in Section IV below.

Electronic weft stop motions, which act faster than mechanical ones and can, if required, monitor the passage of weft in more than one place, are used in most shuttleless machines to stop the machine as soon as the weft breaks or runs out. To prevent unnecessary stoppages from weft running out, magazine weft creels are used. Most magazine creels hold one reserve package for each type of yarn being woven but for single colour machines a weft creel carrying four reserve packages is available.

In jet machines a premeasured length of weft has to be supplied through the nozzle. The weft can be measured either on an adjustable drum or with interchangeable measuring discs and it can be stored, ready for insertion, either on the drum or in a storage tube or chamber where a loop of yarn is held by light air pressure. For high speed weaving, even when yarn does not have to be pre-measured, it is often useful to withdraw it at a steady speed from the cone and store it for one to three picks on the drum of an accumulator or constant tension winder from which it can be withdrawn rapidly with minimum tension variation.

TABLE V. Weft insertion systems.

---

1   SINGLEPHASE (Single or superposed sheds)
   1.1 Shuttle
      1.1.1 Shuttle changing
      1.1.2 Pirn changing
          fed from rotary battery or stacks or box loader or pirn winder on
            machine
          single or multiple boxes on one or both sides
   1.2 Projectile
          single or several projectiles;
          unilateral or bilateral weft supply;
          without or with weft selection
   1.3 Rapier
      1.3.1 Rigid       ⎫  unilateral or bilateral rapiers;
      1.3.2 Telescopic  ⎬  unilateral or bilateral weft supply;
      1.3.3 Flexible   ⎭  tip or loop transfer (with single or double pick);
                     without or with weft selection
   1.4 Nozzles
      1.4.1 Air jet
         with single nozzle or with main and additional nozzles;
         without or with weft selection
      1.4.2 Water jet
         without or with weft selection

2   MULTIPHASE
   2.1 Wave shed
      2.1.1 Flat
         single or multiple weft supply
      2.1.2 Circular
   2.2 Linear shed

---

Terminology based on draft international standard ISO/DIS 5247.2

## 7. Selvedge motions

Selvedge motions are not needed on shuttle machines weaving single width fabrics if the woven selvedge, which is generally slightly thicker than the body of the cloth, is acceptable. If two or more cloths are woven side by side in a shuttle machine or if the weft is cut every pick or every other pick in shuttleless machines a special selvedge motion has to be used to bind in the weft and to prevent warp ends from pulling out during finishing and handling. A wide range of selvedge motions are available and they form either a fringe selvedge, where the edge threads are held in position by a half cross or a full cross leno motion, or a tucked selvedge where, after cutting of the weft, the outside 10 to 15 mm of weft are tucked into the next shed and bound in. Whilst the tucked in selvedge is stronger than most leno selvedges it is also thicker and may therefore require special attention during finishing. Chain stitch selvedges, which are widely used on narrow fabric looms, have found little application on wide machines but could become of greater interest as machine speeds increase. When thermoplastic yarns are woven, electrically heated wires or ultrasonic cutters mounted on the loom will cut off the selvedge and, at the same time, seal the edge.

## 8. Temples

Temples are used to prevent uncontrolled contraction of the fabric caused by tension generated in the weft during weaving. They are fixed on the cloth delivery side of the fell and generally extend up to 15 cm from the selvedge inward. For cloths which are very difficult to hold or which crease easily full width temples may be used. A wide range of temples are available to suit different yarns and different constructions and to ensure that fabrics are not damaged by cutting or tearing.

## 9. Take-up and cloth winding

Take-up motions "F" with widely differing configurations are available to suit the types of fabrics to be woven (for alternatives to "F" in Fig. 2 see Fig. 6a, b). The take-up roller is positively driven and covered with a material such as carborundum, metal filleting, or embossed plastic which will grip the cloth but does not damage it. The cloth is held against it by a system of rollers with adjustable pressures which ensure slip-free take-up and, therefore, even pick spacing. To change the pick spacing the speed of the cloth take-up can be varied either by change gears or by a variable speed gear box. The take-up motion and the pick finding motion should be synchronized so that yarn breaks requiring the removal of a pick can be repaired as quickly as possible and so that the machine can be restarted without making a starting place due to variations in pick spacing.

To keep the cloth clean and to protect it the take-up roller on the cloth delivery end of the machine is covered with a breast plate.

After passing through the take-up the cloth is wound on to a circumferentially or axially driven cloth roller "G" (Fig. 2). If an axially driven roller is used a friction device has to be incorporated to ensure that the cloth winding does not interfere with the take-up. It should be possible to doff the cloth which has been wound on to the cloth roller whilst the machine is running.

## 10. Signal lights and monitoring

The stop motions of high speed weaving machines are generally connected to lights which signal to the weaver that the machine is stopped and different

W. S. SONDHELM

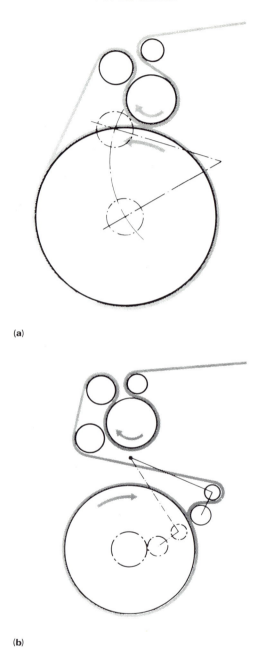

(a)

(b)

FIG. 6. Examples of take-up and cloth winding motions available with Ruti weaving machines, (a) and (b).

coloured lights are often used to indicate the cause of the stoppage. The sensors which note that the loom is stopped can be connected to a micro-processor which monitors the speed of the machine and the cause, frequency, and duration of a stoppage. The computer will process this information according to cloth sorts, weavers sets, or in any other required grouping. This information is immediately and continuously available and enables management and operatives to diagnose faults more rapidly and effectively, increase efficiency, and eliminate unnecessary work.

## B. Machines with inclined or vertical sheds

In most weaving machines the warp sheet is nearly horizontal when the shed is closed. This arrangement has the advantage that fly and size, which separates from the warp yarns during weaving, drop below the shed line and are less likely to accumulate in areas where they can contaminate the cloth. Experience has, however, shown that a horizontal warp sheet is not essential.

The Elitex "P" type machines, which were the first industrially successful air jet looms, have a shed line inclined at about 45°. In this machine (Fig. 7) the warp beam "A" is located above the cloth roller "G". The back rest "B" is at the highest point and from there the yarn moves down an incline through the warp stop motion "C" and the shedding motion "D" over the breast beam "H" to the take up motion "F". This arrangement results in a very compact weaving machine with easy access to the warp sheet, but restricts the size of the beam and of the full cloth roller which can be accommodated within the frame of the machine.

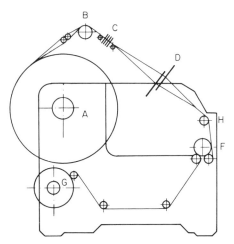

FIG. 7. Section through Elitex air jet machine. For explanation of letters see text.

Most circular wave shed machines with shuttles use a horizontal shed. They are mainly used to weave bags from polypropylene or polyethylene film, which is clean, or from jute. Flat wave shed machines with the warp beam at the top or at the bottom and a horizontal shed line have been tried but have not proved commercially successful so far.

The conventional concept of how a shed is formed does not apply to linear shed machines where each warp thread forms a number of sheds simultaneously. In a recently released machine successive sheds are formed by traversing the warp yarns round grooved rollers mounted on a rotating drum. Sheds are formed round 340 degrees of the circumference of the drum.

## C. Machines with batching motions

When the bearings carrying the cloth roller ("G" in Figs 2 and 7) are accommodated within the framework of the weaving machine the diameter of the full cloth roller cannot conveniently exceed 500 to 600 mm. The diameter can be increased if the bearings are mounted outside the main framework but this makes the work of the weaver difficult because he has to lean across the cloth roller to reach the reed and healds. The cloth, which protrudes into the weavers alley, can also be damaged more easily.

When it becomes desirable to produce large rolls of fabric either to reduce the cost of subsequent processing or because the cloth is required in long lengths without seams, a special batching unit (Fig 8) has to be used. Most batching units have an independent motor drive which can be linked to the weaving machine to ensure that cloth is always taken up. The batching machine can be fitted with an inspection table (which may be swung out of the way for cloth removal) and an accumulator so that a patrolling cloth inspector can inspect, and if required mend, the cloth which has just been woven.

Batching motions, which can produce rolls of cloth of up to 1500 mm diameter, are most frequently placed on the far side of the weaver's alley and the weaver patrols on a platform placed between the cloth delivery side of the weaving machine and the batching motion. This arrangement increases the space required by the weaving machine by up to 100% and also increases the walking time of the weaver. The weaving area can be maintained at standard size if the batching motions are located on a separate floor below the weaving area and the cloth is passed through a slot in the floor located below the take-up rollers.

Batching motions increase the cost of weaving because of the cost of the batching unit and of the additional floor space and because they have to be maintained and the extra space has to be lit and air-conditioned. This extra cost has to be balanced against the savings which can be obtained from long

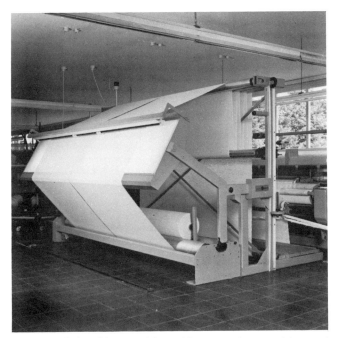

FIG. 8. Hagemann's batching machine with accumulator and inspection table.

pieces and from improved quality obtained by inspection in the shed soon after the cloth has been woven.

## D. Machines with more than one full width warp beam

Two or more full width or four or more half width beams may be required because yarns of widely differing linear density (count) or extensibility cannot be warped satisfactorily on to one weavers beam or because widely differing lengths of warp yarn may be required for ground and pile yarns in terry, velvet, double plush, and similar constructions. When the extra beams are required to cater only for variations in linear density the modifications which have to be made to the weaving machine are simple and consist mainly of the provision of a second beam stand with its own let-off.

In modern terry weaving machines with shuttles (Fig. 9) the pile warp is generally mounted above the ground warp which makes it easily accessible for warp changing and allows a pile warp beam of large diameter to be used without increasing the height of the level shed. The feed roller draws the required length of pile warp from the pile beam, maintaining constant

L

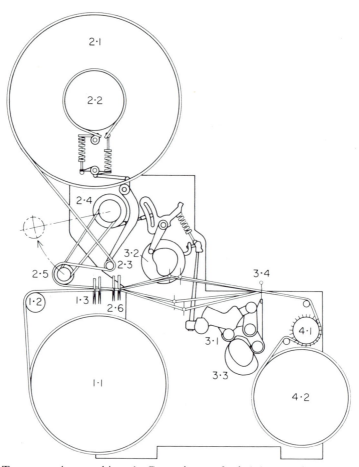

FIG. 9. Terry weaving machine. 1. *Ground warp feed:* 1.1 ground warp beam; 1.2 back rest roller; 1.3 ground warp stop motion. 2. *Pile warp feed:* 2.1 pile warp beam; 2.2 brake drum; 2.3 oscillating roller; 2.4 clutch; 2.5 feed roller; 2.6 pile warp stop motion. 3. *Pile formation and beat-up:* 3.1 control gear for sweep of sley; 3.2 cams for pile control; 3.3 cams for sley drive; 3.4 reed. 4. *Cloth take-up:* 4.1 take-up roller; 4.2 cloth roller.

tension. The terry loops are formed with the help of a double sley. The basic sley positions the reed so as to lock the picks used for binding the terry warp and the terry sley then beats in the 2 + 1 or 3 + 1 pick groups to the fell thus forming the loops. The difference in the distance between the two movements of the reed determines the depth of the pile. In some shuttleless looms the action of the sley has been replaced by a horizontal movement of the breast beam and temples (Fig. 10). Each time they are moved two or three

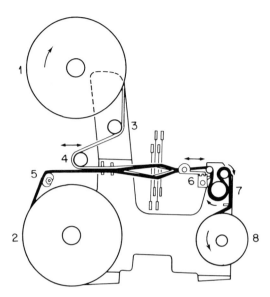

FIG. 10. Projectile terry weaving machine. (1) Ground warp; (2) pile warp; (3) deflection roller for ground warp; (4) whip roller for ground warp; (5) whip roller for pile warp; (6) movable breast beam and temple; (7) needle-type take-up beam; (8) cloth beam.

picks of weft are pushed to the fell and form the pile; the loop height is governed by the distance the breast beam is moved.

Some of the heaviest pile fabrics are woven on rapier machines operating with superposed rapiers and four beams (Fig. 11). The pile is fed positively from two pile beams "P", which have a larger diameter than the ground beams "G", and a pile length of up to 70 mm can be maintained with precision. The pile is cut on the machine and the woven fabrics are rolled on to two separate cloth rollers "R". The number of picks inserted over the two fabrics is equal to twice the revolutions of the machine.

## IV. Selected Weft Insertion Systems of Shuttleless Machines

The weft insertion system is the core of the weaving machine and in Table V some of the available weft insertion systems have been listed. Since Hanton[9] wrote about the automatic loom more than 50 years ago the rotary battery has been replaced by the box loader and the Unifil[10] and some of the mechanical weft control systems have been replaced by faster acting electric

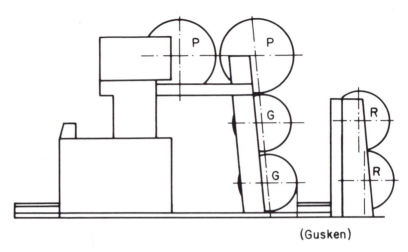

(Gusken)

Fig. 11. Velour weaving machine with four beams (Velourmat GMV-80 made by Gusken).

and electronic ones. The modern automatic shuttle weaving machine (Fig. 12), due to these improvements and to better engineering, looks more streamlined, operates much faster and needs considerably less labour in spite of the fact that the basic principles have changed little. In this section, therefore, only a few of the more interesting weft insertion systems of shuttleless machines are being considered.

## A. Projectile

Projectile machines can use either a single projectile or a series of projectiles to pull the weft through the shed. If only one projectile is used, as in the Investa OK4 or the Crompton and Knowles 400, the projectile is fired from each side alternately using a bilateral weft supply. If a number of projectiles are used, as in the Sulzer PU and PS machines, the weft supply is unilateral and the projectiles are returned on a conveyor outside the shed.

### 1. Single projectile machines

Single projectile machines have identical picking units at each side. The units receive the small projectile, rotate it through 180°, position it in the picking position, insert the weft into it, and fire it pneumatically. The power is provided by double acting air pumps at each side of the machine. The machine requires at least one weft supply at each side but up to four colours

FIG. 12. Automatic shuttle weaving machine with electro-magnetic brake and electronic shuttle flight control (President MDC).

can be fed to the projectile from each side with equal ease. The picking units are affixed to and move with the sley. This makes projectile flight adjustment less critical but, together with the need to turn the projectile after each pick, imposes limitations on the weft insertion rate. So far these machines have achieved little commercial success.

## 2. Weaving machines with several projectiles

Since Sulzer commenced series production, their weaving machines (Fig. 13) have sold more units than any other shuttleless machine and have been continuously developed to cover an ever increasing range of fabrics. At the same time, design improvements have enabled Sulzer to increase the weft insertion rate from 660 m to 1100 m/min and the maximum reed width to 5450 mm.

In the Sulzer machine the weft is carried through the shed by a series of 90 mm long projectiles which weigh only 40 g in the standard version or 60 g in wide machines or machines intended for coarse and heavy yarns. The energy required for the picking operation is stored through the tensioning of

Fig. 13. Sulzer PU four colour weaving machine with a maximum working width of 3340 mm weaving two fabrics side by side.

the torsion bar "R" of the picking unit (Fig. 14). After the lock is released the bar returns quickly to its original position, accelerating the projectile "P" without shock through the lever "L". The velocity of the projectile depends on the turning angle of the torsion rod and can be adjusted depending on the reed width, the machine speed and the type of weft being processed. During its flight through the shed (Fig. 15) the projectile runs in special guides "T" and the warp is never touched by the projectile or the weft. After the projectile has been stopped in the receiving unit it is returned to the picking position by a conveyor located underneath the shed. After insertion every pick is cut off and the length of yarn projecting beyond the edge of the cloth on each side can be tucked into the shed of the next pick by a tucking unit. One or more centre tucking units can also be fitted if more than one width of cloth is to be woven side by side.

## B. Rapier machines

Of the 51 weaving machinery makers at ITMA 1979 and ATME-I 1980, 35 showed rapier weaving machines and some were selling more than one

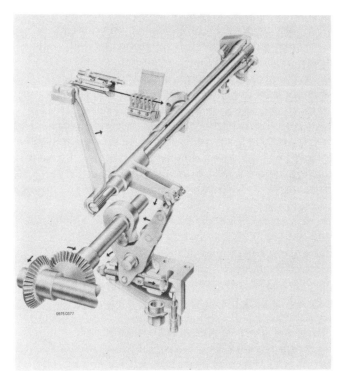

FIG. 14. Picking unit of Sulzer PU weaving machine.

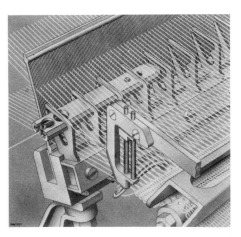

FIG. 15. Sulzer projectile running in special guides.

system. The available machines fell into three main groups: rigid, telescopic and flexible. Whilst most of the systems use bilateral rapiers some rigid rapier systems intended for coarse yarns insert the weft with one or two superposed unilateral rapiers. When a unilateral rapier is used, the weft yarn does not have to be transferred in the middle of the shed and there is less danger of a component of a ply yarn being dropped and pulled back to form a fault. Rapier machines with unilateral weft insertion also need less mechanism and are therefore cheaper to build, but they take more space because the whole length of the rapier, which has to be greater than the maximum reed width, has to be withdrawn on one side. They also tend to be slower because the long rapier arm has to be stiffer and therefore, generally, heavier and because of the extra time required to complete the longer movement by the arm.

Most bilateral systems originally used loop transfer which was also known as the Gabler system. With this system the cut end of the weft is held in a gripper whilst the inserting rapier pushes a loop of yarn forward. When the loop is transferred to the receiving rapier the cut end is released and as the rapier begins to move back the released end of the loop is pulled through to the receiving side. Whilst this system makes transfer easy it imposes considerable extra tension on the yarn which has to be withdrawn from the yarn package at twice the speed of insertion and, in the case of hairy yarns, can result in rubbing up. As yarn speeds have increased there has been a steady change to tip transfer systems (Dewas) which are more gentle on the yarn but requires more accurately made and controlled rapiers. The clamp of the entry rapier is generally positively opened and then allowed to close on to the weft and grip it. The weft is also positively released from the exit rapier after weft insertion has been completed. Positive opening of the rapiers for yarn transfer in the middle of the shed is more difficult because the clamps of the rapier head are surrounded by the warp sheet and have to be operated either through the warp sheet or through the rapier arm from either side of the machine. For most yarns negative transfer between the rapiers is satisfactory but for delicate yarns and yarns which can be easily split or corkscrewed positive transfer is advisable.

Most rapier machines use a unilateral weft supply but for biphase machines and for bilateral machines where each half of the machine operates as a separate unit a bilateral weft supply is needed. The bilateral weft supply does not have to be located on the outside of the machine and, in the case of a biphase machine with a single central rapier, location of the weft supply above the rapier drive saves a considerable amount of space. With most rapier systems the fitting of a four or even six or eight colour device adds little to the cost or complexity of the machine although considerable extra space may be required to accommodate the larger creel.

## 1. Rigid rapiers

Rigid rapier machines have established a good reputation for weaving heavy and fancy fabrics. The Dornier machine (Fig. 16) is now available in reed widths of up to 400 cm. With positive control of the clamps during transfer (Fig. 17) the Dornier can weave nearly all types of fabrics and can handle

FIG. 16. Dornier rapier weaving machine GTV8/SD with a reed width of 320 cm and eight colour pick at will weft insertion.

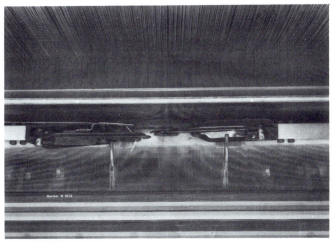

FIG. 17. Rapier heads of Dornier weaving machine approaching centre of shed. Right hand control lever opens right hand rapier head in anticipation of transfer of weft.

weft yarns from less than 1 tex to more than 3000 tex linear density. With Dornier rapier heads it is possible to insert into one shed with one stroke of the rapier two picks which remain parallel even after transfer. When the width in reed utilized in a machine is changed the rapier stroke is adjusted to the new width. This reduces waste to a minimum and enables the full machine speed to be used irrespective of pick width.

Another machine with an interesting and very simple rapier motion is the MAV by SACM (Fig. 18). The two rapiers are driven by a cam, a counter cam, and a lever system. The rapier drive is linked directly to the sley and the length of the rapier stroke can be adjusted by simply moving a lever. The width of the machine is limited to 225 cm because with the existing arrangement of the rapier drive for wider machines the rapiers would be above a convenient working level. There are no guide teeth in the shed, rapiers can be exchanged quickly and easily and different rapier heads are available for different weft yarns. The standard machine has been successfully used to weave a wide range of fabrics. For heavy fabrics a reinforced model is available and there are also models with superposed rapiers for weaving velvets and other double pile fabrics.

Rigid rapier machines of more than 200 to 240 cm reed width are wider than flexible rapier machines because tubes protecting the rapiers extend beyond their other motions.

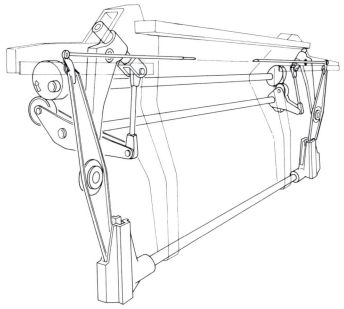

Fig. 18. Sley and rapier drive of SACM's MAV.

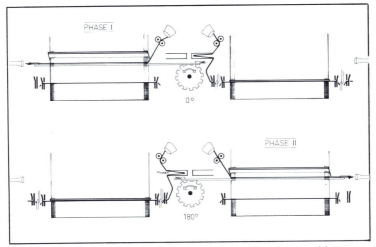

FIG. 19. Saurer 500 two-phase rigid rapier weaving machine.

Whilst the principle of the double acting single phase system (sometimes called the two-phase rigid rapier system) has been known for a long time, the release in 1979 of the Saurer 500 (Fig. 19) marked a new stage in the development of this principle. In this machine a rapier with a yarn clamp at each end is driven from the centre of the machine. The two sides of the machine are displaced from each other by half a cycle (180°) so that one half will be at front centre when the second half is at back centre. In one machine cycle the rapier inserts one pick into both sides and the picks are beaten up on opposite phases. So far this machine is only available in 2 × 185 cm reed width.

## 2. Telescopic rapier

The telescopic rapier combines many of the advantages of the rigid rapier with the space saving of the flexible rapier. In the rapier movement of the Saurer 400 (Fig. 20) a connecting rod eccentrically attached to a fly wheel on the mainshaft actuates an angle lever at its lower end. The angle lever drives the rapier slide which moves on a guide rail. The endless rapier belt is attached to the inner rapier and moves it in and out of the shed completely free from guides or other support. The machine operates at weft insertion rates of above 600 m/min and is available with a simple four colour mechanism. It is limited to a reed width of 205 cm and it is unlikely that, for the present, an unsupported telescopic rapier will operate well at high speed in much wider widths.

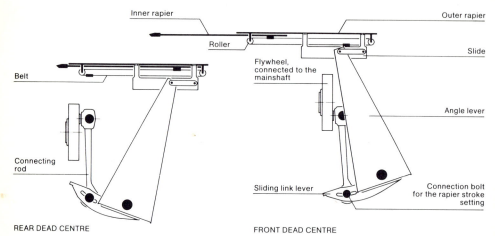

FIG. 20. Telescopic rapier of Saurer 400 fully retracted and fully extended.

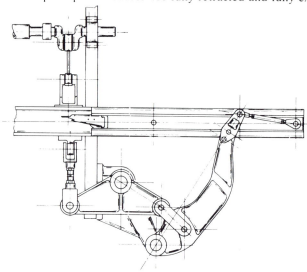

FIG. 21. Drive of telescopic rapier of Galileo's Telematic weaving machine.

The rapier system of the Telematic[11] (Fig. 21) needs no gears and relies on a complex lever system and an equalizer to drive its telescopic rapiers.

### 3. Flexible rapier

The largest number of machine builders have adopted the flexible rapier system. In most models when the rapier ribbons, which carry the rapier heads, are withdrawn from the shed, they are turned with the help of guide rails through 180° (Fig. 22). In the case of very wide machines the ribbons

FIG. 22. Nuovo Pignone's TP300 showing guide rail of rapier ribbon.

can be passed underneath the main loom framing as long as the rapier guides do not interfere with the cloth delivery mechanism. Machines are available in widths of up to 6500 mm although most machinery makers limit their widths to between 3600 and 4100 mm. Draper (DSL) and Ruti (F2001) wind the rapier tapes round tape drums which simplifies the rapier drive. With both systems of rapier guidance the rapier system can be either connected to and mounted on the sley or the weft insertion and the beat-up can be separated. If they are separated, as has become the practice in the latest high speed machines, the sley is stopped at back centre whilst the weft is being inserted.

To reduce weft breaks and obtain good transfers it is essential to accelerate and decelerate the rapiers as smoothly as possible. When the guide rail principle is used the rapier ribbon is generally driven by a reciprocating tape driving wheel. Figure 23 shows how the tapes are driven by positive cams in the Somet UV 770.

Another method of driving is used in the Propellor[12] weaving machine where a crank arm fitted at its end with $2 \times 4$ rollers reciprocates a long worm arm (Fig. 24). The thread of the worm arm has been designed to reduce vibration and no gearings, cams, or levers are required.

The drive of Ruti's F2001 (Fig. 25) represents a breakthrough in the

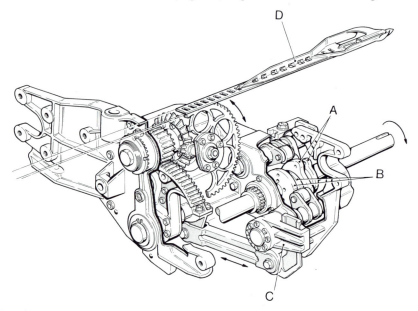

FIG. 23. Positive cam drive to rapier tape in Somet's UV 779. (A) Reed movement cams; (B) weft insertion cams; (C) fabric's width variation; (D) left hand weft insertion gripper.

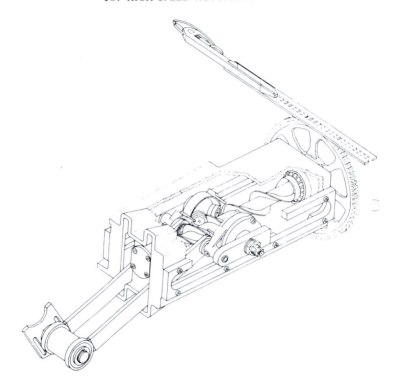

FIG. 24. Rapier drive of Vamatex's Propellor 201.

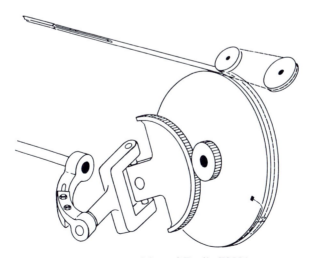

FIG. 25. Tape drive of Ruti's F2001.

method of tape driving and guiding. The tapes carrying the grippers are moved by oscillating tape wheels which derive their motion from a special crank gear. The tapes are guided on the tape wheels by endless guide elements which oscillate with them. This reduces the friction to a minimum and thereby the amount of heat generated by the tapes. This system, by preventing overheating, has enabled Ruti to increase greatly the weft insertion rate of medium width machines. Rates of more than 600 m/min can be obtained with machines with a reed space of only 180 cm, and wider machines (up to 280 cm) have a higher weft insertion rate. The machine is fitted with a rigid sley and short sley sword and has a precise beat-up. It can weave a wide range of fabrics with high cover factors and up to eight weft colours can be used in any sequence.

## C. Jet machines

During the past ten years weaving machines with weft insertion by nozzle have made the most rapid progress. They have reached the highest weft insertion rates of any type of machine in bulk production. Weft mixing and limited two colour facilities have been developed but so far no machines with multiple weft supplies picking in any desired sequence have become commercially available and no machines with superposed sheds have been marketed.

The reed width of the early jet machines was very limited. In air jet machines the air spreads out rapidly as it emerges from the nozzle and as it spreads its speed, and its ability to propel the yarn, decrease. This effect can be mitigated by constricting or guiding the air. Improvements in jet design, better means of air guidance, and the use of booster jets, have made it possible to weave much wider fabrics and there seems no reason why, in the long run, it should not be possible to project most types of yarn as far with the nozzle as they can now be carried with projectiles or rapiers.

### 1. Air jets

The Maxbo[13] air jet loom, which was in production during the 1950s, was the first commercially available weaving machine to reach 400 picks/min. It could, however, only weave fabrics of up to about 100 cm in width and its shedding and beat-up were limited and it was, therefore, much less versatile and could only operate at much lower weft insertion rates than the then also new projectile weaving machine.

The first commercially successful air jet weaving machine was the Elitex[14] which, after more than a decade of development, is now capable of weaving a wide range of filament and staple yarns. More than 40 000 of these

FIG. 26. Confuser of Elitex air jet.

machines have been installed and they are probably the second most frequently used shuttleless weaving machine at present in operation. The width
of the single jet machine has been steadily increased from 105 cm to 190 cm
with an operating speed of 400 to 450 picks/min. The unconventional
arrangement of the machine, which has already been referred to earlier (Fig.
7), gives easy access to the whole of the shedding area. The weft, after
passing through a tension device, is wound with the help of a feed wheel on
to a measuring drum. After most of one pick has been wound on to the drum
picking commences and a blast of air blows the measured length of weft
through a channel of closely spaced confuser blades (Fig. 26) to the receiving
side where it is held tight by a suction tube. While picking takes place the
reed is at back centre with the shed fully open and the confuser, through
which the yarn is blown, in the middle of the shed (Fig. 27a). After picking
has been completed the reed and confuser, which are in close proximity to
each other, move forward through an arc and the weft yarn slips through the
gap in the confuser into the shed ready for beat-up. To prevent damage to
the cloth, the confuser, like the guide teeth in a projectile machine, have to be
below the cloth when beat-up takes place at front centre (Fig. 27b). The

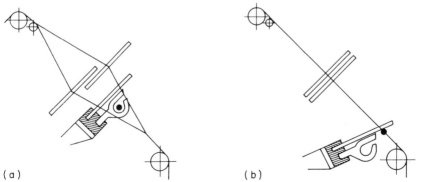

(a)                                        (b)

FIG. 27. Healds, reed and confuser of Elitex air jet. (a) Shed fully open; (b) at
beat-up.

mechanical weft stop motion of early models has now been superseded by a
photo-electric stop motion with the checking unit built into a confuser blade
near the receiving side.

The distance the weft can be thrown depends on the air speed, the design
of the jet, the type and count of weft, the confuser design and diameter, and
the density of the warp sheet. The amount of air consumed in propelling the
yarn accounts for a large proportion of the operating cost and depends on the
width of the machine and its design features. A very fine or a very coarse
yarn will need more air than a yarn of medium count and an open warp sheet
will also need more air because there will be more gaps for the air to escape.
The distance over which weft can be thrown can be extended by adding
booster jets to the confuser. The additional jets continue to propel the yarn
when the power of the main jet becomes insufficient.

The need to penetrate the shed at every pick with a large number of fairly
bulky confuser blades makes it difficult to weave low twist filament yarns and
imposes limitations on the warp and weft cover factors of the fabrics which
can be woven with the Elitex. This problem can be overcome with the te
Strake reed which has been shaped to form a tunnel along which the yarn is
blown. In the weft insertion system which is shown in Fig. 28 the yarn is
accelerated into the weft tunnel by the main nozzle and then carried forward
by air streams from relay nozzles spaced 74 mm apart. The weft, after being
drawn off the supply package, is measured either on an adjustable feeder
drum or between a roller and a measuring disc. The choice of measuring
motion depends on the type of weft to be processed. When a measuring disc
is used the weft, which has been measured, is blown into a tube where a loop
is formed and held under light tension by an air stream from an ancillary
nozzle. The disadvantage of this method is that a disc of different diameter is
needed for each pick length. The drum feeder is used both for measuring and
storage. A number of wraps of yarn are held on the drum prior to picking
when a pin is withdrawn to release the yarn needed for the next pick. The
leading end of the yarn ready for picking is held by a clamp placed just in
front of the main nozzle. Just before the clamp opens the main nozzle begins
to blow and as soon as the weft is released by the clamp it is blown into the
shed where the relay nozzles take over. At the end of the insertion cycle the
clamp is closed and the pick is stretched tight by a suction nozzle on the
receiving side. Individual groups of nozzles are actuated one after another
and closed after 75° to 100° rotation thus reducing the amount of air
required. The L5000 can be fitted with two measuring drums to give weft
mixing. When the weft mixing is used the unwinding speed from the supply
package is reduced to half.

Toyoda and Picanol use an ordinary reed, booster nozzles and a tunnel
formed by special guide blades facing the reed. Whilst this system combines

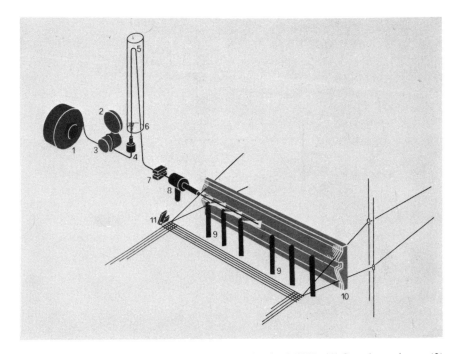

FIG. 28. Weft insertion system of the Ruti-te Strake L5000. (1) Supply package; (2) measuring disc; (3) rollers; (4) auxiliary nozzle; (5) yarn loop; (6) storage tube; (7) clamp; (8) main nozzle; (9) relay nozzles; (10) reed with weft tunnel; (11) cutter.

some of the advantages of the te Strake reed with the ability to use ordinary reeds, it suffers from the disadvantage that the guide elements have to be brought into the shed and again withdrawn before beat-up. The guides can also disturb the warp yarn especially when low twist filament yarns are used.

## 2. Water jets

Water jet machines (Fig. 29) are in many ways similar to air jets but have to be made of stainless steel or protected against rust. Picking has always to be separated from beating up because the water jets have to be permanently fixed and therefore cannot be mounted on the sley. As it is easier to control a jet of water than a jet of air, the width of water jet machines increased more rapidly and a single nozzle will carry weft 280 cm. Neither confuser nor special reed tunnels are needed but nozzle design and pump pressures are critical and jets have to be regularly maintained. To obtain constant performance the temperature and viscosity of the water has to remain steady.

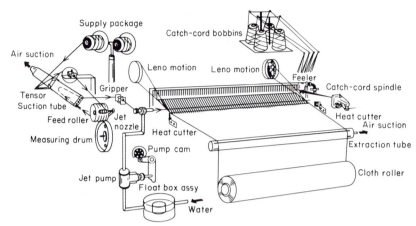

FIG. 29. Nissan water jet machine.

The water has to be filtered to prevent blocking of the nozzles and it must be of the correct hardness and pH. The growth of algae in the feed water can cause serious problems because chemicals used to prevent their growth must not attack the skin of the weaver, the machine or damage the yarns being woven.

When polyester or polyamides are woven, heat cutters instead of scissors are frequently used to trim off the selvedge and, at the same time, seal the new edge. Provision has to be made to shield operatives against the water released in the picking area and to suck away surplus water near the fell of the cloth and to dry the cloth if necessary. Two colour machines are available from Elitex and Nissan but there are limitations on patterning.

## D. Multiphase machines

In all multiphase machines a number of picks are inserted simultaneously. In machines with wave sheds (multiple shed formation weft-wise) a number of weft carriers travel across the warp sheet at the same time but otherwise the basic concept of shedding and of weft insertion of single-phase machines has been retained. In linear shed machines a novel method of shed formation is used and a number of open sheds are maintained in parallel right across the warp sheet and the weft is inserted by a series of parallel rapiers.

### Wave sheds

The weft insertion systems of single-phase machines considered in Sections IV,A–IV,C above require an open shed to be formed from side to side of the warp sheet before the weft is carried or blown through the shed. In wave shed machines a number of sheds are, at all times, in different stages of development. The weft is inserted with the help of small weft carriers which contain one measured pick. The weft carriers are supplied either from one or a number of winding units. After the carrier has been filled with weft the carrier is propelled into the shed which has just been formed commencing at the selvedge on the picking side. As soon as the carrier is completely in the warp sheet the shed is closed behind it and the shed for the next carrier is formed. This carrier immediately locks the weft from the previous carrier into position and for this reason it is practically impossible to repair weft breaks which occur after a carrier has entered the warp sheet. Each shed, in its open position, contains a weft carrier and as each shed moves forward the appearance of a wave motion (Fig. 30) is created. This has given this system its name.

In the TPC 1330[15] and the Kontis[16] the weft carriers are pushed mechanically by rollers mounted on a driving chain below the warp sheet. For shedding the warp is divided into small sections each controlled by a pair of separate narrow heald frames which are moved up and down at high speed to form continuously changing sheds. As the sheds change the weft carriers move slowly forward between a dummy reed and a rotary reed built up from specially shaped reed discs which traverse the weft to the fell and push it

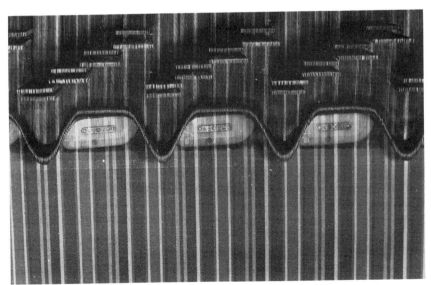

Fig. 30. Shedding elements and reed wires forming a wave around the weft carrier of the Ruti R6000.

home. The empty weft carriers are then returned to the weft winding unit on a conveyor in preparation for their next traverse of the warp sheet. Although the machines achieve high weft insertion rates the weft is unwound slowly as the carriers move forward and this reduces the number of weft breaks.

In the system developed by Ruti the reed dents press against the specially shaped end of the light weft carrier and push it through the shed (Fig. 30). The reed is composed of dents, separated by guide plates, pivoted on an axis and actuated by two spiral cams rotating in opposite directions causing one dent after another to beat up.

Weft carriers can also be driven electro-magnetically. This offers the advantage that no physical penetration of the warp sheet by the driving elements is needed and reduces danger of damage to the warp.

A method of forming the warp sheds by air currents has been developed and it has been shown that sheds can be changed rapidly in this way. Unfortunately no satisfactory way of inserting weft into such sheds in looms of standard width has so far been developed.

## 2. Linear sheds

A linear shed weaving machine, the "Orbit", based on the Bonetti patents has been developed by Bentley Weaving Machinery Ltd.[21] In the Orbit shedding occurs round 340° of a large diameter drum which carries 18 rollers equally spaced around its circumference.

Each roller is built up of a large number of discs—one for each end of the warp. For plain weave, alternate discs of each roller are of large and small diameter (Fig. 31). On alternate rollers discs are so arranged that the yarn goes from a large disc to a small disc and vice versa. In this way 17 sheds are formed around the circumference of the machine. Whilst the warp moves slowly from the point where the warp sheet comes into contact with the first roller to where it leaves the drum, the rollers move with the drum at a much

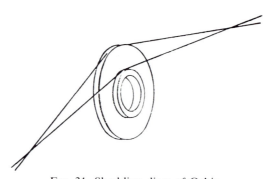

FIG. 31. Shedding discs of Orbit.

faster speed in the direction in which the warp sheet moves. As the rollers move through the warp sheet they move each shed forward. Whilst the sheds are open a series of 18 rigid rapiers and 18 supply packages, which are also mounted on the drum, are rotated with it. The rapiers travel through the shed empty, pick up a weft supply and pull it through the shed. When weft insertion has been completed the warp with interlaced weft is taken off the last roller, the shed collapses, and the weft is beaten into the fell by a beat-up roller with six collapsible reeds. During the traversing of the sheds the spacing and alignment of the warp sheet is maintained with false reeds. The action of the rollers on the warp yarn is more gentle than the action of normal healds and this should either reduce warp breaks or reduce the cost of warp preparation. One set of 18 rollers is needed for each thread spacing in the warp direction and as these rollers have to be precision made they are expensive. So far it has proved difficult to produce discs of very small pitch which are needed for fabrics with a high number of ends/cm.

## V. Conclusions

### A. Objective and summary of requirements

The objective of a weaving machine is to produce fabric of the required design, weight, width and quality for a specified end use at minimum cost.

The cost of a fabric is built up from a large number of interacting items and, in practice, many compromises have to be accepted to enable the machine builder and the weaver to benefit from the introduction of modern methods of processing and control. The most important cost elements in the production of fabrics are;

the cost of the warp and weft yarns;
the cost of preparing the yarns for use in the weaving machine;
depreciation and interest charges for the weaving machine and for the space required for it;
the cost of spares and maintenance;
power for driving, lighting, and air conditioning; and
direct and indirect wages and salaries including all related charges.

To complicate the equation, allowance has to be made for the effect of the various possible combinations on cloth quality, the cost of rejects, sales appeal, and on the cost of further processing.

Some of the underlying principles have, in general terms, been dealt with by Grant and Ireson,[17] and Pennycuick,[18] and, in relation to textiles, by Ormerod[19] as well as by many other authors. In this chapter only factors affecting weaving machines directly have been considered.

In the second section we have seen that within living memory the productivity of single-phase machines has increased fourfold and that of the weaver at least twentyfold and that further increases are likely. Without these improvements it is unlikely that weaving would have retained its dominant role and it is therefore important to understand the factors which have helped to make the modern single-phase weaving machine more productive and more cost effective.

The amount of cloth produced by a weaving machine depends on the revolutions per minute, that is the picking speed, the number of threads inserted per unit length of warp, and the width of the machine. Assuming the threads per unit length remain the same, output is increased if either the revolutions or the width are increased and the efficiency remains the same.

The output of the weaving machine can be doubled if either two picks are introduced at the same time or if two cloths are woven on top of each other by using superposed sheds.

Whilst the maximum possible output of a machine can be calculated from the picking speed and the reed width, the actual output, and therefore the cost per unit of production, is affected by the percentage utilization of the reed width and the frequency and length of machine stops for whatever reason. Such stops can be caused by yarn breaks, mechanical failures, the time required for cleaning, maintenance and sort changes, and for replacing warps. The cost of such stoppages will not only depend on the length of time required by the operative to carry out the work but also on the average time a machine has to wait before the operative can come to it to do the work, which depends on the method of staffing and operating the weaving machines.

## B. Effect of method of pick insertion on picking speed

The first automatic shuttle machines weaving simple fabrics worked at speeds of about 190 picks per minute for 100 cm wide fabrics. Speeds hardly increased during the first 40 years after Northrop had invented the automatic loom. Then cast gears were replaced by cut gears, belt drives from line shafts by individual loom motors and the automatic loom began to change to the weaving machine. During the past 20 years substantial further speed increases were made possible by improved engineering, improved bearings and lubrication, by fitting electro-magnetic brakes and by replacing the mechanical shuttle protection mechanism by an electronic shuttle flight monitor. In this way speed increases of up to 30 or even 40% have become possible, although the maximum speed of the shuttle machine will always remain limited because the large shuttle carrying the pirn of yarn has to be accelerated and decelerated quickly at the beginning and end of each pick and because it has to be possible to stop the machine quickly in case of emergency without damage to the loom, the shuttle, or the yarn inside the shuttle. The time between consecutive picks using the same shuttle has to be sufficient to allow for the automatic replenishment of the shuttle or its replacement whenever the yarn package inside the shuttle is exhausted.

The limitations imposed by the weight of the shuttle can be reduced by replacing it with a smaller and lighter weft carrier holding only one pick or with a projectile which holds the tip of the yarn and draws one pick through the shed. In this way picking speeds have been increased substantially and further limited increases are likely with the help of electronics.

Instead of drawing the yarn through the shed behind a projectile, the weft can be taken through the shed by rapiers. Insertion rates obtainable with bilateral rapiers tend, however, to be lower than those obtainable from projectiles because the weft has to be accelerated and decelerated twice per pick to allow for its transfer between rapiers in the middle of the shed.

When the weft is propelled by a fluid, either liquid or gaseous, the limits imposed by shuttles, projectiles, or rapiers are removed and much higher picking speeds can be obtained. The limiting factor in jet machines will probably be the speed with which sheds can be formed and the weft beaten into the fell without damage to the yarn or excessive generation of fly. As filament yarns tend to be more regular and to contain fewer weak places than staple yarns it is likely that higher picking speeds will be possible when filament warps are used.

So far water is the only liquid which has been used successfully in jet machines and this has generally restricted the application of machines using liquids for weft propulsion to yarns with low water absorption like nylon or polyester although it has been found possible to weave some cotton weft yarns. It may be possible to protect other yarns against water by special coating treatments, but cost may render such treatment uneconomic. It is unlikely that liquids which do not wet cellulosic yarns but are cheap enough for use in jet machines will be found in the near future.

No such limits apply to air jet machines. Compressed air can be used to propel a wide range of yarns as long as they are not too uneven to pass through a nozzle or so fine or heavy as to require an excessive amount of air. Whilst air is free, expensive compressors and a great deal of power is needed to compress it sufficiently to propel the weft. The altitude of the installation and the operating temperature also have a big effect on power consumption. For some applications the success of air jet looms may well depend on the efficiency with which air is used by different types of machines and on the cost of power, which varies greatly from country to country.

## C. Effect of machine width on weft insertion rate

It has already been shown that the weft insertion rate of a weaving machine can be increased by increasing the machine width whilst maintaining the machine speed. In a single-phase machine shed formation, picking and beating-up have to be performed in sequence. The pick cannot be inserted until an open shed has been formed to allow the weft and, where applicable, its carrier to pass through freely. Weft insertion has to be completed and, in the case of rapier machines, the rapiers have to be withdrawn before the weft can be beaten up. Shed formation can, to some extent, coincide with beating up. This applies equally to shuttle and shuttleless machines. In relatively narrow machines the time taken for weft insertion, which is the only time during which the machine actually produces cloth, can be less than half the total time required for one revolution of the machine.

The curves in Fig. 32[20] show the relation between reed width and picks/min and the weft insertion rate of the weaving machine. Figure 33 shows the effect of increasing the reed width of a Ruti "C" shuttle weaving machine. In this model, as the reed width is increased from 100 to 350 cm the picking speed is nearly halved but the weft insertion rate is roughly doubled. Table VI shows the effect of reed width on the performance of the Sulzer weaving machine. Because of the high speed of operation of these machines two limiting factors apply: there is a maximum weft insertion rate governed by the ability of the various motions to handle the weft yarn and a maximum speed of the machine governed by its technical specification. Because of limiting technical factors the widest of their machines has a slightly lower weft insertion rate than their second widest machine. It is also noteworthy

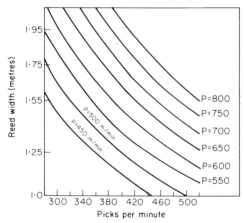

Fig. 32. Relation between output of weaving machine, reed width and machine speed. (P = weft insertion, m/min).

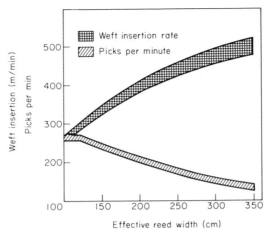

FIG. 33. Relationship between reed width, picking speed and weft insertion rate of a Ruti "C" weaving machine.

that their latest model (PS), which contains many modified motions and electronic controls and is better balanced, can operate at a speed about 20% greater than their latest standard (PU) machine.

TABLE VI. Performance of Sulzer weaving machines with tappet motions.

| Machine type | Maximum reed width mm | Performance | | |
|---|---|---|---|---|
| | | Maximum weft insertion m/min | Maximum speed rpm | |
| PU85 | 2200 | 715 | 330 | |
| PU110 | 2830 | 830 | 330 | |
| PU130 | 3340 | 900 | 295 | |
| PU153 | 3930 | 940 | 270 | |
| PU183 | 4690 | 975 | 240 | |
| PU213 | 5450 | 965 | 225 | |
| PS3600 | 3600 | 1100 | 320 | |

## D. Utilization of reed space

In recent years there has been a steady tendency for cloth width to increase as this results in savings in finishing and making up. Loomstate or grey cloth is always narrower than the warp sheet in the reed and further contraction generally occurs during finishing. The reduction in width depends on the weave, the type of fibres and yarns used, the method of weft insertion, the weaving tension, and the finish. Allowance has also to be made for selvedges or dummy selvedges if these are cut off or if the cloth width is specified between selvedges. The width in reed required to produce the finished cloth of specified width will decide the minimum reed width of the weaving machine in which the cloth can be produced.

Most machinery makers offer their machines in a number of reed widths and most machines can accommodate fabrics considerably narrower than the maximum possible reed width. Some makers produce machines in width steps of as little as 100 mm whilst others have steps of 600 mm or more arguing that the benefits of increased standardization during machine production more than outweigh the additional fixed and operating costs of machines which are slightly wider than necessary for the fabrics to be woven in them. As it is nearly always uneconomic to widen weaving machines once they have been produced it is essential to give the most careful consideration to the machine width before the weaving machine is ordered.

As the width of the machine increases the cost of the machine increases, the picking speed decreases (see Fig. 33), the cost of power, space and maintenance increase, and it may be necessary to reduce the number of machines which can be installed in an existing building. Whilst it is essential not to buy machines which are too narrow, maximum reed space utilization will result in reduced operating costs.

In Section V,C it was shown that a substantial increase in the reed width generally results in an increased weft insertion rate. It can, therefore, be attractive to have a weaving machine which can weave two or more fabrics side by side. Wide machines, as distinct from bilateral machines (see Section V,F below) also have the advantage that the number and widths of fabrics which can be woven in a given reed width can be varied (see Fig. 34) as long as they fit into the total available reed width and there is sufficient space to accommodate the required selvedge motions.

If a number of fabrics are woven side by side and the stop rate per unit area of fabric produced is not reduced, the stop rate per pick inserted will increase in proportion to the increase in width, and the efficiency of the weaving machine will decrease. If there are many stoppage-related faults in the fabric the cloth quality will also decrease when a number of fabrics are woven side by side. To obtain the lowest production cost and good quality from wide machines operating at high speed it is, therefore, necessary to use yarns of good weaving quality, to have well prepared warps, good machine maintenance and to reduce stopped time for warp and sort changes to a minimum.

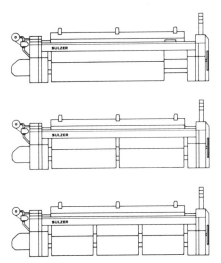

FIG. 34. Weaving of cloths of different widths in a Sulzer PS machine.

## E. Superposed sheds

Superposed sheds have been used for a long time for velvets and other special fabrics where two cloths can be woven face to face with the pile ends interchanging from one cloth to another. The pile ends are cut whilst still on the loom giving two separate pieces of velvet.

With the development of rigid rapiers it became evident that the same principle can be used for weaving two independent fabrics using separate superposed sheds. A number of manufacturers, including IWER and SACM, build such machines and they may have unilateral or bilateral weft supplies. Most motions are identical to those used in single shed machines. The superposed rapiers are operated from the same mechanism and insert weft into two superposed sheds which are formed on top of each other. Whilst a single warp beam can be used, it is generally more attractive to have two separate beams as this makes it easier to adjust warp tensions and, at the same time, doubles the length of warp yarn on each beam and reduces down time for knotting or gaiting.

Weaving machines with superposed sheds can operate at about 90% of the speed of similar single shed machines and therefore have very high weft insertion rates. They need little extra space and their cost per unit of output is lower than that of similar single shed machines. Their main disadvantage is that it is often difficult to repair yarn breaks, especially in the lower shed, and machines with superposed sheds should therefore only be used for weaving fabrics where yarn breaks are infrequent.

## F. Bilateral weaving machines

It has already been shown that weft insertion rates can be improved by increasing the reed width or using superposed sheds. Instead of weaving more than one fabric within the same reed width or weaving two fabrics on top of each other, two fabrics can be woven side by side in two interconnected weaving machines. In 1975 Gardella showed a right hand and a left hand unilateral rigid rapier machine with separate drives and weft supplies making two cloths but working as one unit. The only advantage of such a machine over two separate machines is a slight reduction in capital cost and a saving in space because no passage is needed between the two halves of the machine.

On the other hand, Saurer's two-phase weaving machine (see Section IV,B above) only requires one rapier with a gripper at each end and has resulted in a well balanced and compact machine. Investa, by placing a pair of swivelling nozzles between the two bilateral weaving units, have produced a water jet machine which can weft mix or produce two-colour weft designs. If the nozzles are turned through 180° after each pick weft mixing results and if they are swivelled less frequently two fabrics with two colour weft designs with a mirror image are woven. An air jet machine using a similar principle has also been designed. If the two jet machines side by side are considered to constitute a single weaving machine they give the highest weft insertion rate of any commercially available single-phase weaving machine.

Both halves of a bilateral weaving machine have to stop at the same time. This increases the time required for warp and style change, makes low stop rates important and reduces machine efficiency.

## G. Weft insertion rates of multiphase machines

Whilst it has been found that, with the help of better materials and engineering and with redesigned motions which react more rapidly, the insertion rates of single-phase machines could be increased above the assumed upper limits, it appears likely that weft breakage rates will fix new upper limits for each type of weft. In multiphase machines, where a number of picks are inserted at the same time, this problem can be overcome and, theoretically, very high weft insertion rates should be obtainable in a machine of limited size.

In wave shed machines individual warp threads still have, however, to form a shed for each pick and the shedding speed for a given weft insertion rate has therefore to be as high as for a single-phase machine of the same width. Whilst weft insertion rates of 2000 m/min have been obtained with the TPC1330,[16] normal single-phase jet machines may soon approach this insertion rate when weaving some weft yarns. Beating-up of the individual weft threads immediately they have been inserted also causes problems because this has to be done in a confined space at high speed.

In linear shed machines a succession of sheds are formed and traversed along the warp towards the beat-up point and whilst the sheds are traversed the weft yarns are inserted. There is, therefore, no need for rapid jerking movements of the warp yarn during shed formation or the high speed acceleration of the weft, and the only operation which has still to be performed at high speed is beating-up of the weft. These machines hold out the hope of very high weft insertion rates without imposing excessive strain on either the warp or the weft yarn. Means of replenishing the weft supply whilst the machine is in motion have still to be developed.

## H. Manpower

In previous sections it has been shown how weft insertion rates have increased and are likely to increase further and how the capital element of the weaving cost can be reduced. Weaving would, however, have priced itself out of the market if labour productivity had not been increased even more than machine productivity.

Modern weaving machines have been designed to reduce the work of the overlooker, weaver, and of ancillary labour by automating many operations and by replacing moving parts, mechanical motions, and links by electronic controls. On shuttle machines electronic shuttle flight control, optical weft feelers, and electronically controlled weft stop motions are simpler and faster than their mechanical counterparts. Electromagnetic brakes which stop a loom quickly in any desired position and electrical inching and coupled reversing motions help the weaver.

The battery fillers job on shuttle looms becomes redundant if box loaders are fitted, and pirn winding, battery filling, empty pirn collection and stripping all become unnecessary if a pirn winder and stripper (Unifil[10]) is fitted to each machine. Loom winders also reduce the accidental mixing of yarns because only full cones are brought to the loom. They create, however, additional work at the loom because additional equipment has to be serviced by a mechanic and the weaver or a special Unifil attendant has to repair Unifil related faults. Considerable extra work has to be done at the loom if very coarse yarns are woven and it may be difficult to justify the equipment on grounds of economy for very fine yarns.

On shuttleless machines automatic lubrication, preventive maintenance, and the use of electronics have resulted in similar reductions in manpower. For example on the latest Sulzer weaving machine the arrival of the projectile on the receiving side is monitored without contact, and therefore without wear, by means of sensors. The projectile brake is adjusted automatically and accurately by a servomotor. Functional irregularities are determined automatically and signalled to a control panel.

In jet machines a fluid propels the yarn through the shed and it becomes unnecessary to thread-up, position, propel, brake, or return a weft carrier and thus all the parts needed for it become redundant. As the nozzle or nozzles need few moving parts the weft insertion system and, therefore, the work of the overlooker becomes simpler. Work done by te Strake more than a decade ago suggests that in air jet machines the automatic repair of most weft breaks may become feasible one day.

Whenever a warp beam has woven out it has to be replaced. The time required will vary depending on whether the new warp is knotted back or gaited, the ancillary equipment available, the number of ends in the warp, and the time required for routine checks at each warp change. It is likely to take 4–6 man hours and may require 2–3 clock hours in addition to waiting time if more than one machine becomes empty at the same time. The frequency with which such changes have to be performed depends on the length of warp on the weavers beam, the picking speed, the picks inserted per unit length, and the number and length of stops. When weaving coarse yarns the life of a beam may be short—less than 100 hours. To reduce stopped time the diameter of weavers' beams has been increased from 600 mm to 1000 mm and even to 1200 mm for special applications. Whilst

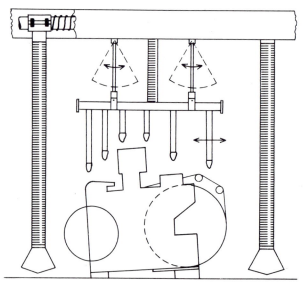

FIG. 35. Parks-Cramer high velocity loom cleaner with vacuum floor cleaning.

this reduces the stopped time it increases the cost of the yarn on beam. Warp yarns which do not need sizing can be drawn direct from a cone creel but, because of the space required, warp creels for weaving machines can only be justified rarely. As beam weights have increased, machine frames and let-off motions have had to be redesigned and beam barrels and flanges have had to be strengthened to withstand the great pressures exerted by synthetic yarns.

The labour required for handling and cleaning can be reduced by automating these operations as far as possible. For example for loom and floor sweeping high velocity travelling cleaners (Fig. 35) are fitted. The gantries used for the loom cleaners can sometimes also be used to transport warps and rolls of cloth.

## I. Space

Space is expensive even if existing buildings can be used when re-equipping. The building has to be paid for and then it has to be maintained, lit, heated, and air-conditioned. The size of the weaving machine in relation to its weft insertion rate therefore affects the cost of production.

Machines having the same output may require greatly varying amounts of floor space. A flexible rapier is likely to require less space than a rigid rapier machine of the same reed width and a machine with rigid bilateral rapiers will need less space than a machine with unilateral rapiers because of the positioning of the rapiers in relation to the machine frame. A machine with a four or six colour motion will need more space for its creel than a similar machine weaving from only one weft supply. If large diameter warp beams and cloth rollers are fitted they require more space because they project further from the loom frame. Allowance has also to be made for passages around the weaving machine. The minimum spacings recommended in the United Kingdom are listed in Table VII. Allowance has also to be made for

TABLE VII. Minimum recommended machine spacings.

| Description of alley | Minimum spacings |
|---|---|
| Main alley | 2000 mm |
| Weavers alley | 600 mm |
| Cross alley without unprotected moving parts | 600 mm |
| Cross alley with moving parts | 750 mm |
| Back alley | 1.75 × maximum beam flange diameter with minimum of 1200 mm |

M

space needed for transporting beams and for ancillary equipment like auto-matic loom cleaners.

A comparison made in 1979 showed that for fabric with a reed width of 165 cm weaving single width the weft inserted per unit of floor space was nearly 2.25 times as great for a modern air jet than for the latest type of automatic shuttle or rigid rapier machines. Two-phase rapier machines and projectile machines weaving two fabrics in a width were second and third best for space utilization.

## J. Machine specification

It is as unsatisfactory to purchase a machine which cannot weave some of the cloths for which it is intended as it is to purchase a machine which has been overspecified and is therefore an unnecessarily expensive producer of cloth. If we are sure we only need six shafts there is no justification for buying a machine with 14 shafts assuming one with 10 shafts is available. If only plain weave fabrics are to be produced there is generally no justification for a dobby. If we have neither to weft mix nor to weave colour check designs we cannot justify a four colour motion which adds to the cost of the machine and the space required for the creel.

Before a weaving machine specification is prepared it is, therefore, essential to know the types of cloths and the cloth widths which have to be woven.

## VI. Postscript and Acknowledgements

This chapter is based on information obtained by the end of 1980. At a time when technology is advancing rapidly, and when we are beginning to learn how to utilize electronic control systems, improvements are frequently made, and methods and machines become obsolescent quickly—this, however, does not justify inaction.

I am grateful to machinery makers and their agents who have supplied me with information and have provided illustrations for publication. These include:– Dornier (Figs. 16 and 17), Galileo (21), Güsken (3), Hagemann (8), Investa (5, 7, 26, 27 and 32), Nissan (29), Nuove Pignone, Divisione SMIT (22), Orbit Weaving Machinery (31), Parks-Cramer (35), Picanol (12), Rüti (6, 9, 25, 28, 30 and 33), SACM (18), Saurer (5, 19 and 20), Somet (23), Sulzer (1, 2, 4, 10, 13, 14, 15 and 34), Vamatex (24), and Zellweger Uster (3). My thanks are also due to my friends and colleagues who made most valuable suggestions on how to improve this chapter.

# References

1. "Malimo" named after machine produced by Textima, G.D.R.
2. "Sulzer Weaving Machine Bulletin" No. 26/E (1968).
3. For definition see "Textile Terms and Definitions" published by the Textile Institute.
4. Draft International Standard, ISO/DIS 5247.2
5. British Standard 3225: Part 2: 1979 (ISO 5241). "Beams for warping, sizing and weaving of textile yarns. Part 2. "Specification for main dimensions of and terminology for weavers beams". (ISO title: Textile machinery and accessories—weaver's beams—terminology and dimensions.)
6. British Standards 2609: Part 1: 1979 (ISO 441) and 2609: Part 2: 1979 (ISO 1150). "Warp stop materials".
7. British Standard 4977: 1973 and AMD 2797. "Terminology and dimensions of serrated bars for warp stop motions".
8. British Standard 4802: Part 3: 1978 (ISO 5243). "Textile machinery and accessories: numbering of heald frames in a loom".
9. Hanton, William A. (1929). *Automatic Weaving*. Ernest Benn Ltd.
10. "Unifil" is the registered trade mark of the Leesona Corporation for their loom winder.
11. "Telematic" weaving machines are made by Officine Galileo Meccanotessile.
12. "Propellor" weaving machines are made by Vamatex.
13. "Maxbo" air jets were made by A. B. Maxbo in Sweden until 1962.
14. "Elitex" weaving machines are made by Investa.
15. "Kontis" is made by Investa.
16. "TPC 1330" is made by Nuovo Pignone SMIT.
17. Grant, E. L. and Ireson, W. G. (1960). *Principles of Engineering Economy*. Ronald Press.
18. Pennycuick, K. (1961). *Industrial Diagnosis*. The English University Press.
19. Ormerod, A. (1979). *Management of Textile Production*. Newnes-Butterworths.
20. *INVESTA Czechoslovak Export Magazine*. January 1980.
21. Name of company has been changed to "Orbit Weaving Machinery Ltd".

# Chapter 11

# Heavy Fabric Manufacture— Carpets and Needlefelts

J. INCE

## I. Introduction

When one surveys engineering advances in the soft floorcoverings industry which have taken place in the last few years, it is immediately obvious, from the tremendous number, that one is considering a very dynamic industry. Whilst other papers[1,2] survey all such developments fairly briefly, this chapter will cover, in a little more detail, only the more commercially important of the engineering advances or alternatively those which have successfully stood the test of time.

## II. Fibre and Yarn Engineering

Fibre and yarn engineering is an integral part of the textile floorcoverings trade and in recent years all fibre and yarn producers have lived through a traumatic period. Ten years ago the situation for all producers looked assured and the future for synthetic fibres looked particularly good. However, the oil situation has had such a dramatic effect on both fibre prices and the world economy that an oversupply situation now exists. This has had the effect of lowering prices and profitability and the resulting shortage of capital has been reflected in the level of new fibre developments. Although many hundreds of fibre modifications are announced annually, these are merely different lustre levels, cross sections etc.; significantly, no new fibres have been developed and major fibre improvements have been surprisingly slow.

The main pre-occupation of many synthetic fibre producers has been in properties particularly important in contract locations, e.g. soil hiding and static electricity.

Figure 1 shows how this has been approached by one producer (Du Pont). The hollow nylon fibre, which has a "light-scattering" effect, gives a soil hiding property and the nylon containing the conductive carbon core gives a permanent anti-static effect. Similar effects have been achieved by other producers by using different fibre cross sections and fibre additives or coatings, e.g. Anso. IV Nylon from Allied Chemical.

Because of their unique scale structure, wool fibres have the property of feltability and this has been used to advantage to produce felted yarns for carpets. The felted structure has the advantage of producing chunky textures in coarse gauge carpets as well as giving less fibre shedding in cut-pile constructions. The use of batch felting machines (solvent and aqueous) is currently the most popular method, although a recently developed continuous yarn felting machine[3] is likely to replace such systems as well as having more far reaching possibilities such as producing yarns directly from roving or sliver.

The principle of operation of the Periloc process is shown in Fig. 2. The yarn (or other fibrous structure) is passed through a flexible tube together with a felting medium. The tube is continuously flexed and compressed in a similar manner to a peristaltic pump and the resulting mechanical agitation produces the felting action on the yarn. The degree of felting is governed by the nature and diameter of the tube, the yarn blend, count and twist, rotor speed and yarn speed and also the felting medium itself.

One outstanding development in the floorcoverings industry in the last

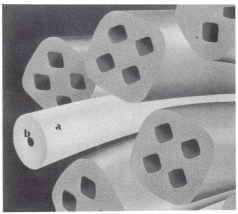

FIG. 1. Antron III HF (a) nylon filament, (b) carbon core.

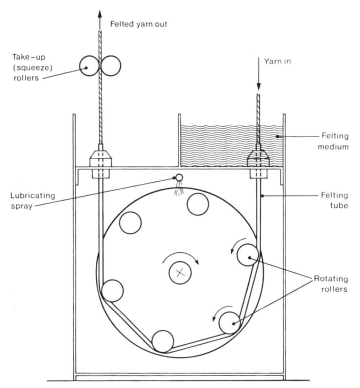

FIG. 2. The Periloc process.

few years is the trend towards natural or berber yarns. Initially these were produced using blends of undyed wools and achieving the desired effect by using a minimum carding action. More recently in order to widen the range of yarn types available, several machines[4] have been developed for introducing effect material (slubs, neps, flames etc.) either at the carding or spinning stage.

Figures 3 and 4 show two such developments. The PA-FA machine is intermediate between a spinning frame and a twisting frame. In one operation it produces a coarse count yarn with long injected flames. Double woollen rovings are twisted together and effect material is injected at the front rollers. Two injection units are fitted to introduce two different coloured rovings. The Wünsch attachment fits above a woollen card in front of the condenser stage. The effect material in the form of pre-prepared roving on four independently driven rods is fed onto the web either mechanically or electro-magnetically.

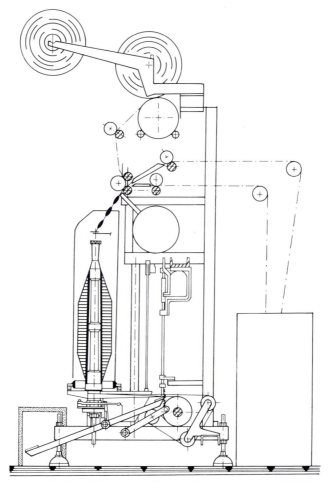

FIG. 3. The PA-FA novelty woollen frame type PFSC.

Two vitally important aspects of carpet yarn engineering are the introduction of yarn bulk and the fixation (or setting) of yarn twist. Various machines for the introduction of fibre crimp (particularly in continuous filament yarn) are described in earlier chapters. Other machines for the concurrent heating and relaxing of yarns have been available for many years. It is only more recently, however, that continuous autoclaves such as the Superba or Gilbos Perfecta Set, have been developed and a typical installation is drawn schematically in Fig. 5.

FIG. 4. A Wünsch attachment.

## III. Backing Fibres and Fabrics

Jute and cotton are still the dominant backing fibres used in weaving, although the use of heat-set polypropylene is steadily increasing. However, in the field of tufted backing cloths the situation is completely different and partly as a result of greater research and development, the use of polypropylene is now more widespread than the use of jute. It accounts for over 80% of primary backing cloths and a steadily increasing percentage of secondary backing.

At this point it is opportune to describe the fairly simple principles involved in the production of polypropylene fabrics.[5] Firstly the polypropylene granules and additives are fed from silos into a temperature-controlled extrusion head. This head may be a slot for film or a smaller rectangular orifice for tapes. The extruded fibre is cooled by water, air or chilled rollers. Extruded film is slit into tapes of approximately twice the final required

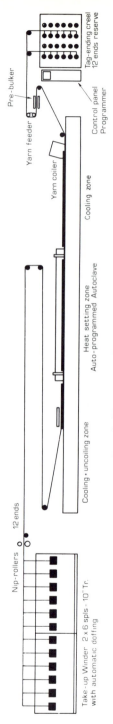

Fig. 5. The Gilbos Perfecta-set setting line.

width. Molecular orientation for fibrillation is achieved by heat stretching the tapes approximately six to one and then heat relaxing them to achieve better stability. The tapes are finally wound onto packages for weaving into wide tufting fabrics in the usual way. Warp and weft densities are determined by the carpet producers' requirements. The above is clearly an overmodification of what is a highly specialized industry and tufted backing cloths are engineered in many ways to satisfy the end users' requirements. Heat stability during carpet finishing is particularly important and many fabrics are given an additional heat setting treatment to ensure stability. One manufacturer[6] produces fabric with an additional fibrillation process to give less needle deflection, better tuftbind, improved latex adhesion and more level pile surfaces. Other producers additionally needle a small quantity of fibre (usually nylon) onto the polypropylene fabric. This results in a backing which is useful in carpet piece dyeing or printing where the needled fibre dyes the same colour as the pile thus obscuring the undyed polypropylene.

Spunbonded fabrics have been widely adopted in tufting as they satisfy many of the requirements outlined above. The basis of this process, which integrates spinning and fabric formation, is shown in Fig. 6.

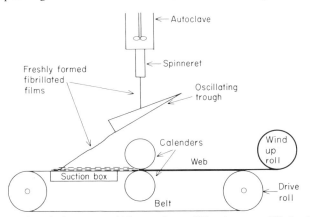

FIG. 6. The spun-bonded process: (A) autoclave; (B) spinneret; (C) freshly-formed fibrillated films; (D) oscillating trough; (E) web; (F) suction box; (G) calenders; (H) belt; (I) wind-up roll; (J) driving rollers.

## IV. Carpet Weaving Machinery

Although many tufting machines are being sold, there is still no sign of a corresponding fall off in the sale of carpet weaving machines. In fact, although there may have been a drop in woven production in certain areas, this will be more than compensated for by increases in production in other areas. For example, countries in the Middle East and low labour cost countries such as India, are buying many traditional weaving looms. These are to produce mechanically the styles of carpets which have traditionally been produced by hand. In some such areas legislation has prohibited the use of child labour for hand weaving and the ensuing drop in production, coupled with growing demand, has meant that the demand can only be met by the installation of modern machinery.

The tufting machine is capable of high rates of production and it is this which has been the main stimulus for the manufacturers of weaving machines. Consequently the main emphasis has been on increasing speed of weaving.

## A. Spool Axminster

The spool Axminster loom is capable of weaving unlimited colour combinations by the use of single or multiple pile yarn spools carried on endless pattern chains. These chains are supported on a gantry and during the weaving operation, they are brought mechanically to the correct position so that the spool (or spools) of yarn are sequentially transferred from the chain, into the weaving position, and back onto the chain, one such sequence representing the insertion of one weft row of tufts. It is this complicated transfer mechanism that limits the output of a spool Axminster loom and in spite of some improvements to the loom (e.g. marginally higher speed as a result of mechanical changes) this type of machine has become less competitive and there have been virtually no sales in recent years.

## B. Gripper Axminster[7]

Whilst the gripper Axminster loom does not have the same pattern potential as a spool Axminster, nevertheless in recent years many significant developments have taken place with this machine and as a result, there is still a wide demand for them.

In the late 1960s, David Crabtree Ltd. (Bradford), realized that an improved gripper loom was essential to meet the growing demand and consequently they embarked on what they termed their Stage III and Stage IV development programme. The Stage III was a general re-designing of the machine. This consisted of a new weft needle mechanism and a modified frame at the drive end. Bevel gears were eliminated and a rope mechanism was introduced. At the same time the pile cutting mechanism was re-designed because it became troublesome at the higher speeds. Accordingly, the number of knife boxes was increased to one every 18″ and the rope device was replaced with a rack and pinion. This cutting improvement meant that the grippers did not dwell at the draw-off position for as long as had been the case previously. The arc of travel of the grippers was also reduced marginally and the sum total of all improvements gave a speed increase of some 80%. The Stage IV machine underwent further modifications—more robust frame, chain driven jacquard, further reduction in arc of travel of gripper, incorporation of weft feeder device etc. An additional important modification was to the pile draw-off mechanism where the yarn/carriage draw off system was replaced with a gripper draw off system. This was operated by compressed air cylinders mounted on either side of the gripper shaft (for machines up to 3 m). For larger machines a centre cylinder was also incorporated. The system is economical in air (8 ft³/min) and obviates the use of thousands of carrier springs. It is also particularly useful for 12/16 frame looms and looms weaving long pile (rye type) carpets. By 1976 the Stage IVA machine was available. This incorporated all the earlier improvements together with others such as a double eccentric needle drive and was capable of speeds up to 50 picks/min for a 4 m loom weaving three-shot carpet (i.e. one row of tufts for every three weft insertions).

Brintons Ltd, have also modified their gripper shuttle loom. In one patent they improved loom speed by having the warp yarns just below horizontally placed carriers. The grippers come from underneath and pass between the warp yarns to take the tufts. This very short arc of gripper movement not only increases the speed of production but the layout has the additional advantage that the weaver can replace missing tufts/yarns without moving from the front of the loom.

## C. Spool gripper Axminster

This machine is also produced by D. Crabtree Ltd, and this company have incorporated many of their gripper loom improvements to this type of machine. The spool gantry itself has been completely re-designed to allow for large capacity spools and to make it more mechanically efficient. The spools are not removed from the chain as is the case during spool Axminster weaving and consequently speeds of 60 weft insertions per minute are obtainable for 4 m wide looms. This is equal to 20 pile rows a minute in three-shot operation, but when operating at a two-shot construction, a rate of 30 pile rows per minute is possible.

## D. Wilton looms

Weft insertion by shuttles is still the most common operation in both single face wire looms and face to face looms. Therefore, one of the most significant developments in this sector has been the ALL 60 face to face machine from van de Weile (Belgium). The problems associated with weft insertion using shuttles were completely overcome when they introduced their flexible rapier system (Fig. 7). The weft is inserted by a "giver" rapier from a large cone on the left of the machine after which it is picked up half way across the shed by a gripper or "taker" rapier. Top to tail linking of the weft yarn prevents all unnecessary loom stops. The machine can also be set up to produce "through to the back" carpet squares. Such carpets with the face design reproduced on the back are particularly popular in many Oriental countries, because they closely resemble knotted carpets.

Fig. 7. The van de Weile flexible rapier system.

## V. Tufting Machines[8]

To describe fully all the recent engineering developments in tufted carpet machinery that have occurred in the last few years would form a fairly large text book in its own right. Indeed many developments are now so firmly established that it seems hard to believe that they have only been used for a few years.

Obviously the greatest limitation in tufting is the narrow scope for varying design and colour and this is the field which has received the greatest attention and which is the most interesting.

The earliest systems were of course the fairly simple devices such as sliding needle bars or jute movers to give a simple zig-zag effect when space dyed or different coloured yarns were creeled into the machine.

One of the next most successful systems for obtaining more sophisticated tufteds was the high/low or "buried end" technique. In this system two different coloured yarns were creeled alternate end and end across the machine and when a particular colour was required in the pile pattern, it was tufted high pile and when not required tufted low pile. The low pile being achieved by braking a particular yarn and "robbing back" a previously formed loop. Typical of such development was the "Clutch in Roll" system devised by Cobble and launched in 1971 (Fig. 8). This was an electro-mechanical system in which an opaque pattern was prepared on a sheet of acetate film. Lights shining (or not) through the sheet onto photocells activated the tufting/braking mechanism as desired. Buried end or simple high–low sculptured loop pile patterns were very popular for many years. Perhaps the ultimate in such systems was made by Cobble and utilizes two separate needle bars; the first is threaded with a single colour of yarn and has a plain yarn feed system; the second needle bar (same gauge) is threaded

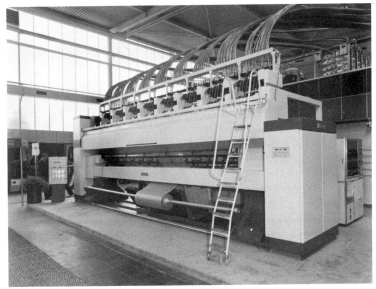

Fig. 8. A Cobble ST80 tufting machine fitted with "clutch in roll" pattern attachment and remote control pattern drum.

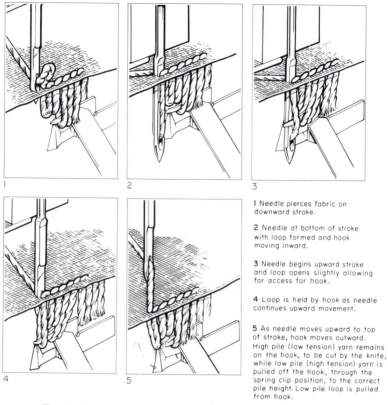

1 Needle pierces fabric on
downward stroke.

2 Needle at bottom of stroke
with loop formed and hook
moving inward.

3 Needle begins upward stroke
and loop opens slightly allowing
for access for hook.

4 Loop is held by hook as needle
continues upward movement.

5 As needle moves upward to top
of stroke, hook moves outward.
High pile (low tension) yarn remains
on the hook, to be cut by the knife,
while low pile (high tension) yarn is
pulled off the hook, through the
spring clip position, to the correct
pile height. Low pile loop is pulled
from hook.

FIG. 9. Sequence of operation of cut-loop tufting machine.

alternately with two colours of yarn. These yarns are controlled by an attachment giving three heights of pile. The medium pile height of the scroll is set to give the same pile height as that of the pile yarn feed on the first needle bar. With the machine threaded in this way, seven different colour combinations can be devised. If we call the colours A, B and C these possibilities are A, B, C (plain colours), and AB, AC, BC and ABC (mottled colours).

The preparation of patterns for such carpets is exactly the same as for more usual carpets incorporating three heights of pile. In addition to colours obtained in the normal operation, all the combinations of permanent pile height and invert switching* can be used to produce many patterns from the same design. Singer Cobble have produced a demonstration carpet showing 21 different patterns. This had been produced merely by using the invert and

* Invert switching reverses the instructions from the photocells to the tufting mechanism.

permanent pile switches and by switching the lights on and off either pattern drum.

The use of high–low patterns was eventually extended to cut–loop constructions (i.e. high cut pile and low loop pile). These machines were meant to copy the traditional carved Wilton textures. In early machines the density of the cut-pile was frequently rather open, e.g. in two colour carpets the cut-pile area of one colour in $5/32''$ gauge carpet would effectively only be $5/16''$ gauge. This disadvantage has now been overcome with the introduction of finer gauge machines, e.g. $5/64''$ gauge. The mechanism for the production of cut-loop carpets is shown in Fig. 9.

A very recent development is the production by Cobble of their LCL machine (level cut-loop) (Fig. 10). This is aimed at copying the typical small geometric design Wilton carpets, the design effect being achieved by having level loop pile and cut pile areas which, because of different light reflection, appear to be two different shades. The principle of operation is as follows: (a) when the looper gate is in the closed position, yarn is not admitted to the cutting throat of the looper and is dropped when the looper withdraws, remaining as loop pile; (b) when the gate is open, yarn is admitted to the cutting throat and operation is identical to conventional cut-pile.

In the mid-1970s a $9/64''$ cut-pile machine was introduced specifically for the imitation of contract Wilton qualities. A typical product is shown in Fig. 11.

As such patterns are achieved with a double needle bar (using various cams) and various creeling arrangements, a small number of geometric designs could be produced. At the time, this development seemed to attract

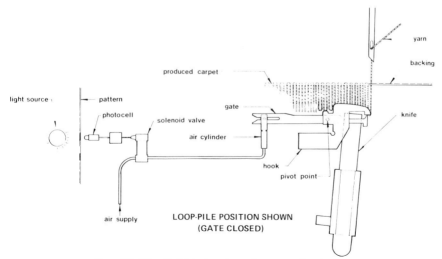

FIG. 10. The Cobble level cut loop mechanism.

FIG. 11. A ⁹⁄₆₄″ gauge cut pile tufted design.

very little interest and few machines were sold. Customers were presumably waiting for further developments which would give greater patterning potential or perhaps an additional reason was the various problems associated with cam operated needle bars. Such difficulties include wear on the cam necessitating frequent adjustments, a limited pattern repeat (usually 16 stitches), lack of co-incidence of needle and looper at different positions etc. These difficulties have largely been overcome in a very recent machine developed by Tuftco Corp. in conjunction with MOOG Inc. They have devised a hydraulic system to replace the cam action. Such a machine has only one needlebar and is called the "Hydrashift". It uses an hydraulic cylinder (linear actuator) controlled by a programme memory card which can supply repeats of up to 512 stitches and change over in 10–15 sec. Wear in the component is eliminated by having lamina seals in the piston and cylinder heads of the actuator. Accurate machining gives a diametrical clearance between cylinder and piston of 0.0254 mm. The movement takes place in 56 milliseconds and is independent of machine speed. The control system is a "closed loop" system which provides a constant feedback of information telling the system exactly where the needles are. Any forces which might tend to move them are immediately counteracted and the needles move to successive positions with an accuracy of ±0.1524 mm. As indicated, the programme card can be changed rapidly. It is in the form of a small block 3″ × 4″ × ¼″ with an infinite shelf life and is merely plugged in when required.

The GRAPHICS system is a further development of Tuftco's Hydrashift. The Graphics machine has two needle bars (both having Hydrashift type patterning) both of which are capable of being shifted laterally. These two needle bars are mounted in such a way that the two needle rows are on a normal ¼″ stagger, and co-operate with normally staggered hooks. The type of fabric produced is shown in the photograph (Fig. 12). Programming of the shift of the two needle bars is entirely independent. They can move together, both remain stationary, move in opposite directions etc., or they can do all these at different points in the programme.

The amount of shift is up to 3″ on each needle bar, so that in making a diamond type pattern the total width can be 6″. Depending on creel threading, the pattern figures may overlap, touch or have a space in between. Patterns can have two, three or more colours. Cobble have recently introduced their version of the above concept. Their pneumatically operated

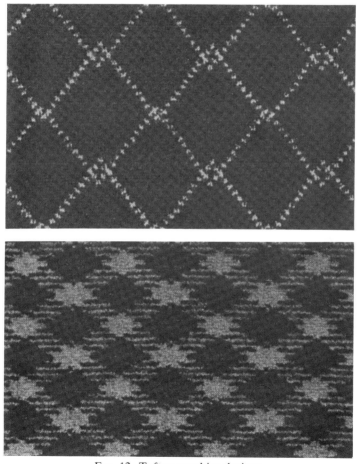

Fig. 12. Tuftco graphics designs.

single needle bar machine is called the PNEUMOVER and the two needle bar machine has been termed MOSAICS.

One promising development in the late 1970s was a tufting machine with independently controlled needles. The machine, produced by Singer and called the ICN machine was developed to produce true cut-pile patterns with no yarn wastage (Fig. 13). The needles were controlled by photo-electric impulses and compressed air. The machine was too coarse (³/₁₆″ gauge) for conventional wall-to-wall carpet although several machines were sold for producing small rugs and bath mats. Two pre-dyed yarns are creeled end and end on the machine; if a yarn is in the design an open ³/₈″ gauge construction will result. So far no machine has been developed for finer gauges although it is believed several attempts have been made.

Over recent years there have been two concurrent trends in the carpet industry. Coarse gauge machines have been produced to manufacture chunky berber type structures which have now become popular; carpets in ³/₈″ gauge and coarser are now quite common, particularly in loop pile constructions. At the other end of the scale, moves towards finer gauges continue. This is being encouraged by the growing popularity of dense plain velour textures. Gauges such as ¹/₁₆″ and ¹/₂₀″ are now available although the majority are used for upholstery fabrics. The practical commercial development of such fine gauges has undoubtedly been encouraged or indeed made possible by the introduction about three years ago of modular gauge parts.

The Cobble Modular Tufting Machine was fitted with gauge components fixed in 1¼″ wide die cast blocks (Fig. 14). These were mounted precisely

Fig. 13. The Cobble ICN machine. (a) In front of needles; (b) behind needles.

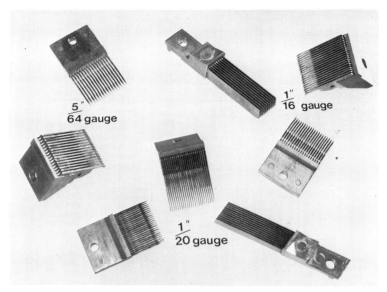

FIG. 14. Modular gauge parts.

side by side on the needle and looper bars respectively to form the required working width. For example, in a ⁵⁄₆₄″ gauge machine each block contains either 16 loopers or 16 needles. Being die cast, precise fitting is both accurate and straight forward. The needles are raked within the module, so obviating the need to do this manually on the machine.

The speed of tufting machines has gradually been increased over the years and even standard machines such as the Cobble ST80 are now capable of speeds up to 1100 rpm.

Whilst such speeds are themselves desirable to obtain increased output, they can in fact have the reverse effect when it comes to overall efficiency. To this end Cobble introduced their MCS (Machine Control System) (Fig. 15). This can be used to monitor and record the performance of a complete department of 16 tufters.[9] There are five main elements in the system.

1. Pile yarn feed. Problems (particularly with spun yarn) occur frequently between the creel and the needle. This element takes the form of a highly sophisticated creel stop motion. Indicator lights inform the operator of the exact location of particular problems. The system can distinguish between tight ends and breaks and automatically slows or stops the machine.
2. The fabric quality in the vicinity of the needling area is constantly checked using a number of photo-electric scanners which form the second

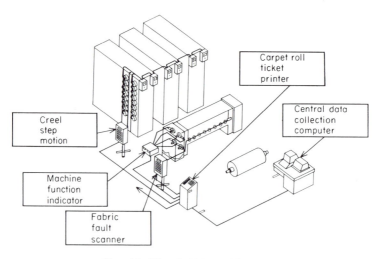

Fig. 15. The Cobble MCS system.

elements. Up to three faults can be identified simultaneously and the machine is slowed or stopped and the fault announced on a digital display which identifies offending needles by number.

3. Machine operation is displayed on the machine function indicator. This gives visual readings of fabric speed, yarn input, running times, machine speed, cloth tension, stitch rate and yarn per stitch. Tolerances can be built into the system and the machine will adjust itself within the set limits.

4. A ticket printer details all carpet specifications once the roll has been tufted.

5. Finally a central computer monitors and records all the data from up to 16 machines and can give production personnel all necessary information for the accurate running of their plant.

It seems likely that the use of these individual or complete monitoring systems will grow in coming years as the use of micro-computer technology becomes more widely accepted. Tuftco[10] have also recently announced a similar highly sophisticated computerized system for controlling stitch rates/pile weights etc. during tufting.

It is often the case that some of the simplest ideas turn out to be the best and in hindsight one wonders why they were not thought of before. An example of this could be the recent modifications to needles and loopers announced by Wool Research Organisation of New Zealand (Fig. 16).[11] It had long been realized that knots, joints and unlevel yarns were the main

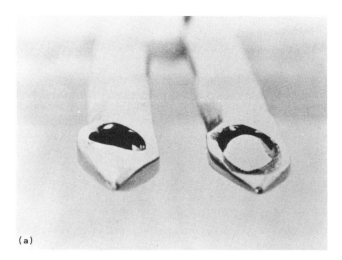

(a)

(b)

FIG. 16. WRONZ needles (a) and loopers (b).

causes of machine stoppages when using spun yarns. High speed photography revealed that joints or irregularities usually jammed in the needle as it was being withdrawn, causing a yarn break and a machine stoppage. The efficiency of tufting was increased tremendously by modifying the yarn path in the needle so that slubs etc. did not get caught in the needle eye.

Premature cutting of the yarn on the looper when a knot becomes jammed in the needle can also be reduced by rounding off part of the looper. This modification does require an additional manufacturing operation although this is more than justified from resulting increases in tufting efficiency.

## VI. Needle Punching

Needling as a method of textile fabric production is at least 100 years old, but it is only recently that it has been widely used for making surfaces with good wear properties.

The general method of production is to form a carded web, needle the fibres in some way to consolidate it and then bond or fix the fibres in some way. There is a vast amount of literature on the subject of needling and this has been reviewed by Crawshaw[1,2] and more recently by Barounig.[12]

Although needle punching (or needle felting) is an old process, it was not until about 1967 that needle felting started to become widely used for floorcoverings as distinct from underlay. At this time the NL12 machine from Fehrer was introduced. At the time the machine was considered remarkable in that it ran at 1000 strokes/min and had a double needle board. The latter innovation meant that all the drive elements could be sectionalized, balanced or counterbalanced, and mounted in an enclosed oil lubricated gearbox. By putting these elements in a single row, the machine could be built to any required width (Fig. 17). Not long after the introduction of the NL12, speeds of 1500 strokes/min were more usual and machines were available in widths up to 6.6 m (for blankets, leather substitutes, paper makers and industrial felts).

The introduction of the double needle board meant that all the various actions could be counterbalanced, and this in turn meant that machines no

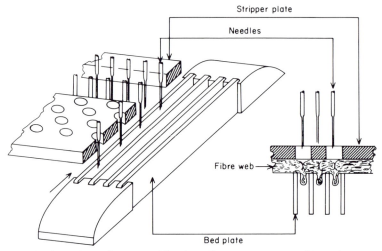

FIG. 17. Principle of needleloom with a lamellar arrangement of lengthwise slots in bed plate.

longer required massive floor foundations. This also undoubtedly contributed to the growth of needle-punching.

Currently machines are available with a choice of 4000, 6000, 10 000 and 15 000 needles/linear metre. With this range, and by selecting an appropriate throughput speed, felt producers can select the best needling density for their particular material.

It is sometimes necessary to build a suction unit into certain machines. This is so in the case of the production of spun bonded products where it is almost impossible to stop the fibre filament production in front of the needling machine. Fehrer can supply a combined discontinuous air jet and suction arrangement which effectively removes all dust and fibre from the bed plate perforation, thus minimizing stoppages from this cause. Lightweight fabrics are also another frequent cause for concern and specially designed boards are often necessary to minimize turbulence caused by air currents.

Figure 18 shows a double needle board machine. This has been devised by

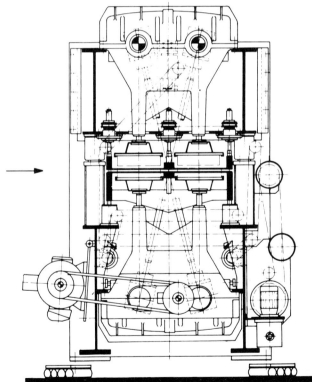

FIG. 18. Section through the NL 42 Quadropunch needleloom of Fehrer showing two needleboards above and two below the web of material.

Fehrer for materials which need high consolidation, i.e. needling from both sides and/or a large number of penetrations. This machine can needle from top to bottom or vice versa if required.

After needling, most fabrics are bonded with resin solution or emulsions. Impregnation may be total or partial depending on the manufacturers' requirements. Sometimes webs may be resin impregnated and subsequently heat moulded, e.g. car carpets. Impregnated felts may also be calendered to produce embossed carded or similar textures.

An alternative to adhesive or resin bonding is to have two different fibres (usually both synthetic) in the felt. After needling, when the web is heated up to the temperature of the lowest melting component, point to point bonding occurs. Several thermobonding machines have been developed for such applications.

## VII. Unconventional Methods

Although tufting and weaving, and to a lesser extent, needling, are the major methods of carpet production, a few unconventional methods are worthy of consideration.

Bonded carpets are probably the most important in this category and they are of particular interest because they give engineers and machine designers the opportunity to implant yarn directly into an adhesive backing so producing a finished carpet in one operation at high speed. There are many bonding systems but all have a basically similar production technique i.e. a sheet of yarn or fibre is made into a corrugated form (using grooved rollers, oscillating blades etc.), and implanted into an adhesive coated fabric. The most commercially successful techniques have been those in which sheets of yarn have been corrugated by pairs of oscillating blades moving reciprocally in conjunction with each other. Two types of carpet can be produced; (i) loop pile, by backing one side only of the pleated yarns, (ii) cut pile by backing both sides and slitting the cloth in a similar way to face to face carpet. The Titan II machine made in Belgium is an example of the latter. This is shown diagrammatically in Fig. 19.

There have been many attempts to produce patterned bonded carpets but the majority of these have been abandoned for commercial or economic reasons. One technique which has had some success is the BONDI Machine[13] (now owned by Fieldcrest Mills in the USA). In this fully computerized system the pattern is produced by pre-assembling the pile yarns as the weft across a temporary warp of a flat woven fabric. This fabric is then folded or lapped backwards and forwards in concertina fashion to form a stack or "bale". When the bale is rotated the ends of the yarn can be seen to

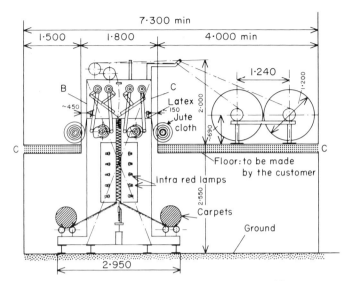

FIG. 19. The Titan II non-woven carpet machine.

form the desired pattern. The yarn ends are then brought into contact with adhesive coated backing fabric. The bale is then sliced in such a way as to form a carpet of the required pile height. The backing fabric is then moved forward the distance of the pattern repeat (comparable with movement in flat screen printing) and the process repeated.

As this process does not depend on depositing single tufts or even rows of tufts, but rather deposits a complete pattern repeat in one operation, it is clear that production rates can be very high. Admittedly a yarn preparation stage is involved but it is claimed that this is far more mechanized than for Axminster spool setting and therefore pattern and quality changes are extremely quick and simple.

## VIII. Electrostatic Flocking

This is a slightly different form of bonded carpet production. There are about 10 or 12 manufacturers of suitable machinery. Most of the fabric produced is used for floorcoverings, particularly in cars, although in Japan, flocked materials are widely used in garments and for wallcoverings. Essentially, the process consists in drawing a backing material (coated with an adhesive on its upper surface) over a plate which is one electrode of a high potential system. A hopper containing the cut flock is placed over the plate. Fibre is sifted through a metal grid in the base of the hopper and this grid forms the opposite electrode. As the flock falls, it receives an electrostatic charge. Because all the fibres are given a charge of the same intensity and polarity they repel each other and distribute themselves uniformly and are projected end on into the adhesive. During the last few years there have been no real changes in this basic technology, although, of course, flock cutting, machine design and adhesive technology have all been improved. Electronic curing and di-electric heating are now more widely used to give improved adhesives and higher productivity.

## IX. Knitting

One unconventional method not falling into the bonded category is knitting. Circular, Raschel and sliver knitting have all been used to produce floorcoverings, although none has been used to any large extent. Knitting has been most widely used in Comicon countries where the limited design and texture possibilities have not been such a deterrent in the market place.

## X. Coloration

There have been a vast number of developments for imparting colour to carpets (particularly tufted). It would be impossible to describe fully all these various machines in this chapter and for full description the reader is directed to the various books on the subject. However, it was felt that a chapter on carpet machinery would be incomplete without at least indicating current trends in this area, particularly in the field of piece dyeing and carpet printing.

Piece dyeing[14] in winches, which was used widely in the late 1960s for producing either plain carpet or simple multicoloured designs using variable dye nylon, has declined. Continuous dyeing has grown and it is undoubtedly in this and also in the field of carpet printing that the majority of developments are taking place.

If we ignore transfer printing which has not as yet been widely adopted in carpet coloration there are several types of carpet printing machines available.[15]

### A. Relief printing

The Stalwart-Pickering (Fig. 20) and, more recently, the similar Tuftco machines consist of a series of rollers upon which a raised pattern is produced using either rubber or flocked material mounted on wood or rubber

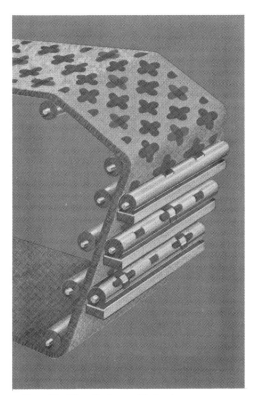

FIG. 20. The Stalwart-Pickering system.

blocks. The rollers are each immersed in a trough of dye liquor and the raised portions transfer the liquor to the passing goods. This essentially simple machine is relatively inexpensive and usually produces three or four colour designs at speeds of up to 12 ft/min. The main disadvantages of these machines are the limited design definition and colour possibilities.

## B. Flat screen printing

Flat screen machines were amongst the first to be produced for carpet printing. Of these, the Peter Zimmer machine achieved the greatest commercial success (Fig. 21). The print paste is applied by means of twin magnetic roller squeegees, as shown, or a hydrostatic pressure slot. The carpet is carried on endless belts with protruding pins which penetrate the carpet backing and hold it firmly in place. This type of machine can print up to eight colours and operates at speeds between 4 and 10 m/min.

It is clear that a machine of this type occupies a considerable floor space and for this reason it was felt that a smaller, more compact machine could have potential. For this reason, Mitter and Co. developed their SEMI-MATIK machine (Fig. 22). This machine has only one printing station and all the screens (up to 20 in number) are stacked above it. Each screen is fed sequentially to the station, removed after printing and returned to the top of the stack. When the last screen has been printed and the design is complete, the carpet is moved forward one repeat length. This machine uses a slot

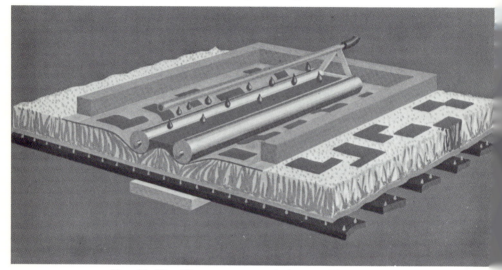

FIG. 21. The Zimmer carpet printing machine TDA.

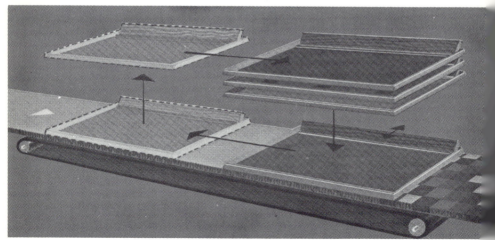

FIG. 22. Semi-matik compact flat screen carpet printing machine.

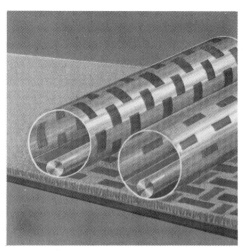

FIG. 23. Rotaprint rotary screen carpet printing machine.

squeegee and a vacuum slot to achieve a good penetration with minimum pile distortion. The machine has been widely used for producing "Oriental type" carpet squares, although it is also possible to produce wall-to-wall designs.

## C. Rotary screen printing

This was developed to give greater production capacity and to make possible the printing of large design repeats. Two successful rotary screen printers are the Mitter Rotaprint machine (Fig. 23) and the Peter Zimmer Rotary Printer. In such machines the print paste is applied with roller squeegees with a large circumference. With the Mitter system (Fig. 24), penetration of

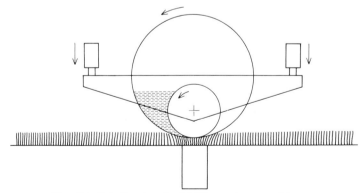

FIG. 24. Printing head of Mitter rotary machine.

the dye liquor is controlled by independently varying the squeeze roller speed and the screen can speed. In the Zimmer system, penetration is achieved with a "Hydroslot squeegee". For extra long repeats (up to 4 m), the rotary screen can be replaced by a continuous screen in triangular form (see Fig. 25). An additional type of screen assembly equipment is required.

FIG. 25. "Rotafilmprint" unit.

## D. Drop printing

Most carpet printing machines were designed to simulate the woven carpet patterning systems, but there are some machines that produce less precise effects that are similar to using space dyed yarns.

The KUSTERS TAK system is one such development. A uniform ground shade is first applied to the carpet using a Kusters continuous dyeing machine. Another colour is then applied using a special dye applicator (see Fig. 26). In this applicator the doctor blade, which has serrations along its lower edge, transfers dye liquor to the carpet in a series of thin streams. These streams fall on a laterally moving ladder-like belt and produce droplets which fall on to the passing carpet. With various combinations of roller, belt and carpet speeds, different patterns can be obtained.

Even more sophisticated effects can be obtained by using MULTITAK (2 applicators) or GUMTAK, i.e. a negative effect achieved by displacing a thick padded ground colour with drops of thinner colour. The latter technique is widely used in the USA for producing the popular "shadow" prints. (Shadow prints can also be produced on conventional printing machines.)

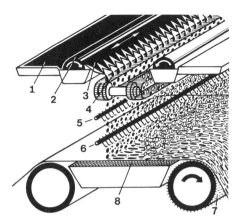

FIG. 26. TAK dye applicator. (1) dye trough; (2) dye applicator roll; (3) doctor blade; (4), (5) and (6) drop cutters; (7) conveying pin roll; (8) fabric table.

## E. Jet printing

Undoubtedly one of the most spectacular developments in carpet technology over the past few years has been "jet printing". This technique has several advantages over screen printing, the main one being the capability of operating "short runs". In the MILLITRON computer injection dyeing system (Fig. 27) the carpet is transported on an inclined table under several banks of dye jets (one bank for each colour). The jets, which are spaced ten to the inch, run continuously. If the colour is not required in the design, the jet is deflected into a collector system by means of an air jet. The latter have electromagnetic valves which are activated by magnetic tape on which designs are stored. Other jet printers are the Titan (Godfrey Hirst), Chromotronic (Peter Zimmer) and the Wadsworth Greenwood machine.

## F. Yarn printing

In the Pickering–Crawford machine (Fig. 28), yarn warps were printed in such a way that a pattern was formed when the yarn was subsequently tufted into carpet. The system was devised by Allan Crawford in the USA, developed by Mohasco Industries and manufactured by Edgar Pickering (Blackburn) Ltd. After many initial problems, several machines came into use and about 1976 many beautifully designed carpets were produced. These were mostly in semi-shag constructions and free flowing designs. The

N

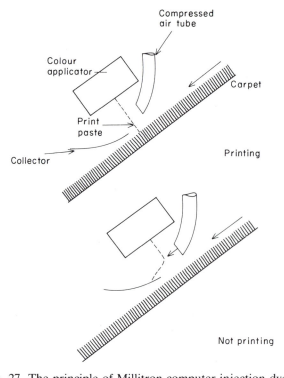

FIG. 27. The principle of Millitron computer injection dyeing.

machine was never fully developed to the stage where it could be used to produce the design complexity of a typical Axminster on a short cut-pile construction. The machine was later completely re-designed (and re-named Kalaedoscopic)[16] and computerized using a microprocessor control system.

Yarn was fed from a creel to the printing head in warp sheet form. A group of yarns (four or five in early machines and eight in later models) was directed across a 1″ square plunger on a printing roller. After printing, fixation etc., each of these identically printed yarns was diverted to a separate take-up beam. Up to eight printing heads (one for each colour) consisted of a top roller covered in ribbed rubber pads and a lower drum covered with spring-loaded pads which carried the print-paste. The pads were moved into the printing position by means of aluminium programme bars contoured according to the design required.

A group of index yarns were also printed black periodically and wound onto each beam. These were scanned by a photocell device during subse-

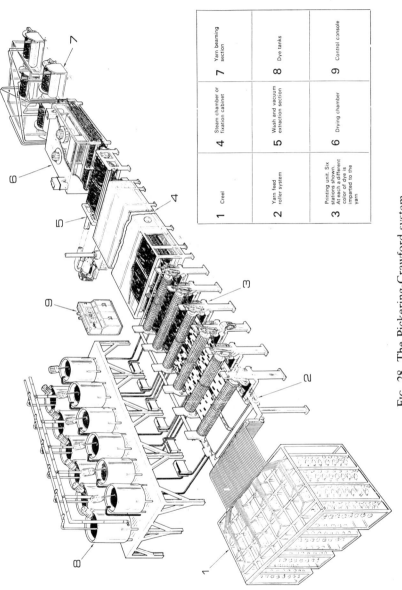

| 1 | Creel | 4 | Steam chamber or fixation cabinet | 7 | Yarn beaming section |
|---|---|---|---|---|---|
| 2 | Yarn feed roller system | 5 | Wash and vacuum extraction section | 8 | Dye tanks |
| 3 | Printing unit. Six stations shown. At each a different color of dye is imparted to the yarn | 6 | Drying chamber | 9 | Control console |

Fig. 28. The Pickering-Crawford system.

quent tufting to ensure that the individual beams were tufted at identical rates thus keeping the design in register.

Of course the actual printing machine is only part of a carpet printing system and in recent years there have been many improvements and modifications to the processes which follow the printing machine. A feature common to all printing systems is a steamer. This follows the printing directly without intermediate drying. Steaming times vary between 5 and 15 min and the temperature is generally about 100°C. Carpets can go through a steamer horizontally or vertically (festoon) or sometimes in a combination of the two. Horizontal steamers are preferred giving better dyestuff penetration and no danger of dyestuffs contaminating carrier rollers. Frequently, however, in order to save space and money, a compromise combination of horizontal steaming followed by festoon steaming is used.

After steaming, printed carpets are washed to remove unfixed dyestuffs and auxilliaries. Again, several types of washer are available, e.g. from spray and squeeze types to perforated drum washing. Washing is frequently followed by a suction slot to remove as much surplus water as possible after which the carpet is finally dried before going to a conventional carpet finishing plant.

## XI. Computerized Patterning[17]

Before leaving the section on carpet coloration, the recent developments in computerized patterning must be mentioned.

There are several time consuming processes necessary between the production of patterns in the design studio and before the production of printing screens etc., and recent developments have enabled these to be very effectively speeded up. A typical example is the Sci-Tex system in which an electronic camera transfers the design (in tonal groups) to a computer memory bank. From here it can be recalled and displayed on a TV monitor in selected colours. Using an electronic pen, changes in design or colour can be made at this stage. Once the design is acceptable to the designer, apparatus using laser beam techniques makes colour separations on photofilm for making into screens. A similar system is used for the production of the jet controlling tapes on the Millitron machine.

The Scan-Plan system (Fig. 29) also uses a computerized scanner to transfer an artists design (up to 15 colours) onto a TV monitor. Again after any necessary modifications with an electronic pen the design is transferred to an eight channel tape. The tape can be used to feed a card punching machine punching as many as 2,400 cards an hour. The Hill TDP system provides similar facilities.

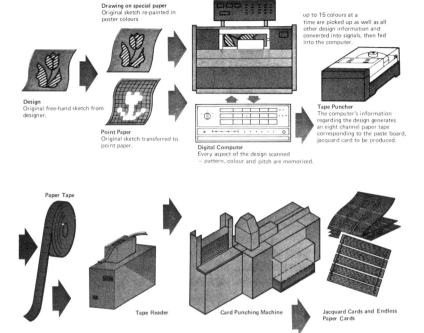

**Design**
Original free-hand sketch from designer.

**Drawing on special paper**
Original sketch re-painted in poster colours

**Point Paper**
Original sketch transferred to point paper.

up to 15 colours at a time are picked up as well as all other design information and converted into signals, then fed into the computer.

**Digital Computer**
Every aspect of the design scanned — pattern, colour and pitch are memorised.

**Tape Puncher**
The computer's information regarding the design generates an eight channel paper tape corresponding to the paste board, jacquard card to be produced.

**Paper Tape**

**Tape Reader**

**Card Punching Machine**

**Jacquard Cards and Endless Paper Cards**

FIG. 29. The Hardaker Scan-Plan system.

## XII. Carpet Finishing

After weaving or tufting all carpets require finishing before they can be despatched to the ultimate user.

### A. Woven carpets

Most woven carpets are now latex backed rather than starched or sized. Rubber latices have been found to give a more durable finish as well as making them non-fray during fitting and laying. The application of such latices is carried out in exactly the same way as sizing, i.e. by using a lick roller and trough and doctor blade, followed by oven drying.

### B. Tufted carpets

All tufted carpets have to be latexed in or finished in some way to ensure dimensional stability and tuft anchorage. Over recent years, many changes have taken place in techniques of application as well as in materials used. Initially a single coating of latex was applied (i.e. primary backed only). However, it was soon discovered that this procedure gave inadequate dimensional stability and it is now more common practice to apply a secondary backing. Foam backed carpets (both plain and embossed) have been extremely popular in Europe, although in recent years, consumer demand has reverted to a preference for secondary backed carpets and a separate underlay. After latexing all carpets (except low level loop pile products) are sheared to remove surface long fibres after which they are rolled and measured before despatch.

Many companies offer a cut-length service to smaller shops and customers and in recent months several labour saving cut-roll and wrap machines have been built to cater for their requirements. A few machines cutting carpets automatically to room shapes are also available. Some are fully computerized and after being fed with information from the sales department, will latex the selected roll, and cut it so that all orders are cut, rolled, wrapped and labelled and also ensure minimum carpet wastage.

## References

1. Crawshaw, G. H. and Ince, J. *Text. Prog.*, **9**, No. 72, 1977.
2. Crawshaw, G. H. and Ince, J. *Text. Prog.*, **4**, No. 2, 1972.
3. Pitts, J. M. D., *Wool Sci. Rev.*, **56**, 1980.
4. Smigielski, E. *IWS Techn. Inf. Letter*, **68**, 1979.
5. Shealy, O. L. *Text. Res. J.*, **39**, 254, 1969.
6. Smigielski, E. *IWS Techn. Inf. Letter*, **80**, 1980.
7. D. Crabtree Ltd. (Bradford). *Technical Catalogue.*
8. Ward, D. T. "Tufting—An Introduction". Published by Texpress, Manchester 1975.
9. *Tufting Yearbook 1980.* (Textile Business Press Ltd). p. 34.
10. *Tufting Yearbook 1980.* (Textile Business Press Ltd). p. 104.
11. Carnaby, G. A. and Ross, D. A. *Text. Inst. Industry*, Sept., 1979.
12. Barounig, J. *Text. Inst. Industry*, Aug., 1980.
13. *Wool Sci. Rev.*, **51**, 12 (1975).
14. Junker, G. *Textilveredlung*, **4**, 1969.
15. *Bayer Farben Revue*, No. 30.
16. *Tufting and Needling News*, **119**, July, 1980.
17. J. Hardacre Ltd (Bradford), *Technical Catalogue.*

## Chapter 12

# Advances in Stitch-Bonding, Warp- and Weft-Knitting Systems, and Automated Knitwear Manufacture

G. R. WRAY and R. VITOLS

Research and development in textile machinery have been pursued in the Department of Mechanical Engineering at Loughborough University of Technology since 1966 and Fig. 1 is a chronological flow-chart summarizing the extent of the main research avenues investigated.[1,2] The editor of this volume has requested us to review briefly advances in warp- and weft-knitting, and in stitch-bonding, in order to put our own machinery development contributions in context with contemporary advances in these wider areas.

## I. Stitch-Bonding and Pile-Fabric Machinery

Between 1966 and 1970 a significant commercial textile machine and process[3,4] was devised, researched and designed in the University Department. This was the Pickering Locstitch pile-fabric machine and it owed its origins to stitch-bonding technology as is described below.

In the late 1950s the Arachne stitch-bonding process was developed in Czechoslovakia[5] to produce a non-woven fabric by stitching through a fibre fleece using parallel warp yarns for the stitched reinforcement, while in East Germany the Malimo process stitched parallel warp yarns through a bed of cross-laid weft yarns.[5] Since then a substantial range of stitch-bonding processes has evolved as shown in Table I which lists several different types

375

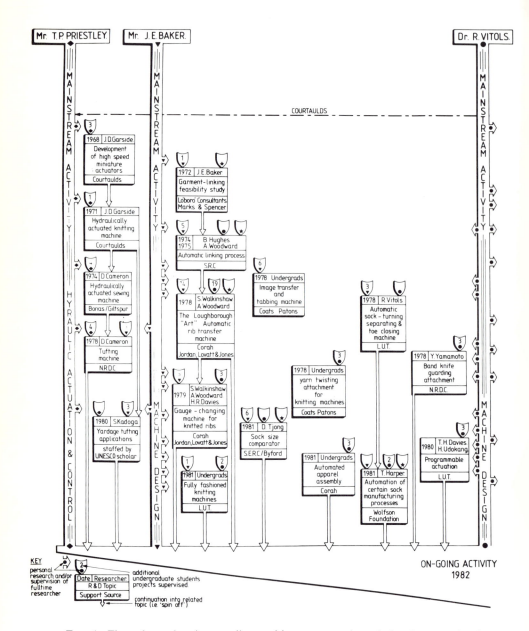

FIG. 1. Flow-chart showing textile machinery research and development in the Department of Mechanical Engineering, Loughborough University of Technology 1966 to 1982.

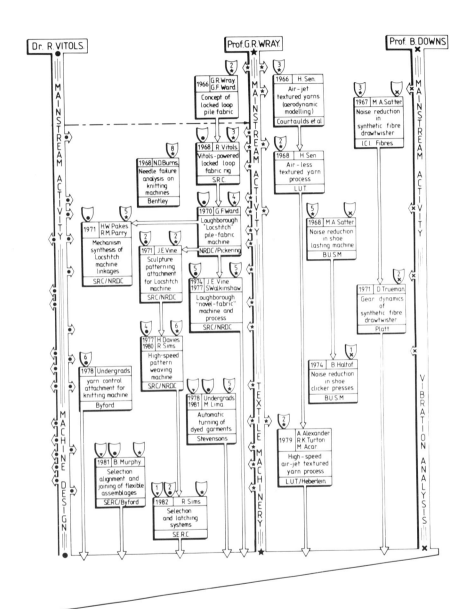

TABLE I. Some commercially-available stitch-bonding and pile-fabric machines and processes (based on Burnip and Marsden[6]).

| Machine | | Nature of process |
|---------|---|-------------------|
| Arabeva | Malivlies | Loose staple-fibre web only |
| Arachne | Maliwatt | Loose staple-fibre with stitching warp or warps |
| Araknit | | Stitching warps only |
| Arutex | Malimo | Stitching warps binding weft-way yarns with extra pile warp |
| | Schussopool | Stitching warps binding weft-way yarns with extra pile warp |
| Araloop | Malipol | Stitching warp as pile into a base cloth or web |
| | Voltex | Loose staple-fibre web as pile stitched into a base cloth |
| Bicolour Araloop | | Stitching two non-interacting warps into both sides of a base cloth on a double needle bed |
| | Liropol | Triple warp machine forming double-sided pile |
| Kraftamatic | | Stitching warp into a base cloth to form unsymmetrical double-sided pile |
| Locstitch | | Stitching two interacting warps into both sides of a base cloth to form locked-loop double-sided pile |

of commercially available machines.[6] It is to be observed that early variants of the two major stitch-bonding processes were the Araloop and Malipol processes for producing pile-fabrics by stitching through either a base cloth or a fibre web.

Traditional types of pile-fabrics are woven on special looms at low speeds (around 10 cm/min) and typical examples are terry-towelling, velvets, carpets, blankets and upholstery fabrics. The underlying principle is the interlacing of the pile warp yarns with the ground warp and weft yarns. Thus, the conventional woven three-pick terry fabric illustrated in Fig. 2a is formed in this manner. The three weft threads (1, 2 and 3), shown in cross-section, are beaten into the fabric so that they interlace with the ground warps (4) which are held tightly stretched to form the basic woven structure of the fabric base. The slack pile warps (5) are at a much lower tension, and during manufacture they are allowed to form pile loops in the vertical plane, as shown by the arrows. As well as being formed by a slow mechanical operation, the resultant fabric is not very stable, because the pile loops are held in place merely by simple coil-frictional forces produced by interaction with the weft yarns. A modest pull on any one of these pile loops could cause the

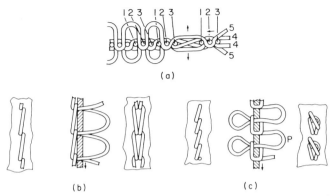

FIG. 2. (a) Woven-terry structure. (b) Araloop (and Malipol) pile stitch. (c) Kraft-amatic pile stitch.

structure to "run". Moreover, if the fabric is cut for making into garments, its structure will easily fray.

The basic Araloop/Malipol pile stitch (shown in Fig. 2b) is a single yarn chain stitch having upstanding loops on one side of the base fabric. In the Araloop system (Fig. 3) a parallel row of tubular compound needles penetrates the ascending base fabric (2) to receive pile yarns (3) from the lapping guides (4) and form them into pile loops over loopers (5). The needle hooks are closed by the sliding closing wires or latches (6) during the pulling-through operation, the other warp yarns (7) being used merely to reinforce the web. If a preconstructed material is used for the base fabric, these warp yarns are not essential.

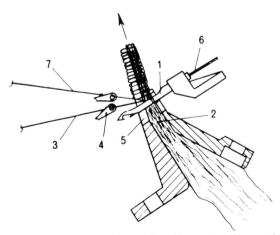

FIG. 3. Araloop pile formation (through a fibre-web).

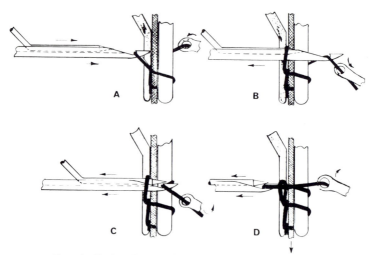

FIG. 4. Cycle of operations in the Malipol process.

The action of the stitch-forming elements in the Malipol machine (Fig. 4) is similar to that of the Araloop process, although the spatial orientation of the elements and their actuators is completely different. At position A the parallel row of compound needles is advancing to penetrate the base cloth. Each closing wire has its point hidden within the groove of the compound needle to permit the yarn loop to ride onto the shank of the needle so enabling the point to pass through the base fabric clearly, the base fabric having been advanced one pitch distance from the previous needle with-drawal point. At B, the needle has reached the fully penetrated position and the pile yarn has been laid over the looper and into the open needle hook by the yarn guide. At C, the needle has partially retracted with the pile yarn securely gathered in the hook. The closing wire, whilst returning, is guiding the hook through the fibres of the base fabric and the previously formed loop. At D, a new loop has been pulled through the previous one and is held in the hook. The base fabric must now be advanced to permit a new penetration.

When compared to double-faced pile-fabrics, such as the terry weave (Fig. 2a), the Araloop/Malipol stitch has limitations, in that the pile loops are produced on only one side of the base fabric, and for each pile loop, there is a complete loop of "wasted" yarn lying flat against the base fabric. Both systems utilize a complex stitching system, involving compound knitting needles and guide bars, similar to those used in warp-knitting, and this limits their potential production rates, even though, at approximately 1000 stitches/min, they easily surpass those of conventional pile-weaving.

In 1966, a British development, the Kraftamatic Machine[7] produced the stitch shown in Fig. 2c. This was an improvement on the Araloop/Malipol stitch, in that it had pile loops on both sides of the base fabric, although they were unsymmetrical in appearance and character. There was less yarn wastage because only half-loops of yarn lay against the base fabric compared to the full loops of wasted yarn in the Araloop stitch. However, there was a tendency for the row of stitches to run if a force was continuously applied, for example, to loop P. The fabric manufacturing process was also complex, comprising a row of double-latch needles (2) which interacted with rows of sewing needles (1), as shown in Fig. 5, in such a manner that the former required three separate accelerations per stitching cycle, which of course formed a severe dynamic limitation to high-speed operation. Nevertheless, this technique operated at some ten times the speed of conventional pile-fabric weaving.

The practical limitations of these earlier processes led the University researchers to seek improvements. It was considered that (i) a more idealized pile-fabric would be run-resistant, with uniform pile-loops symmetrically arranged on either side of the base fabric, whilst preferably using less "wasted" yarn; and (ii) simpler mechanical methods should be devised to obtain higher production speeds. From these considerations the Locstitch

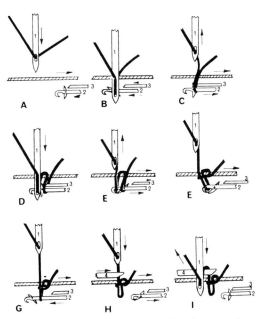

FIG. 5. Cycle of operations in the Kraftamatic Process.

pile-fabric and its manufacturing process were both evolved in the Department of Mechanical Engineering at Loughborough University of Technology between 1967 and 1969.[3] From Fig. 6 it can be seen that the "Locstitch" fabric advances the chain stitch concept of the Araloop/Malipol fabric but with the subtle difference that one set of yarn chain is used to lock the other, and vice-versa. It also uses less yarn per pile stitch because the half-pitch spacing advantage of the Kraftamatic fabric is incorporated without its two major disadvantages of unsymmetrical appearance and running tendencies. Two sets of pile yarn warps penetrate the pre-constructed base fabric such that the upstanding pile loops of the yarns protruding on one side of the fabric are secured by the lassoing effect of prone loops in the yarns protruding on the other side. If a loop is severed or snagged, a pull in any direction only serves to tighten the structure, so avoiding undesirable "run" features. The "Locstitch" fabric will not fray when cut and is ideal for brushing or cropping for blankets, rugs, furs, etc. The pile on each face of the base fabric may be constructed from yarns having separate characteristics and/or colours to enable novel reversibility effects to be produced, e.g. clothing fabric with a self-contained lining.

This new stitch concept enabled the stitch-forming operations to be simplified to obviate the complicated knitting mechanisms which give dynamic limitations to the other unconventional pile-fabric machines. Indeed the first stitching was hand-done with two domestic sewing-machine needles through a piece of cardboard. This formed the basis for an exceptionally high-speed process using simple stitch-forming elements.

The development of the Locstitch process has been detailed elsewhere.[4] Initially it involved undergraduate projects as feasibility studies followed by a 1968 SRC grant for undertaking basic research leading to a machinery rig

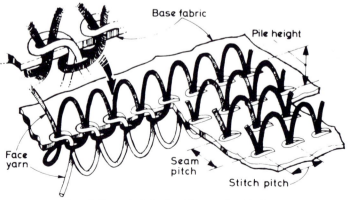

FIG. 6. Locstitch locked-loop pile-fabric.

which produced sample fabrics at 1300 double stitches/min. This was a remarkable achievement considering that all the loopers and needles were individually hand-made.

The University patents were assigned to NRDC and the rig and sample fabrics were shown to several British textile machinery companies, and in 1970 Edgar Pickering Ltd, Blackburn, was offered a licence to produce Locstitch machines. In June 1971, a prototype "Locstitch" machine was exhibited at the International Textile Machinery and Accessories Exhibition in Paris, but the major innovatory step of designing this full-width industrially-sized (3 m length) machine was undertaken by the University team between September 1970 and March 1971.

The smooth orbital motion of the main moving parts, i.e. rows of lightweight sewing needles, permitted the design of an extremely quiet prototype having low inertia forces due to the absence of high accelerations. Figure 7 is a sectional view through the fabric forming region; parallel rows of 780 needles ($N_1$) on one side of the base fabric (F) cast off yarn loops to the corresponding needle row ($N_2$) on the other side, and vice-versa, the tensioning of the yarn causing the locking effect to be attained. The height of the pile was independently adjustable by pre-setting the height of the loopers

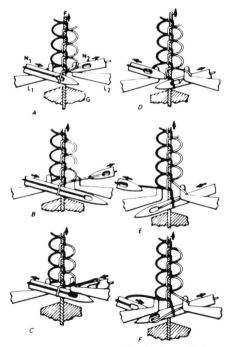

FIG. 7. Cycle of operations in the Locstitch process.

(L) which laterally traversed to engage the yarn during stitch formation; by removing one set of loopers, the height of the loops on one side of the fabric could be reduced to zero to give a single-sided pile if required. The stitch pitch was also adjustable to accommodate the possibility of widely differing fabric constructions.

The prototype "Locstitch" machine, running at 750 cycles/min, produced a variety of locked pile-fabrics at speeds of the order of 2 m/min. Commercial "Locstitch" machines, as shown in Fig. 8, were built by the Pickering Company, which has since been acquired by Cobble-Blackburn Ltd. These were largely based on the University designed prototype, and are currently operating in various parts of the world.

A further SRC grant in 1974 was directed towards scientifically-based improvements to the Loughborough process and J. E. Vine developed a sculpture-patterning system[8] (see Fig. 9) in which individual looper elements were pre-selected according to a pattern read from a continuous transparent film by photo-electric cells, these signals being electro-mechanically amplified so that high and low pile height patterns were produced at the same fast production rates. Changes of pattern were easily accommodated since the painted transparent film carrying the pattern design could be replaced by another in a few seconds, as shown in Fig. 10.

Mrs. R. M. Parry undertook linkage mechanism synthesis research aimed

FIG. 8. The "Locstitch" Pile-Fabric Machine as produced commercially by Edgar Pickering Ltd. (now part of Cobble Blackburn Ltd.).

FIG. 9. Sculpture-patterned pile-fabric emerging from the Loughborough University research rig.

FIG. 10. The preparation of painted pattern designs for the sculpture patterning pile-fabric rig.

at simplifying the costly conjugate cam mechanism drives used for the prototype Locstitch cycle. A family of mathematically devised mechanisms was eventually derived, fuller treatments of the synthesis and optimization techniques being given elsewhere.[9,10] A model of a typical synthesized linkage mechanism is illustrated in Fig. 11.

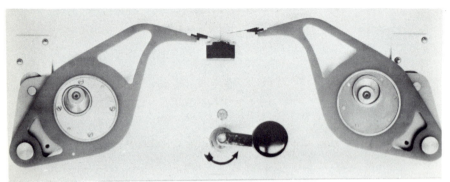

FIG. 11. Model of the synthesized needle orbit linkage mechanism intended for the Locstitch pile-fabric machine.

## II. Warp-Knitting and Similar Fabric Forming Processes

Knitting is a process whereby fabrics are formed by the interlacing of neighbouring yarn loops. The knitted structure may be formed either by weft-knitting, in which one or more individual weft supply yarns are laid across beds of needles so that loops of yarn are drawn through previously made loops, or by warp-knitting, in which case the fabric is formed by looping together parallel warp yarns as they are fed collectively from a warp beam.

In warp knitting, each needle can be supplied with either a separate yarn or with several yarns. The simplest type of single-bar fabric is shown in Fig. 12 where it can be seen that each warp yarn is caused to zigzag along the length of the fabric; this forms a loop at every change of direction as individual yarns are intermeshed with adjacent yarns which have similarly contorted paths. Each row of loops running along the fabric is known as a wale, and each row of loops running across the fabric is a course, the loops in one course being produced simultaneously. This means that each yarn usually traverses a horizontal distance of only one or two needle spaces during the formation of a knitted course making warp knitting highly productive when compared with weaving where wefts have to be inserted by traversing across the width of all the warp threads.

Fig. 12. Single-bar warp-knitted structure.

The simple fabric shown in Fig. 12 serves to illustrate the principle, but would be too unstable for practical use. In order to make more stable fabrics, two or more sets of warp yarns are used, resulting in a straight wale structure if at least two yarns forming each loop are wrapped around the needle under suitable tensioning conditions. The warp yarns are supplied on beams, there being usually one beam per guide-bar assembly; tension devices must be provided to absorb the excess lengths of yarn which may occur at certain stages of the knitting cycle, as well as to provide the tension necessary for the formation of stable loops. As with weaving, the let-off of warp threads may be based on negative methods, in which case the yarns themselves pull the beams around, or alternatively a servo-controlled positive drive to the beams may be used to feed the yarns at the correct speed for knitting.

The needles used in warp-knitting may be either of the bearded or latch types, but a compound tubular type of needle has also been employed in some machines. Bearded needle warp knitting, in its simplest form, is illustrated in Fig. 13. In position (a) the guide bars, G, are moving away from the front of the machine to pass the warp threads through the spaces between needles. Each needle, N, is still rising, having just knocked over the previous stitch and having left it held by the throat of the sinker, S. An overlap has been formed over the needle beard when position (b) is reached, as the guide bars have been shogged slightly sideways before moving forward. The needle descends to position (c), the presser, P, having closed the beard to trap the newly lapped thread. The sinker retracts to allow the fabric to be lifted from its throat so that the previously made stitch is landed on the outside of the needle beard. The cycle is completed at position (d) when the needle has descended to its lowest position, having caused the previously formed fabric loop to be knocked over the needle heads as the fabric is again brought into the sinker throat, which has moved forward appropriately. The guides then take up a new lateral position ready to perform a similar operation on another needle.

Thus one can see that, although the needles remain in the same vertical plane, the guide bars are able to be shogged sideways by one or more pitches. When the shogging motion occurs on the beardside of the needle, as between positions (a) and (b), it makes an overlap; when it occurs in front of the needles, as between positions (d) and (a) an underlap is formed. The provision of as many as 48 different guide bars, each capable of independent lateral shogging motions, together with an additional needle bar in some cases, enables very intricate patterns to be knitted. In such cases a visual

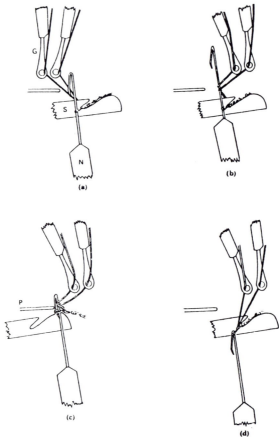

FIG. 13. Cycle of operations in warp-knitting using bearded needles.

representation of the required fabric design has to be translated onto either pattern chains or pattern wheels to actuate the guide bars. The scope for variation in fabric design is wide indeed, but is limited by the physical problem of fitting so many mechanical parts into the knitting region.

The reader will appreciate that the simple cycle of operations shown for producing the basic stitch using a bearded needle could just as easily have been represented using either a latch needle or a compound needle, as the principle remains unchanged. The only points of detail are that, instead of a presser being used to close the beard, the latch is closed by the action of taking the previously formed loop over it, similar to the action shown for weft-knitting (see Fig. 24): in the case of the compound needle, a mechanically operated tongue, positioned inside the hollow needle stem, bridges the gap between the needle stem and the hook, thus obviating the need for a negatively closed latch, but introducing an extra positively driven element in the same way that the presser is an extra positively driven element when it closes the latch needle.

End-uses of warp-knitted fabrics include sheetings, shirtings, fancy lace, nets, curtains, thermal cloths, packaging bags, upholstery and other loop-raised fabrics, simulated suede, as well as many types of lingerie and dress fabrics. Hence warp-knitting is not only a competitor in many fields traditionally associated with lace making and weft-knitting, but it also rivals weaving in some of its outlets, principally because of its very high rates of production which can be in excess of 1000 courses/min. Machines suitable for many fabric types are often over 4 m wide, and widths can be as great as 6 m on some curtain net machines; such machine widths also increase their competitiveness on the basis of output/machine.

One disadvantage of warp-knitting is that it is mainly suitable for filament yarns rather than for spun staples. The fabric structures produced by warp-knitting, although at first glance often similar to woven fabrics in certain constructions, e.g. filament nylon shirtings, cannot as yet simulate woven fabrics entirely. No weft is used and, therefore, the high stability of a closely interwoven fabric is hardly realized; moreover, the varied weft effects which are cleverly exploited in many types of weaving are not possible by the very nature of the process. However, it is interesting to note that warp-knitting machines have emerged in recent years that incorporate the laying in of weft threads during the process; there is no doubt that such developments will continue to increase the versatility of the warp-knitting process. Some idea of the vast range of commercially available warp-knitting machinery can be gained from Table II which shows how this is sub-divided into generic groups.[11]

It was against the dual backgrounds of warp-knitting technology and of the Locstitch process that the authors developed a novel high-speed sewing–

knitting fabric manufacturing process at Loughborough University of Technology.[8] If the base fabric of the Locstitch pile-fabric (Figs. 6 and 7) is imagined to be removed then a double chain of yarns remains. However, the base fabric constrains both sides of the loop whilst the picking-up of the loop by the complementary needle is accomplished. A modification to the process would be needed to satisfactorily intermesh the yarns into a "pillar" seam comprising a two-yarn chain structure of the type shown in Fig. 14 which shows six stages in the formation of two chain loops, the relative needle tip locations in simplified planar orbit being indicated at the top left in 60 degree intervals of one operational cycle.

Stage A is an arbitrary starting point where the needle $N_L$ is at its left extremity from the central fabric plane C and is threaded with the lighter shaded yarn $Y_L$. This yarn is in the form of a loop $M_L$ around both the needle $N_R$ and its darker yarn $Y_R$. The needle $N_R$ has passed its left extremity and is

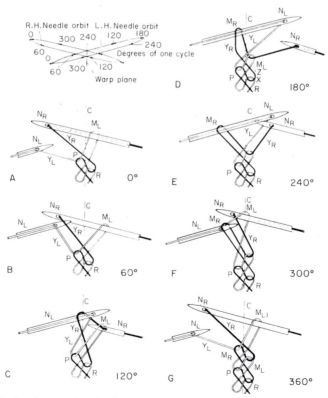

FIG. 14. Cycle of operations in the "novel" high-speed fabric manufacturing process developed at Loughborough University of Technology.

returning towards C, the previously looped chain being tensioned down-wards along this central plane.

At stage B, 60° later, needle $N_L$ in advancing towards C passes behind the suspended yarn $Y_R$. It should be noted that needle $N_L$ is on a collision course with needle $N_R$ and therefore a transverse "shogging" action has to be provided to pick-up the suspended yarn $Y_R$ to form a new loop. Needle $N_R$ is returning axially to the right but at relatively lower speed than the advancing needle $N_L$. Both running-yarns must be tensioned to prevent the elongation of the loop $M_L$.

Stage C, at 120°, shows that the left-side needle $N_L$ has progressed to the right and has shogged transversely (i.e. towards the reader) to avoid colli-sion with needle $N_R$ and to allow the suspended yarn $Y_R$ to form a loop around the yarn carried by needle $N_L$. Needle $N_R$ has also progressed away from the central plane C and it is about to cast-off the loop $M_L$ which envelopes the running yarn $Y_R$.

Stage D indicates that needle $N_R$ has retracted to its right extremity (180°) and has fully cast-off the loop $M_L$ which in turn envelopes the running yarn $Y_R$. Needle $N_L$ is advancing further to the right whilst supporting loop $M_R$. This forward progression of $N_L$ may suitably reduce or "set" the length of the loop $M_L$ between points X and Z.

Stage E is at 240° when the tip of needle $N_R$ has just passed in front of the suspended yarn $Y_L$. As with the complementary needle in Fig. 6B, a trans-verse "shogging" motion is again required to avoid collision of the needles and to pick up the suspended yarn $Y_L$ to form a new loop. Needle $N_L$ is retracting to the left, supporting the loop $M_R$ which must be cleared away from the approaching point of the needle $N_R$.

Stage F shows that at 300° the needle $N_R$ has further advanced to the left, passing between the suspended running yarn $Y_L$ and the far side of the needle $N_L$, thus forming a loop $M_L$ around its projected body cross-section. Needle $N_L$ is just about to cast-off the formed loop $M_R$ which in turn envelopes the running yarn $Y_L$.

Stage G (360°) shows the two needles and their associated yarns in identical positions to those depicted at stage A thus completing one cycle of operations. However, during this *one* cycle, *two* new loops $M_L$ and $M_R$ have been formed. This formation of two loops/cycle gives rise to the prediction that at least a doubling of the production rates of warp-knitting should be possible with this novel fabric production technique. Therefore, a compari-son must be made with a typical warp-knitting cycle to establish the relative dynamics between the two processes.

A precise representation of a typical warp-knitting cycle is hardly practic-able as the types of warp-knitting and their development variations are numerous, as has been shown in Table II. However, in all warp-knitting

TABLE II. Generic grouping of warp-knitting machines (after Reisefeld[11]).

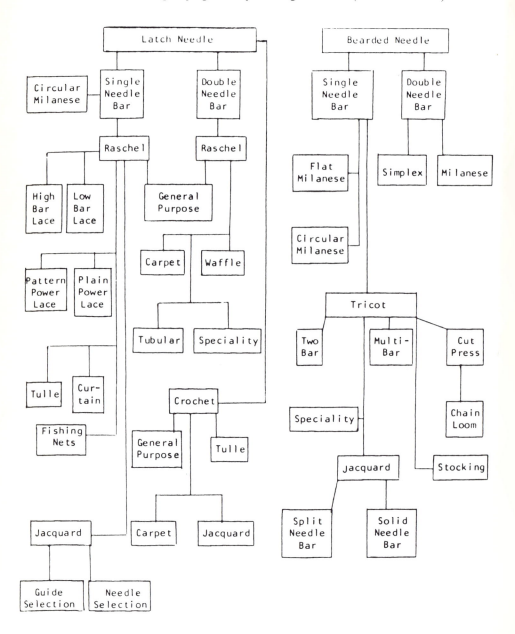

processes, the knitting cycle is a recognizably repetitive part of a continuous process in which a loop of yarn is pulled through a similar loop that has been made during a previous cycle. This entails: controlling the previously made loop; gathering and orientating the warp yarn; pulling a loop of this yarn through the previously made loop; and controlling the new loop for the start of the next cycle. Eyed guides and hooked needles of various designs are the most typically used yarn manipulating elements.

The bearded needle tricot process (Fig. 13) is a somewhat unfair basis for comparison with the authors' novel process because it entails additional presser bar movements and a more complex locus of the needle hooks than does the simpler latch-needle raschel-type process. This latter process would therefore be a more reasonable basis for comparison since the number of elements is small and the manipulating paths are relatively simple and dimensionally similar to those of the novel process. Figure 15 illustrates the approximate loci of (a) a latch-needle hook and (b) its associated guide-bar. The needle's rise and fall paths are coinciding as in pure reciprocation, but the path of the eye of the guide needs to be orbital and in a plane which is transverse to the needle path. It may be further assumed that the motions are generated by cranks (i.e. modified SHM) without dwells, thus resulting in the lowest accelerations and their associated forces. Figure 15c shows the approximate curves of needle and guide displacements and accelerations for a warp-knitting machine running at 1000 cycles/min.

Similar assumptions for needle motions may be made for the novel method, except for the transverse or shogging motion which has to be of relatively shorter duration than a full cycle, thus necessitating a cam motion. The approximate loci of the two needle tips are shown in Fig. 16a,b and the

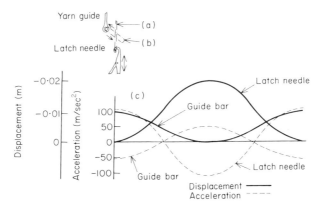

FIG. 15. Warp-knitting: (a) latch-needle orbit; (b) yarn guide orbit; (c) displacements and accelerations.

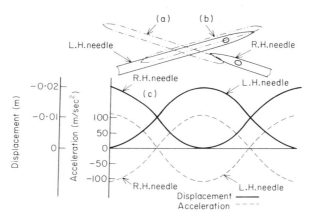

FIG. 16. "Novel" process: (a) RH needle orbit; (b) LH needle orbit; (c) displacements and accelerations.

curves of displacements and accelerations of the needle tips at 1000 cycles/min are plotted in Fig. 16c.

It is reasonable to assume that the reciprocating and oscillating masses would be of similar magnitudes for both processes. Therefore the acceleration curves in Figs 15c and 16c may be compared directly. Their magnitudes are similar and hence the inertia forces will also be of similar magnitude. The dynamic behaviour of the knitting needle latches is complex and is often a limitation to the optimum running speeds of latch-needle type machines. Eyed needles are envisaged for the novel method, thus removing this potential limitation to running speed. This comparative assessment is used to predict that the cyclic running speeds of the projected designs of machines, based on the novel method, should be at least as fast as the fastest warp-knitting machines, whilst producing two stitches/cycle instead of the conventional single stitch. However, the directions of the forces are diverse and careful balancing of the machines will be essential.

The optimum productivity of such machinery can also be limited by the yarn stresses during fabric forming. Although this is subject to continuing investigation, it appears that, in the novel knitting method, slightly more yarn should be back-robbed for control of the stitch length and its uniformity than is necessary in the warp-knitting process. However, during warp-knitting, successive loops of yarn are subjected to full rubbing friction between themselves on both sides of the hook of the needle. On the novel knitting it is feasible to shield, at least, the feed-side yarn with a suitable groove in the face of the needle. Thus, on balance, no reduction in loop-forming rates should result. In productivity comparisons of knitted goods, the machine speeds, or courses produced/unit time, are effective for a particular type of process only. A more representative comparison of process productivity is obtained by relating the number of similar loops produced/unit time. Consequently, the relative productivity of the novel knitting should be at least double that of comparable warp-knitting since two loops are produced per cycle. Based on the authors' previous experiences with the Locstitch machine and process, a powered research rig (see Fig. 21) was developed at the University in order to produce several envisaged new fabric types. The primary needle motion was provided by a linkage mechanism to produce an elliptical orbit (Figs 14 and 16). The required needle shogging motion (Figs 14B and E) was cam-driven. Because the needles had to shog 2 mm to pick-up the loops, both needle banks were arranged to move in opposite directions so that only half this displacement was needed, the acceleration characteristic was dynamically sound, and one needle bank tended to dynamically balance the motion of the other.

In order to pick-up the yarns with a high degree of reliability an extra element was needed and this was termed the "grabber" (Fig. 17). This grabber element needed to push the picked-up yarn loops down the shank of the needle and it positioned the base corner of the pick-up triangle at a known point in space independently of the formation of the previous stitches. In so doing, the seam take-off tension could be minimized because all corners of the pick-up triangle would be formed by mechanical compo-

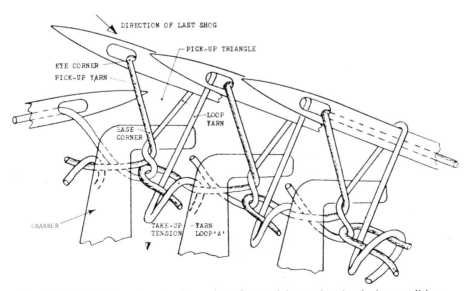

FIG. 17. Pictorial view showing formation of yarn pick-up triangles during conditions of seam interaction in the Loughborough "novel" fabric process.

nents rather than by yarn and seam tensions. To perform the above-mentioned dual-function the nose of the grabber had to lie across the shank of the needle and therefore the needles had to be cranked (in their non-working shank area) to allow the grabber nose to come up through the spaces between the adjacent needles (Fig. 17). Thus as the grabber moved over the working length of the needle its nose lay transversely to the seam and pushed the pick-up loops down the shank of the needle on the com-plementary bank. Having performed this function the grabber moved on to a point in space to hold the next yarn to be picked up in a predetermined position irrespective of the correct or incorrect formation of the previous stitch.

Figure 14 shows only the production of a simple pillar chain seam of interlooped yarns without any seam interaction or weft yarn inlays being used. However, even these simple configurations could appeal as potential textile products because they closely resemble some types of speciality effect yarns (e.g. chainette and mock-chenille) which are used because of their increased bulk and novelty in appearance. The bulk of such chains may be varied with the singles yarn counts and with the setting of the loop lengths. The novelty appearance of such two-yarn chains would be enhanced with the freedom to select two very different yarns. Coupling this to potentially very high multiple production rates could create new markets for such novel products. Other speciality types can be produced, e.g. if an appropriately stretched elastic yarn is inlaid and the chain loops are suitably tightened, then a new type of covered elasticized yarn results.

Many fabric types can be manufactured when seam interaction is pro-vided, i.e. when adjacent seams are connected to each other as in warp-knitting systems. Therefore the authors designed their new process to incorporate seam interactions so as to produce a fabric without the need for extra inlaid weft yarns. The needle-bar shogging mechanism was thus arranged as a "series-cam-set", i.e. the total movement could be the sum of the pick-up shog and the interaction shog, the latter being readily change-able to facilitate different fabric constructions. Figure 17 is a pictorial view showing the position of the yarn in the knitting zone for three needles when a single-pitch shog from left to right has just occurred. The base corners of the pick-up triangles are accurately controlled by the grabbers as both pick-up yarns and loop yarns are retained by mechanical components and seam interaction has been attained. Figure 18 shows a fabric with a connected chain structure, where both needle banks have been shogged relatively to one another, by one wale pitch in alternate directions on successive loop forming cycles. Shogging can also occur over two wale pitches as in Fig. 19, or even over three or more pitches.

The technique for inlaying warp yarns can also be combined with inter-

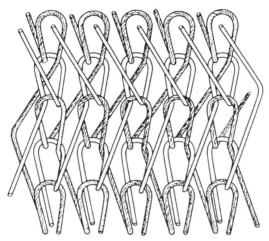

FIG. 18. Seam-interlaced structure over one wale pitch in the "novel" fabric process.

laced structures involving seam interaction. Only one illustration can be given here of the many possible variants. When straight warps are laid in between the wale loops they become distorted into a wave form normal to the plane of the fabric. However, warps may also be laid-in in the form of waves lying within the plane of the fabric. The amplitude of this wave may envelope one, two or more wales as shown in Fig. 20. The frequency of the

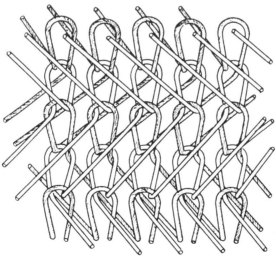

FIG. 19. Two-seam interlaced structures (over two wale pitches) in the "novel" fabric process.

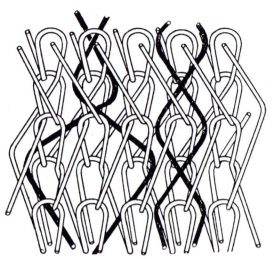

FIG. 20. Seam interlaced structures with selective warp lays in the "novel" fabric process.

wave may be the same or lower than the loop forming frequency. It will also be noticed that this illustration also includes the possibility of a "composite" waveform, comprising two warps in "anti-phase", interlacing with the same pillar seam. This could result in a more stable structure due to the balancing of warp tensions.

Interlaced chain structures can also be intermeshed with weft lays or even with both weft and warp lays. Figure 22 shows one such fabric that contains wefts and the simpler straight warp lays only and this would be expected to possess rather similar weft-wise and warp-wise stability to certain types of woven fabrics.

The above few examples of the many fabric variants typify the wide range of fabrics that are possible with the novel technique. It is hoped that the availability of this new system of fabric manufacture, which performs equally well with spun staple yarns and filaments, will encourage fabric specialists to suggest potential end-uses for the products and that there will be sufficient interest from both machinery makers and users to ensure that it achieves successful commercial exploitation.

## III. Weft-Knitting and its Associated Needle Dynamics

As will be observed from Fig. 1, the University workers have also been active in the fields of weft-knitting machinery particularly in association with

FIG. 21. The Loughborough University research rig for producing "novel" fabric types by a sewing-knitting process.

machinery makers and users in the Leicestershire area. Two "case-study" examples of this work are given after first describing the general area of weft-knitting. The first was sponsored by the Bentley Engineering group to establish the dynamic forces between the cams and needles in circular weft-knitting machinery with a view to decreasing knitting needle failures at high-speed. The second was sponsored by Corah Ltd for the design and development of a prototype system for the mechanized transfer of knitted rib trims from a V-bed knitting machine onto a magazine bar for feeding to the ensuing fully-fashioned knitting process. This has now reached proto-

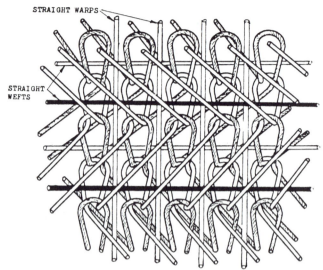

FIG. 22. Two-seam interlaced structures with warp and weft lays in the "novel" fabric process.

type stage as the Loughborough Automatic Rib Transfer (ART) Machine and is described in Section IV.

Figure 23 shows the simplest type of weft-knitted structure, namely, plain (or Jersey) knitting. Horizontal rows of loops are called courses, and the rows of loops down the fabric are known as wales. Thus, if a circular

FIG. 23. Plain (jersey) weft-knitted structure.

weft-knitting machine is producing this type of fabric, the number of wales will be the same as the number of needles in the cylinder of the machine, the number of courses per inch depending on the size of the stitch being produced, i.e. the stitch length.

Figure 24 illustrates how this basic type of knitted stitch is produced by a simple vertical reciprocating movement using a latch needle. It should be noted, however, that this is an end-view of one typical needle only, and that identical needles will be operating in just the same manner immediately before and behind it so that a continuous knitted fabric is produced rather than the simple "one-wale" row of stitches shown. At position (a) the needle is rising, the previously formed loop serving to hold the latch open. When the needle has risen to its top-most position (b), the previously formed loop has been cleared below the latch, and new yarn is fed into the needle hook. The needle then descends to its lowest position (c) so that the newly fed loop of yarn is drawn through its immediate predecessor, which is now cast off to form part of the knitted fabric body. In being cast off, however, it also performs a useful function in closing the latch to retain the new loop. Then the cycle of operations restarts as the needle rises through position (a) again.

In most circular weft-knitting machines the vertical movement is imparted to the needle butts as the needles in their revolving cylinder pass over stationary cam tracks mounted on the machine frame around the cylinder. Hence, if the positions of the tracks are suitably adjusted, or if different types of cams are substituted, more complicated variations of this basic stitch may be produced to give variety to the knitted structure. For instance,

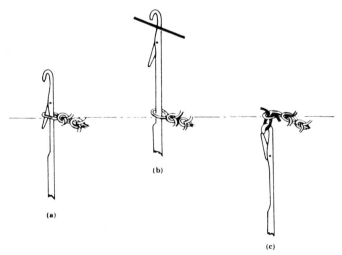

FIG. 24. Cycle of operations in weft-knitting using latch needles.

particular needles need not knit at one or more particular yarn feeder stations; tuck stitches may be formed by using cams which cause the needles at one particular feeder to rise only to the tuck position (Fig. 24a), where the needle is able to receive the new loop without having cleared the old loop below the latch.

Even more variety can be obtained by introducing a second set of needles in a dial which has its needles positioned radially so as to be at right angles to the cylinder needles. Then the yarn feeders can incorporate a yarn-changing mechanism, so that extra yarns may be introduced when required by the particular knitted pattern or design, or alternative colours or yarn types may be substituted during the process. Also, if jacquard selection mechanisms are incorporated, complex patterns can be produced for effect purposes. Machines with two sets of angularly opposed needles can produce a loop structure known as "rib" (see Fig. 25).

The Bentley sponsored University work in weft-knitting needle dynamics has been reported in detail in a 17-part series of papers by Wray and Burns[12-14] which describe how specially designed transducers, incorporating micro-miniature semi-conductor strain-gauges, were used to detect the dynamic forces acting between the cams, needles and yarns in weft-knitting machinery. In particular they identified the impact forces in the stitch cams, guard cams, and in the latch needles themselves, but space precludes the techniques from being re-described here. However, a typical result is interpreted below.

FIG. 25. Rib-type weft-knitted structure.

In Fig. 26 cam forces and yarn forces, as displayed on a storage oscillo-
scope from the strain gauge signals, are shown interposed between successive side-views of a needle head (as it draws a new yarn loop) and the
corresponding front views of the needle head. The needle is moving to the
right as it is carried in a trick within the revolving cylinder, but it is also
moving downwards as the needle butt makes contact with the stationary
stitch-cam. The yarn force is measured by its contact with a verge transducer
since the knitted loops are caused to form over the verge.

The needle impacts the cam at position N1 when the verge transducer is at
V1. Then at position N2 the cam force is largely that required to overcome
the trick's resistance to motion. At N2 this force increases rapidly as the old
loop expands over the widening needle section just below the latch rivet, and
by N3 it reaches a local maximum (Peak 1) when the old loop is at the

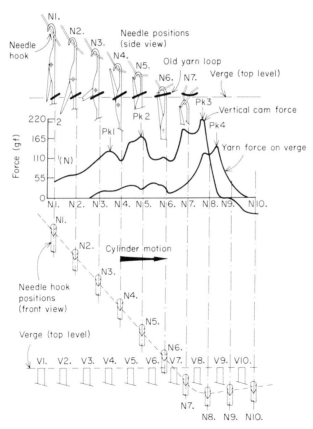

FIG. 26. Cam forces and yarn forces during loop formation in weft-knitting.

thickest section of the needle. It then decreases sharply at N4 as the loop passes on to the decreasing section of the needle above the latch rivet. At N5 the loop rises abruptly on to the latch, the cam force initially increasing rapidly as the yarn expands over a relatively steep latch slope, but as this unfavourable slope decreases the loop expansion becomes more constant and the force stabilizes at a high value (Peak 2). By N6 the needle has reached a position within the knitting cycle where it is possible to rob-back yarn from the cast-off loop one or two needles ahead. As soon as robbing back commences, the cam force rapidly decreases and then remains at a fairly constant value until the needle passes below the verge at V7 and begins to draw the new loop at increasing yarn force. At position N8 a maximum cam force exists (Peak 3) as the yarn is held in the hook at high tension (see also yarn force trace). This tension drops suddenly as soon as it becomes possible to rob yarn from previously formed loops and the cam-force drops to zero as the needle moves from the upper stitch cam to the lower cam but yarn-force is then at its maximum (Peak 4). The cam-force then assumes a smaller constant value until the needle leaves the lower cam.

This research, and that relating to cam/needle impact forces, considerably assisted the Bentley company to redesign their machines so as to minimize needle failure at high operational speeds; indeed the company stated that, as a direct result of the University investigations, which included theoretical studies as well as precise experimentation, their machine speeds were able to be increased by some 20%.

## IV. Automation in Fully-Fashioned Knitwear Machinery

Fully-fashioned outerwear garments are formed by the assembly of separately made knitted garment pieces. The main body panels are shaped or "fashioned", usually in plain-knitted constructions, on straight-bar fully-fashioned machines by increasing or decreasing the number of stitches in the knitted courses. The other parts, e.g. collars, cuffs, and ribbed waist-bands, often collectively known as "trims", are made on flat, or V-bed, machines. Therefore these two types of components have to be joined together. For instance, the collars are usually joined to the neck of the garment by *linking*, i.e. a highly skilled labour-intensive operation in which the knitted loops in the collar are individually stitched to the garment on a separate linking machine away from the knitting machinery that produced the components. The rib trims are normally transferred to the fully-fashioned machine by a protracted rib-transfer sequence of processes, so that the plain-knitted main body panels can be knitted on to the rib-trims within the machine itself. Because such processes are highly labour-intensive, attempts have been made to make them automatic, or at least semi-automatic, in their modes of operation.

Random linking is typified by the Mathbirk[15] system which uses a twin-needle sewing system to insert a multiple seam so that at least one stitch passes through each loop. Mock linking is also provided by the Arndt[16] system which uses a special knitting technique to provide a rolled edge of fine yarn on the trim; this can then be attached to the body of the garment as a "single-linked" operation by a purpose-built sewing machine possessing a mechanically assisted feeding device for the specially prepared trims. The Boehringer Autolinker[17] uses trims produced on a special V-bed knitting machine which transfers the loops on to a comb-like point bar. The garment body is then positioned on this bar against the trim and the whole assembly is passed through a sewing head. This largely automatic process more closely approaches true linking because the trims are attached to the point bar with one knitted loop per individual point.

Before describing the advances in automatic rib transfer machinery the normal "manual" rib transfer process, which is still common industrial practice, should be considered. The rib-trims are knitted in series on V-bed machines, each trim being connected to the other by waste courses and draw threads. They are then removed from the machine and are separated by manually removing the draw-threads from whence they are taken in stacks to the running-on area. Skilled "runners-on" impale the individual loops of a specially-knitted slack course on to the points of a comb-like magazine bar, with one loop per point as with the action of linking described above. The number of trims thus impaled on each bar corresponds to the number of heads on the fully-fashioned knitting machine (usually 8, 12 and 16). The waste yarn courses, provided to facilitate easier handling of the ribs on to the magazine bar, are then unroved by hand. The bars are then taken to the fully-fashioning machine where "bar loader" operatives transfer the trims on to machine bars, one trim for each knitting head. These are then conveyor-fed to each head so that the trims can be "run-on" to the knitting elements. The plain-knitted fashioned body panels can then be knitted on to the trims.

The manual loading of ribs on to bars is expensive in terms of both labour and yarn wastage and consequently automatic rib transfer machines have been developed. In these the basic V-bed knitting principle is used for making the ribs but instead of being connected to each other the individual ribs are automatically transferred from the knitting elements to a magazine bar on the same machine. Typical machines operating on this general principle are the Fabrique Nationale Model R.A. System,[18] the Bentley-Cotton Autorib 3 System,[19] the Scheller Transrobot System,[20] and the Boehringer Ribomat System.[21]

The authors and their colleagues at Loughborough University of Technology have also been active in developing the Loughborough ART Process[22] as a sponsored research and development project with Corah Ltd of Leicester. As will be observed from the second column of Fig. 1, this evolved from the earlier University SRC-sponsored research work in automatic linking, the Corah company having shown considerable interest in furthering the University inventions towards automatic rib transfer.

Whereas the above-mentioned commercial systems are all available only as complete integral machines the Loughborough ART System operates with a unique take-off principle and it is therefore possible to make it as a "bolt-on" attachment which will operate in conjunction with *any* flat (V-bed) knitting machine. A University prototype machine (Fig. 27) is at the industrial trial stage; a secondary technical requirement, namely a "gauge-changing and doubling unit", has also been developed at the University with Corah support (Fig. 28).

FIG. 27. The Loughborough University prototype ART (Automatic Rib Transfer) machine.

FIG. 28. Gauge-changing and doubling unit for use with the Loughborough ART Rib Transfer System.

## Acknowledgements

The authors wish to thank their past and present colleagues in the Department of Mechanical Engineering at Loughborough University of Technology, many of whom are named in the flow-chart (Fig. 1), for their valued cooperation in the textile machinery case-studies mentioned in this chapter. They also gratefully acknowledge the support of the named industrial sponsors, and that of the Science Research Council and the National Research Development Corporation.

## References

1. Wray, G. R. "Contributions of a University Mechanical Engineering Department to Innovation in Textile Machinery". *SRC Survey of Notable Projects in Engineering,* (in press).
2. Wray, G. R. "How can University Engineering Research be Directed to the Needs of Manufacturing Industry?" *Brunel Lecture to the British Association for the Advancement of Science Meeting.* "Engineering Challenges in the 1980s", Vol. 1, ch. 2. Cambridge Information and Research Services Ltd.
3. Wray, G. R., Ward, G. F. and Vitols, R. "A New Development in the Manufacture of Pile-fabrics by a Sewing-knitting Technique". *Studies in Modern Fabrics,* pp. 30–39. The Textile Institute, Manchester, 1970.
4. Wray, G. R. "The Application of Mechanism Theory to a Textile Machinery Innovation". (Nominated Lecture of the Institution of Mechanical Engineers)— *Proc. I. Mech. E.,* **190**, 45/76, 367–78, 1976.
5. Krcma, R. "Nonwoven Textiles". Textile Trade Press. Manchester, 1967.
6. Burnip, M. S. and Marsden, G. "Some Factors Affecting the Production of Patterned Needlepunched Fabrics". *Proceedings of European Conference on Nonwovens.* Paris, 1977.
7. Wray, G. R. "The Kraftamatic Process". *Textile Institute and Industry,* **6**, 155–58, 1968.
8. Vitols, R., Vine, J. E., Walkinshaw, S. and Wråy, G. R. "Sewing-Knitting Fabric Manufacture". *The Fabric Revolution.* The Textile Institute, Manchester, 1981. (Proceedings of the 1981 Annual Conference of the Textile Institute, University of York, March 1981).
9. Wray, G. R., Parry, R. M. and Pakes, H. W. "The Mechanism of a University-designed Textile Machine". *Proceedings of the Fourth World Congress on the Theory of Machines and Mechanisms,* Newcastle-upon-Tyne, published by the Institution of Mechanical Engineers, pp. 1123–29, September 1975.
10. Pakes, H. W., Parry, R. M. and Wray, G. R. "The Application of Optimisation Techniques to Mechanism Synthesis of a Planar Orbit with Constant Attitude". *Mechanism and Machine Theory,* **14**, 171–78, 1979.
11. Reisfeld, A. *Warp Knit Engineering,* National Knitted Outerwear Association, New York, 1966.
12. Wray, G. R. and Burns, N. D. "Transducers for the Precision Measurement of Weft-Knitting Forces". Part I: "A Cam-Force Transducer". *J. Text Inst.,* **67**, 113–18, 1976. Part II: "A Yarn-Force Transducer", 119–22. Part III: "The Experimental Measurement of Cam and Yarn Forces", 123–28.
13. Wray, G. R. and Burns, N. D. "Dynamic Forces in Weft-Knitting". Part I: "Comparison of Measured Non-Knitting Cam-Forces with Mathematically Predicted Values". *J. Text. Inst.,* **67**, 149–55, 1976. Part II: "The Effects of Cam-Cylinder Clearance and the Presence of Yarn on Cam-Forces", 156–61. Part III: "Yarn Tensions during the Loop Forming Process", 162–65.
14. Wray, G. R. and Burns, N. D. "Cam-to-Needle Impact Forces in Weft-Knitting". Part I: "Theory of Stitch-Cam Impact". *J. Text Inst.,* **67**, 189–94, 1976. Part II: "A Stitch-Cam Impact Transducer", 195–98. Part III: "Some Measurements of Stitch-Cam Impact", 199–205. Part IV: "Further Measurements of Stitch Cam Impact", 206–9. Part V: "Guard-Cam Impact", **69**, 229–34, 1978. Part VI: "A Guard-Cam Impact Transducer", 235–37. Part VII: "Influences of Yarn Tension and Machine Speed on Guard-Cam Impact",

238–43. Part VIII: "The Effects of Some Knitting Parameters on Guard-Cam Impact", 244–49. Part IX: "Needle Latch Impact", 301–8. Part X: "The Characteristics of Latch-Needle Breakages", 309–14. Part XI: "The Measurement of Impact-Induced Strains in Latch-Needles", 315–20.
15. Mathbirk Ltd. *Machines for Securing Textile Fabrics One to Another*, BP 1 520 637, 19 December 1974.
16. Anon, "Arndt System Speeds Garment Assembly". *Knitting International*, pp. 90–91, May 1978.
17. Anon, "World-Wide Use for Ribomat-Autolinker". *Knitting International*, pp. 52–53, December 1978.
18. Anon, "Automatic Knitting and Loading Rib Ends". *Hosiery Trade Journal*, p. 67, August 1969, and "Fully-Fashioned Garment Manufacture". *Hosiery Trade Journal*, pp. 108–9, April 1973.
19. Anon, "Reviewing the Latest Cotton Models". *Knitting International*, pp. 95–96, March 1979.
20. Anon. "Transrobot: A New German Rib Loading Machine". *Hosiery Trade Journal*, pp. 104–5, April 1973.
21. Goadby, D. S. "Fully-Fashioned Knitwear Manufacture by Boehringer". *Knitting International*, pp. 56–57, June 1974.
22. Anon, "New Rib-Loading and Gauge-change Units, for Fully Fashioned Knitwear Production". *Knitting International*, pp. 96–97, January 1982.

## Chapter 13

# Microprocessors and Associated Micro-electronic Devices

---

## I. Introduction

This chapter on microprocessors and associated micro-electronic devices has four main aims: (a) to explain briefly the main properties of these devices and show how they can be used in control systems; (b) to discuss the importance of these devices to the textile industry; (c) to survey present and possible future uses in the textile industry; (d) to look in more detail at a few examples of their use in textiles.

Because of the limited space available, much of the treatment especially of the internal workings of microprocessors will be brief. Any reader needing more detail can obtain it from standard texts such as those by Zaks,[1] Ogdin,[2] Morgan[3] and Potton.[4] The choice of textile examples is also limited by space and is biased somewhat towards knitting, rather than other areas of textiles. There are two reasons for this, firstly knitting machine builders have to date probably used microprocessors more than most other sections of the textile industry and secondly the author's own expertise tends to be biased towards knitting. It is not intended to give the impression that microprocessors are going to be confined mainly to knitting. On the contrary they will, if they are not already doing so, soon penetrate into every sector of the textile industry.

Microprocessors and associated micro-electronic devices are products of the 1970s which compared with the long history of textile engineering makes them extremely new devices. The first microprocessor, the Intel 4004 was introduced in 1971 by the American Company "Intel". This was a relatively simple 4 bit microprocessor intended for use in desk calculators. In 1972 the

411

same company introduced the Intel 8008 the first general purpose 8 bit microprocessor. It was about 1975, however, after a large decrease in price, before they began appearing in products such as textile machinery and general purpose micro-computers. Further decreases in price have taken place since, resulting in these small microprocessor chips selling for as little as £5 stg.

Microprocessors have brought with them a language of their own, derived from the jargon of computers and digital electronics. This language may not be familiar to many people in the textile industry, and therefore a brief explanation of what microprocessors are and how they work will first be given. In addition, a glossary of some of the more important terms used is included at the end of this chapter.

## II. What is a Microprocessor

To explain what a microprocessor is, it is best to look first at the basic structure of a digital computer. This can be thought of as consisting of four main parts as shown in Fig. 1.

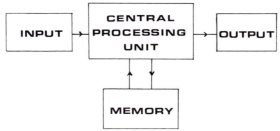

Fig. 1. Main parts of a computer.

## A. Central Processing Unit

At the centre of the illustration is the central processing unit (CPU). This includes the control unit (CU) and the Arithmetic Logical Unit (ALU). The ALU performs arithmetic operations (e.g. addition and subtraction) and logical operations (e.g. "logical or" and "logical and"). The CU manages all the control signals necessary to synchronize operations and the flow of data within the ALU, as well as outside it. It controls the flow of signals within and between the various parts of the computer. These signals flow along groups of conductors referred to as buses. A typical bus could contain 40 conductors, 8 for the data bus, 16 for the address bus and 16 for the control bus. A microprocessor is all or most of a CPU miniaturized on to a single chip of silicon about 5 mm square and 0.4 mm thick (Fig. 2).

FIG. 2. An 8 bit microprocessor.

## B. Memory

The memory is used to memorize both data and instructions (programs). Several different types of memory are possible. The main or working memory consists of a large number of addressable memory locations, one word of information being able to be stored in each location. Some memories have information pre-programmed into them which can be read but not altered, these are often referred to as ROMs (read only memory). Other addressable memories can have information written into them by the computer as well as read from them, these are often called RAMs (random access memory) (Fig. 3). This is something of a misnomer as ROMs also have random access;

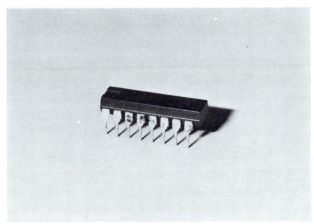

FIG. 3. A 1 kbyte RAM chip.

Read/Write RAM would have been a more precise description. Both RAMs and ROMs can now be made on small cheap integrated circuit chips in a similar manner to that used for microprocessors, a development equally as important as the development of the microprocessor itself. These integrated circuit RAMs have however one important drawback, they are volatile, the information in the memory is lost when the power is removed. ROMs are not volatile and for this reason most microcomputers will have a main memory consisting of both RAM and ROM chips. The ROM will be used for information (usually programs) that is required by the computer all the time and the RAM used for information (data and programs) that is needed temporarily or may require to be altered. The distinction between RAM and ROM is becoming somewhat blurred with the development of special kinds of ROM such as EPROMs (Fig. 4) and EAROMs. The information in these can be altered but only by special means, usually off the computer. Altering them or reprogramming them is very slow compared with writing into RAM and the number of times the information in the memory can be altered before chip failure is limited.

In addition to the fast random access main memory, computers often have backing or serial access memory in the form of magnetic tape, drums or disks (Fig. 5). These are cheaper than the fast main memory but access times are much longer. A relatively new kind of serial access memory which is only just becoming available and is as yet little used is magnetic bubble memory. This is a compact memory, similar to an integrated circuit in form and has faster access times than magnetic tape, drums or disks. It is however relatively expensive, although the price will almost certainly decrease over the next few years.

FIG. 4. A 2 kbyte EPROM.

FIG. 5. A floppy disk backing store.

## C. Input/Output devices

A wide variety of input and output devices are available. One of the cheapest output devices is a visual display unit (VDU) or TV type screen and this is now used on most computers. One of the most popular and reasonably cheap input devices is a typewriter keyboard and this appears on most computers. A computer may have more than one output and more than one input device either of the same type or of different types. Input devices for example could be paper tape readers, punched card readers, photoelectric scanners, graphic tablets, digitizing tables, light pens on a VDU screen or electrical signals from transducers. Output devices could be printers, digital plotters, LED displays, electrical signals to operate control switches, valves, stepper motors etc. Both input and output devices can be at a distance from the CPU but special precautions have to be taken to prevent corruption of signals and the effects of noise in long lengths of cable. A combined input/output device at a distance from a CPU is usually referred to as a computer terminal.

## D. Dedicated microcomputers

A computer built for a special single purpose such as the control of processing machinery is usually referred to as a dedicated computer. Unless the machinery is extremely large and complicated an 8 bit microprocessor would probably be sufficient for the CPU (a 4 bit microprocessor may even be satisfactory). ROM would normally be used for storing the program which would be written in terms of the limited set of instructions that the particular microprocessor can accept (i.e. machine code). A well known 8 bit microprocessor, the Motorola 6800, has, for example, an instruction set of only 81 instructions. Writing programs manually in machine code can be very tedious, time consuming and therefore costly. It is possible however to use a "microprocessor development system" a special purpose computer which allows a higher level language (usually assembly language) to be used for writing the program, then it is converted automatically to machine code and programmed into the ROM. Some RAM would also be needed for storing variables. The main output would be electrical signals to switches, valves etc., for controlling the machinery. There may also be visual outputs, perhaps LED displays, to inform the operator what is happening. The main input would probably be a simple keyboard but others are possible, such as card readers. In addition inputs could come from transducers attached to the machinery for measuring such parameters as temperature, pressure or position.

With this type of computer the program is fixed and cannot easily be altered and the computer cannot easily be used for jobs other than the one job for which it was designed.

## E. General purpose microcomputers

General purpose microcomputers on the other hand are designed so the operator can quickly change the program. This user written program will be stored in RAM and probably written in a high level language such as BASIC. This involves having a "compiler" or an "interpreter" program either stored permanently in ROM or loaded into RAM every time the machine is switched on. This type of program which could occupy about 8k of memory is needed to change the BASIC statements into machine instructions that the microprocessor can understand. The output would normally be a VDU and the input a keyboard, although other output and input devices can often be easily added. This type of general purpose computer usually has a backing store using magnetic tape or floppy disks for the storage of programs that are not at that moment being executed or for storing data in excess of what the main memory can hold. Computers of this kind have become available in the last three years which are very cheap compared with the cheapest computers of five years ago. The Commodore Pet (Fig. 6) for example only costs about £600 stg for a version with 32 kbytes of usable memory (RAM). It also has 16k of ROM for the basic interpreter, control of the input/output and the operating system. A printer can be added for about £350 stg and a double floppy disk unit for a further £700 stg.

FIG. 6. An 8 kbyte Commodore Pet microcomputer with magnetic tape backing store.

## III. Importance of Microprocessors in the Textile Industry.

Microprocessors and associated micro-electronic devices are extremely important to the textile industry, since they present the industry with an opportunity to increase its efficiency and remain competitive in an increasingly competitive world. Those who grasp these opportunities are much more likely to prosper while those who do not are increasing their chances of going out of business. The most obvious result of introducing micro-electronics is that they can automate and speed up various operations. This will lead to less need for labour and perhaps redundancies. On the other hand failure to improve efficiency can lead to whole firms collapsing and even more redundancies. In the long term therefore microprocessors may be doing more to save jobs than to lose them. This effect on the amount of labour required in the textile industry is not likely however to be a very large effect nor is it likely to be the main effect. Much more important will be the improved reliability and accuracy of operations, producing better quality goods and less material waste. Better planning and organization should also result in better utilization of materials, buildings and staff. Significant energy savings could also be made by better control techniques. More flexibility and the ability to respond to market changes will also be of considerable benefit as a result of introducing these devices. The availability of various devices, products and machinery is changing rapidly. New products are appearing almost every day. It is therefore becoming increasingly important for textile technologists to keep abreast of the latest technological developments.

## IV. Survey of Uses in the Textile Industry

Present and possible future uses of micro-electronics in the textile industry can conveniently be divided into four main areas: (a) incorporation in textile processing machinery for automatic control purposes; (b) use in data monitoring; (c) use in instruments and testing equipment; (d) use for ancillary and industrial engineering functions.

### A. Textile processing machinery

Microprocessors are very useful devices for sequencing and timing operations. Any machinery that has to follow a predetermined sequence of events can have this sequence stored in a microcomputer memory. This sort of system is usually smaller, cheaper and more flexible than similar mechanical systems. If also feedback is required from sensors in the system, microcomputers again are ideal. They can look after a large number of inputs from sensors, do any necessary calculations, build in delays if required and issue the required instructions to modify the operation. They will thus find uses in these types of control systems in almost all sections of the textile industry. In the yarn manufacturing section, for example, microprocessors are already being used to control the hopper feed on cards.[5] In 1980, Haigh-Chadwick introduced such a system which is claimed to reduce long and short term variations in output weight.* At the beginning of each weighing cycle the empty weighpan is weighed and the weight fed to the microprocessor. The microprocessor then gives the signal to start filling. Just prior to the correct weight being delivered the microprocessor can actuate a trickle feed to obtain very close control. About one second after delivery has stopped the pan is reweighed and then emptied. The microprocessor then checks the actual material weight against the required weight as set by a thumbwheel and if different, a correction is made to the next weighing. In this way accurate control of the output material is obtained. A speed pick-up also feeds the output web speed to the microprocessor, to enable it to calculate and display the output weight per unit area, or weight per unit length.

Microprocessors have also been used by Benniger of Switzerland on their new sectional warping machine.[6] During warping the actual warp build up is continuously measured and a signal fed to the microprocessor. This is compared with the set value and if there is a difference a signal is sent to adjust the yarn tension. In this way uniform warping conditions are maintained from start to finish and from one warp to another.

---

* See Fig. 10, Chapter 6, p. 188.

The weft knitting sector of the textile industry has probably made more use to date of microprocessors than any other sector. Computer control of knitting dates back to the early 1970s when several knitting machine builders introduced double jersey knitting machines with the fabric pattern information stored in minicomputers. These were not particularly successful, probably due to the extra capital cost compared with mechanical machines and also to the contraction in the double jersey industry that occurred at about the same time.

In 1975 however when Stoll introduced the first microprocessor controlled knitting machine, the ANV Selectanit,[7] it was a great success. Not only did this machine offer more than its mechanical counterparts it was also no more expensive. Since then Stoll have brought out several more advanced microprocessor controlled models and numerous other knitting machine builders have also launched a variety of microprocessor-controlled machines.[8] Monarch,[9] Dubied,[10] Universal,[11] Bentley-Alemannia[12] and several other firms have produced microprocessor V-bed flat knitting machines (Fig. 7). Bentley[13] and Nagata Seiki have produced microprocessor-controlled half-hose machines, Camber their microprocessor-controlled stripper for single jersey[14] and Wm. Cottons have produced their GEMINI microprocessor-controlled fully fashioned machine. In the case of the GEMINI the microprocessor controls not only the fabric design but also the garment shaping and the sequence control. Two microprocessor-controlled knitting machines, one a V-bed and the other a half-hose machine, are described in more detail later.

One would expect perhaps that warp knitting would be a similar story to weft knitting, especially for the raschel lace area where large pattern chains can be very expensive. It was only recently however (January 1981), that

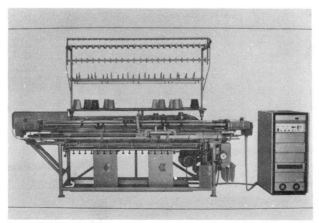

FIG. 7. A Dubied Jet 2 computer controlled V-bed knitting machine.

Karl Mayer introduced their microprocessor-controlled lace jacquard ras-chel machine.[15] This is based on Texas Instruments 9900 series of LSI chips. The CPU is a 16 bit microprocessor and the operating program stored in an EPROM. The pattern information however is stored in magnetic bubble memory, one of the first uses of bubble memory on textile machines. This is a very good use of bubble memory, the pattern information is not lost in the event of loss of power and the fact that bubble memory has serial access only does not matter as the information is needed in sequence. The pattern information is prepared on a specially designed microprocessor-controlled pattern preparation and transferred to the knitting machine on a magnetic tape cassette.

In weaving, one obvious use of microprocessors would be for replacing mechanical jacquard and dobby mechanisms. Development work on these types of mechanisms are being carried out, but very little has appeared commercially yet; it should not be long however, before such devices are being sold commercially. Some of the problems associated with applying microprocessors to weaving are discussed in a recent article by Grigg.[16] He suggests magnetic bubble memory as the most appropriate type of memory for storing the pattern information because of the harsh mill environment and the large amount of data. He also concludes that, "The problem area is the loom interface, the transducers in the system. New techniques are needed of novel nature to make for a viable, reliable and economic system."

Computer control of dyeing and finishing machinery is not new—mini-computers have been used for the control of individual machines or small groups of machines, and main frame computers for the control of large groups of machines. These have not been all that successful due mainly to the cost. The coming of microprocessors and associated micro-electronic devices have brought down these costs considerably and have produced a trend away from multi-machine control to individual machine control. Courtaulds have for example introduced their "Celcon" dye cycle control-lers for the sequence and temperature control of dyeing machines. Bates Textile Machine Co. have also introduced microprocessor control on their new ADT5 perforated drum dryer for the continuous drying of tubular and open-width knitted fabrics.[17] The microprocessor controls the drying temperature and also the percentage overfeed. A more complex system the "Protex" process control system[18] was shown at ITMA in 1979 on a stenter, although it was claimed that it is suitable for any continuous dryer.

The clothing sector of the textile industry has also seen the introduction of microprocessors in the last few years. Several sewing machine manufactur-ers have introduced machines that can automatically stitch a complicated pattern stored in a microcomputer memory. Union Special have for example introduced their 2800TI Memory Stitching Sewing System. A PROM is used to store the pattern which can hold up to 300 stitches. Jones and Brother have also produced a similar machine. An Israeli firm, Beta Engineering and Development Ltd., has produced an interesting system which can convert a standard lockstitch machine of any of the main makes into a microprocessor-controlled memory stitching machine. This stores the stitch pattern in an EPROM which has to be programmed on a separate microprocessor based programming unit.

Multi-head embroidery machines are also now being produced with microprocessor control. One of these, the Barudan BEHJ, can have a plug-in module containing an EPROM which can hold patterns of up to 8000 stitches.

## B. Data monitoring

Computerized data monitoring systems are nothing new in the textile industry. They have been around for more than ten years and have been applied to many different sectors of the industry, e.g. spinning, weaving, knitting, making-up, etc. They have tended to be large, complex, very expensive systems controlled by a central main frame computer and gathering and sorting out large masses of information. They have not been very popular in Britain, mainly because of the initial cost, but also because of the disruption they cause to existing manual systems, the long settling times and, in some cases, the fact that they produce too much information—more than managers can digest.

With the introduction of microprocessors the position has changed and much smaller cheaper systems are becoming available. Systems which are perhaps less ambitious in the amount of data they collect, but systems which can more easily be tailored to suit the individual needs of a particular factory. One of these systems, the Tamarix Data monitoring system which can cost less than £7000 stg was first installed in a making-up factory in Britain in 1980. This will be described in more detail later. Other, somewhat dearer and more specialist systems, were on show at ITMA in 1979,[19] including Uster's "Ringdata", "Rotordata", "Conedata" and "Loomdata". The Ringdata consists of a microprocessor-based central unit connected to travelling sensors, one for each side of a ring spinning frame. The sensor travels backwards and forwards along all the spindles on one side of each ring frame checking for ends down. The central control unit can thus, when required, print out average end breaks per 1000 hours, mean down time per end break and end breaks on individual spindles. In this way defective spindles, defective frames or defective raw materials can quickly be spotted and corrected.

The Rotordata does a similar job for rotor spinning, Conedata for cone winding and Loomdata for weaving.

The monitoring of false twist texturing machinery using microprocessors has been discussed recently by Peacock and Jacques.[20] They describe the development of a system for scanning 24 spindles of a false-twist texturing machine, checking tension transducers and end break detectors on each running end. This system is however not commercially available yet.

## C. Instrumentation and testing equipment.

In the use of many instruments and textile testing equipment several readings may have to be taken, written down on paper and then calculations done to arrive at the final result. In this sort of situation a microprocessor is ideal for recording the results, doing the calculations and presenting the final result on a printer. The microprocessor could also if required compare these results with standards in its memory and print out any differences.

Many such textile testing instruments with microprocessor control have been launched in the last 18 months and no doubt many more will be appearing soon. One instrument, the Fastran metre measurer, is described in detail later. Another is a tensile strength tester by J. J. Instruments Ltd., which has a Commodore Pet microcomputer attached for recording results, calculating averages etc. The Commodore Pet is also used as part of the Hosiery Equipment Ltd. "qualitymaster" system for measuring the width and length of socks. Not only does it print out the results, it calculates the averages and prints out standards. In this way an operator free of the task of writing down the results and looking up standards, can carry out tests more than twice as fast as without the computer.

A much more sophisticated system[21] is the Almeter AL100 fibre length machine launched in 1979 by Siegfried Peyer Ltd, of Switzerland. This is a development of the Almeter machine developed by Centexbel-Verviers in Belgium. The preparation of the fibre sample and the actual testing is the same as the pre-1979 Almeter, but instead of the fibre length distribution parameters being calculated by analogue computing circuits, a microprocessor is used. A printer can print out the Hauteur, the Barbe, CV% of fibre length, histogram and cumulative histogram. It is intended that this system will be combined with a system for measuring fibre diameter and its variation, and an instrument for detecting visible defects in tops or slivers. The results from these three instruments will be fed into a minicomputer which will supply "prediction bulletins" describing the expected performance in processing and the properties of the products at later stages.

Weighing systems especially those for dyestuffs[22,23] seem to be popular for microprocessor control. Several firms have introduced such systems, which can store recipes and prompt the operative, telling him how much of each dyestuff is required. They can also check that the correct amounts are weighed. In this way it is claimed that weighing can be speeded up and errors reduced.

Another favourite is colour measuring instruments;[24-26] both tristimulus colorimeters and spectrophotometers have for a number of years been made with minicomputers incorporated. Now there are several instruments coming onto the market that incorporate microprocessors. These can calculate the colour co-ordinates and predict the dyestuffs required to achieve a correct match.

## D. Ancillary and industrial engineering functions

Under this heading there has been numerous uses of microprocessors introduced in the last year or two. Some of these involve specially designed dedicated microcomputers and some involve using general purpose microcomputers. General purpose microcomputers, such as the Commodore Pet and the Apple II,[27-29] are increasingly being used for such operations as stock control, production planning, costing, work study, printing work tickets, production control, payroll and invoicing. One of the main problems in these areas has been the lack of suitable software. Most of the "off the shelf" programs have been general purpose programs, which are usually not all that suitable for the peculiarities of textile firms. This situation is however changing and reasonably priced programs are beginning to appear specially written for textile firms. The Hosiery and Allied Trades Research Association, for example, has recently started selling programs for Apple II and Commodore Pet microcomputers, written specially for knitting firms.

Another area which is making considerable use of microprocessors is in computer-aided design and pattern preparation. A large number of systems are now available for assisting with the preparation of patterns for weft knitting machines,[10,30,31] warp knitting machines,[15] sewing machines, embroidery machines and cutting jacquard cards for weaving. Most of these systems are special purpose dedicated microcomputers that prepare patterns for one make of machine only. They cannot be used to prepare patterns for other machines nor can they easily be used for any other purpose. This is a great pity as it means that a firm may have to have a number of pattern preparation machines for its different knitting machines, some for sewing machine patterns and a separate general purpose computer for its stock control, costings etc. Each of these microcomputers could be in use for less than half the working week. One exception to this is the Stoll[30] pattern preparation for Stoll V-bed knitting machines. This is based on an Apple II microcomputer and as the programs are stored on floppy disks not in ROM, they can easily be changed and the Apple used for other purposes.

Most of the uses of microprocessors mentioned so far involve some degree of automation or automatic control of an operation. Some operations in the textile industry are however notoriously difficult to automate and are likely to remain manual for some time yet. Sewing and examining are just two examples of such processes. These manual operations contribute consider-ably to the final cost of the product and anything that can help to increase their efficiency is becoming increasingly important. Several firms[32] have recently realized this and produced microprocessor-based instruments for pacing manual operations and continuously displaying to the operative an up to date figure for her efficiency. This type of instrument is claimed to be able to increase production by up to 40% and to decrease training times by as much as 35%. One of these, the Tamarix TEMPO, costs about £190 stg and is based on a single chip computer (Fig. 8). Any target time from 1 to 999 seconds can be set and the instrument continuously calculates and displays the percentage efficiency on a bright red LED display. A more sophisticated instrument, the Dextralog PERFORMER (Fig. 9), in addition to displaying the percentage efficiency can print out at the end of the day the number of stops, waiting time and earned value. This however costs about £1100 stg including the printer.

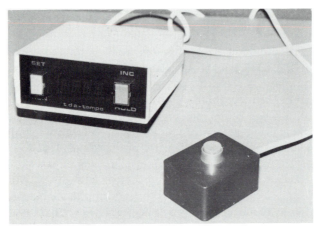

Fig. 8. Tamarix Tempo performance monitor.

## V. A More Detailed Look at Four Examples

### A. Heal's "Fastran" metre measurer

The Fastran metre measurer is an automatic device for measuring yarn

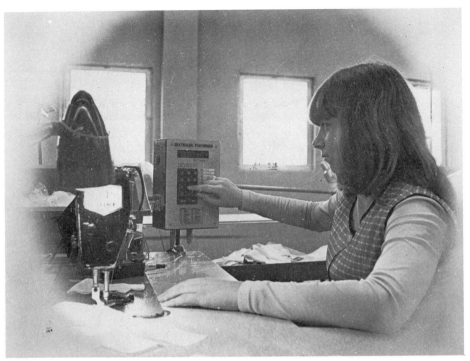

FIG. 9. Dextralog Performer.

linear density and represents a very simple but very useful application of microprocessors to testing equipment (Fig. 10). It automatically reels and weighs a preselected length of yarn and indicates the count in any of ten different count systems. The length of yarn to be measured, normally 100, 50 or 25 m, is preselected on an illuminated digital counter. After reeling the correct length, the measuring rollers automatically stop, the yarn is cut and the spider wheel onto which the yarn is wound, is released for the yarn to be weighed. The spider wheel with the yarn wound on it is mounted on a thin beam which has strain gauges attached to it. These strain gauges are connected to an electrical bridge circuit which produces an analogue signal representing the weight of yarn. This signal is amplified and then digitized by an analogue-to-digital convertor.

The digital signal is fed into a small dedicated computer for calculating the count in any of ten possible count systems. The count system required is selected by a rotary switch and the final figure is displayed on a LED display. The computer is a very simple one consisting essentially of four chips only; one microprocessor, one RAM, one PROM and one interface chip. The

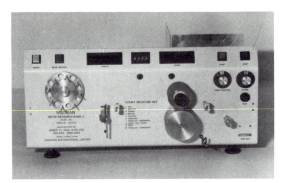

Fig. 10. Fastran Metre Measurer Mk 2.

microprocessor is an 8 bit processor and the RAM is for storing the numbers
involved in the calculations. The program is stored permanently in the
PROM and although it is a relatively simple program, just multiplying or
dividing by a constant when written in machine code, it can fill several pages
of A4 paper. This is because in machine code every tiny step in the working
has to be spelt out in detail. The interface chip converts the parallel digital
signal from the microprocessor to a binary coded decimal signal, for feeding
to the LED displays.

In this example the microprocessor is being, in a sense, under-used, in that
it is only working for very short periods of time and even then, only doing a
simple calculation, when it is capable of much more complex work. Because
of the low price of microprocessors, however, this kind of work is a sensible
and common use for them. The microprocessor could, if required, also
produce the timing signals for measuring the correct length of yarn and this
may happen in future versions of the instrument.

## B.  Komet TC2/E electronic half-hose machine

This machine was launched in 1979 at the Hanover ITMA Exhibition and is
one of the first half-hose knitting machines incorporating microprocessor
control (Fig. 11). It differs from conventional double cylinder jacquard
half-hose machines in that the normal four pattern drums positioned around
the jackbox are replaced by four actuator stations at the same positions
around the main cylinder. Three of these stations correspond with the three
feeders, the fourth is for transferring needles from one cylinder to the other.
Each of these actuator stations has four moveable blades, each of these
blades being operated by an electro-magnet. These blades then operate
directly on to the selector butts as they rotate with the main cylinder, thus

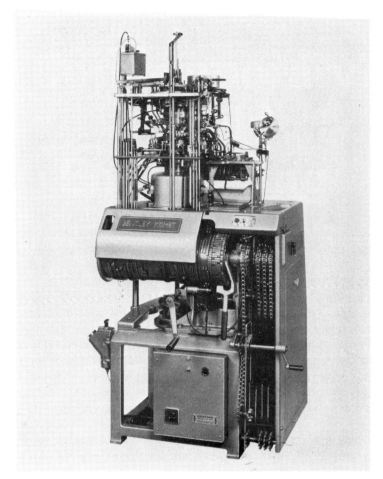

FIG. 11. Bentley Komet TC2/E.

obtaining the needle selections. Four blades are required at each actuator station in order to give the electro-magnets time to act, each blade thus only has to operate on every fourth needle selector. If we consider a machine with a 4″ diameter cylinder, 12 needles/inch and a top speed of 180 rpm, each actuator station will have to make selections at a rate of:

$$\frac{180 \times 12 \times 4 \times \pi}{60} = 452.4 \text{ times/second}$$

Or, each actuator station will have about 2.2 milli-seconds to complete the selection cycle. This is faster than available electro-magnets can be guaranteed to operate and there is thus the need for a bank of four blades, one above the other, with the selector jack butts arranged at four different levels, so that a particular blade only acts on every fourth selector butt. If machine speeds are going to continue to increase, as they have done over the last decade, this will necessitate faster acting electro-magnets or more blades and electro-magnets per actuator station. With this system, the layout of these selectors stay fixed and never require to be changed even when a different pattern is introduced, as is the case with conventional jacquard half-hose machines.

The electrical signals for operating these electro-magnets, are produced by a microprocessor based control box, which is built into the knitting machine. The microprocessor used is the Zilog Z80, a very popular microprocessor which is used in many dedicated computer control systems and in many general purpose computers such as the Sinclair ZX80, Superbrain, Zenith and Research Machines 380Z. It is an 8 bit microprocessor, that is, it has a word length of 8 bits. It also has 16 address lines, so it can address directly up to 64k ($2^{16}$) different memory locations. In addition to the Z80 the control box contains 8 kbytes of PROM (4 IC chips), 4 kbytes of EAROM (8 IC chips), 1 kbyte RAM (1 IC chip) and various interface IC chips. The PROM is permanently programmed in machine code with all the instructions needed for controlling the patterning on the knitting machine and also with instructions for pattern preparation and for putting the pattern information into the knitting machine memory. This is an unusual feature of the machine. Most computer-controlled knitting machines need a micro-processor-based pattern preparation system separate from the knitting machine, with the pattern information, when prepared, transferred to the knitting machine by means of punched paper tape or magnetic tape.

The pattern information is stored in the EAROMs and this again is unusual; most computer-controlled knitting machines use RAM. The main advantage of EAROM is that if the power fails the EAROM still retains the pattern information and it is not lost as with RAM. A disadvantage is that special circuitry using higher voltages is needed for writing into the EAROM. This writing is much slower than writing into RAM and the number of times the contents of the memory can be changed before chip failure is less than with RAM. These disadvantages are not particularly important in the case of a computer-controlled knitting machine where new patterns are not being fed in every minute, but they would be important if EAROMs were used in place of RAM in a general purpose computer.

A pattern area of up to 8000 stitches can be produced allowing for example a pattern of 88 stitches wide by 90 rows. The Z80 microprocessor is capable of addressing more memory so even greater pattern areas could be obtained if required by the use of extra EAROM chips.

To allow the pattern information in the EAROMs to be changed the control box has an input/output connection which allows it to be connected to any computer terminal device which has the standard RS232 Serial Interfree. Thus inside the control box circuits are provided for converting the parallel method of handling data to the serial method and via a Modem to the input/output socket. Although a wide range of terminal devices could be used, the one that most people will prefer is probably the well known Teletype with paper tape punch. This has the advantage of being cheap and already in use throughout the world, so service is readily available in most countries. It also has the advantage that permanent copies of patterns can easily be produced on punched paper tape for long term storage in a form that can easily be re-entered into the knitting machine.

## C. Bentley-Alemannia ACE V-bed knitting machine.

This microprocessor-controlled knitting machine was also launched in 1979 at the Hanover ITMA Exhibition, where several other knitting machine builders were also launching microprocessor-controlled V-bed machines (Fig. 12). In fact every major builder of V-bed knitting machines was showing a microprocessor-controlled model at Hanover.

The ACE is a double system full jacquard machine which has full individual needle selection on both beds. Stitches may also be transferred from

FIG. 12. Bentley Alemannia ACE V-bed knitting machine.

one bed to the other on the same traverse as knit/tuck/miss combinations on both beds. It can handle up to four different yarn colours and has automatic racking, the front bed having three positions and the back bed being able to rack ½ or 1 to 7 in steps of 1, 2 or 3. The control of all these patterning features is obtained from a microprocessor-based control box built into the knitting machine. This box has for its main processor an Intel 8085 microprocessor, a chip which is a development of the very first 8 bit microprocessor, the Intel 8008, produced in 1972. It has a word length of 8 bits and has 16 address lines, thus it can address directly up to 64k different memory locations and is in many other ways similar to the Zilog Z80. The memory consists of 14k EPROM for permanently storing the operating program and 32k dynamic RAM for storing the pattern information. In addition there is a small extra RAM type solid state memory which is battery protected and stores the position of the program in the operating cycle. So if the power fails, although the pattern information in RAM will be lost and has to be reloaded, the position in the operating cycle is known, so knitting can continue where it left off.

Because of the large number of outputs (108) from the control box to control the various patterning features of the machine, a second microprocessor, an Intel 8035, is used to look after the input/output lines. Sixty of these outputs go to the moving carriage itself to control the knitting and transfer cams. These are in banks of six actuators, five banks on the front bed and five on the back bed. Six actuators are required in each bank because of the need to select needles faster than a single actuator could act, in the similar way as on the Komet TC2/E. One bank of actuators is positioned in the middle of the two cam systems and two at each end. This enables each needle to be selected on each traverse for: (a) deferred leading transfer or miss; (b) knit, tuck or miss on leading system; (c) knit, tuck or miss on trailing system; and (d) trailing transfer or miss.

This is a vast improvement on many other electronic V-bed jacquard knitting machines that cannot knit and transfer on the same traverse on the same system. The other 48 outputs control the yarn change, the racking, stitch length changes, etc.

The pattern information is prepared on a separate microprocessor-controlled pattern preparation and transferred to the knitting machine on a magnetic tape cassette. The knitting machine has a built-in high speed cassette reader to enable this information to be read into the knitting machine memory. As special high speed (9600 bits/second) tape recorders are used, special high quality cassettes have also to be used.

The specially designed pattern preparation unit consists essentially of a typewriter keyboard, a VDU type screen and twin high speed tape recording heads. It is controlled by another Intel 8085 chip and has 20 kbytes of ROM for the permanent storage of the pattern preparation program and 32 kbytes of RAM for storing the pattern information. The pattern information is entered by a question and answer technique in which the program asks the question and the operator has to type in the answers. Built in checks are provided by the program, so as not to accept illegal answers. In addition a pattern preparation system based on the Apple II has been introduced recently (Fig. 13).

FIG. 13. Pattern preparation system for the ACE based on an Apple II microcomputer.

## D. Tamarix data monitoring system

The Tamarix data monitoring system is, compared with most data monitoring systems, a relatively inexpensive microprocessor-based system. The first system was installed in 1980 in the making-up department of a British textile firm and tailored to the requirements of the firm's management. The system can however with minor modifications be used for other sectors of the textile industry.

The aim of this first system was to make readily available to the management up-to-date information about the status of all the sewing machinists. At any one time, not all the machinists are working normally, 10 to 20% are often "off-standard" for a variety of reasons, such as machine breakdown or shortage of cut work. As the machinists are the factory's most valuable resource, it is important for the efficient running of the factory that as few machinists as possible are off-standard. It is one of the manager's main jobs to get these machinists back to normal working as quickly as possible. To do this, he needs up to date information about all machinists off-standard, the reason for being off-standard and the length of time off-standard.

This is done by displaying the names of all machinists off-standard together with the reason and the time they went off-standard on the VDU of a microcomputer in the manager's office. This information is also repeated on a factory floor VDU for all to see. The computer used is a 32k Commodore Pet, which is connected to up to 48 terminals placed throughout the factory. This ensures that no machinist or supervisor has more than a few yards to go to enter information. The terminals consist of a simple numeric keypad with keys 0 to 9, a CLEAR key and an ENTER key (Fig. 14). In

Fig. 14. One of the remote keypads used with the Tamarix data monitoring system.

addition the terminal has a four digit LED display to display information as it is entered for checking. This LED display also allows the computer to signal that it has received the information and not rejected it as illegal information. The only information that needs to be entered at a terminal is a three digit number to identify the machinist and a single digit number for the reason off-standard. (Return to standard is represented by a zero.) All other information, such as the machinist's name and the time, the central computer already knows.

Each terminal contains an IC that contains not only the circuits of a simple CPU but also some RAM, some ROM and some interfacing circuits. This sort of LSI chip is referred to as a single chip computer. The RAM is for working memory, the ROM for storing the program and the interface circuits convert the data from parallel 8 bit format to 7 segment coded decimal suitable for connecting direct to the 7 segment LED display. The terminal also has a UART chip to convert the data from parallel format to serial format so that it can be transmitted down a single wire to the main computer. At the central computer is a control box, which contains another UART to convert the data back to parallel format, before entering the Commodore Pet. To save wire all the terminals are daisy-chained in one long line and each terminal boosts the signal as it passes through it, to prevent any corruption of data in the transmission lines. In this way only four wires are required for the transmission lines—one for data and three control wires.

At the end of every day or week, the Commodore Pet can analyse the data and present reports highlighting any trends, or pointing out particular trouble spots. These reports are particularly useful in comparing the before and after effects of any organizational change in the working arrangements.

## VI. Conclusions

We have already, in the first few years of the availability of cheap micro-electronic devices, seen considerable utilization in most areas of the textile industry. As new and more powerful devices become available and as prices continue to decrease, we will certainly see an increasing use of these devices in all aspects of the textile industry. They will not just affect a small number of specialist occupations, but will affect everyone from the managing director to the shop floor.

These changes are already being referred to as a "Second Industrial Revolution", a revolution which could have far greater consequences than the first Industrial Revolution and one which could take place over a much shorter time scale. The textile industry will not be the leader this time, that place belongs to the space industry, but textiles need not be too far behind if the opportunities available are grasped.

## References

1. Zaks, R. *Micro-processors*, Sybex, Berkley 1977.
2. Ogdin, C. A. *Micro-computer Design*. Prentice Hall, Englewood Cliffs, 1978.
3. Morgan, E. *Micro-processors, A Short Introduction*. Department of Industry, London, 1973.
4. Potton, A. *Digital Logic*. Macmillan, London, 1973.
5. Anon. *Textile Institute and Industry*, **19**, 9, January 1981.
6. Anon. *Knitting International*, p. 88, May 1981.
7. Buswell, D. A. *Textile Institute and Industry*, **17**, 362, October 1979
8. Schofield, B. *Knitting Times*, **49**, No. 10, 38, 1980.
9. Anon. *Knitting Times*, **48**, No. 8, 38, 1979.
10. Cooke, W. *Knitting International*, p. 48, November 1980.
11. Tollkuhn, Dieter. *Knitting International*, p. 56, October 1977.
12. Lennox-Kerr, P. *Textile Institute and Industry*, **18**, 87, April 1980.
13. Anon. *Knitting International*, p. 54, October 1979.
14. Goadby, D. R. *Knitting International*, p. 95, May 1981.
15. Wheatley, B. *Knitting International*, p. 74, March 1981.
16. Grigg, P. J. *Textile Institute and Industry*, p. 59, March 1980.
17. Anon. *International Dyer*, p. 129, March 1981.
18. Anon. *International Dyer*, May 1981.
19. Douglas, K. *Uster News Bulletin*, **27**, 8, August 1979.
20. Peacock, G. B. and Jacques, D. *Textile Institute and Industry*, **18**, 63, March 1980.
21. Grignet, J. *Wool Science Review*, **56**, 81, May 1980.
22. Anon. *International Dyer*, p. 59, January 1980.
23. Wolfenden, P. H. *International Dyer*, p. 59, January 1980.
24. Ferguson, J. P. *Textile Institute and Industry*, **17**, 397, November 1979.
25. Dickenson, P. *International Dyer*, p. 224, March 1980.
26. Anon. *Textile Institute and Industry*, p. 99, April 1981.
27. Schofield, B. *Knitting International*, p. 89, June 1981.
28. Gunston, R. *Knitting International*, p. 40, April 1981.
29. Gurrick, H. *The Clothing and Footwear Journal*, **1**, No. 12, 339, 1980.
30. Anon. *Knitting International*, p. 82, March 1981.
31. Tollkuhn, Dieter. *Knitting International*, p. 97, June 1981.
32. Schofield, B. *Knitting International*, p. 48, March 1981.

## Glossary of Terms

*address*    An expression, usually a number which describes a specific location in a memory device.

*ALU*    Arithmetic Logic Unit.

*analogue*    Information represented by a continuous property such as voltage, current, frequency, etc.

*backing store*    A store (memory) of much larger capacity than the working store, but requiring longer access times: e.g. magnetic tape and magnetic disk stores.

*BASIC*    A high level language very popular with microcomputers.

*binary*    Information presented in the form of digits, each digit only having two possible states. The two states are usually denoted by 0 and 1.

*bits*    Binary digits.

*bus*    One or more conductors used for transmitting signals between the different parts of a computer. A complete bus is often made up of a data bus, an address bus and a control bus.

*byte*    8 bits.

*CPU*    Central processor unit.

*clock*    A device that generates periodic signals used for synchronization.

*digital*    Information in discrete or quantized form, not continuous as in analogue information.

*disk*    A flat circular plate with a magnetic surface on which information can be stored.

*EAROM*    Electronically Alterable Read Only Memory.

*EPROM*    Erasable Programmable Read Only Memory. ROM which is not programmed during manufacture but can be programmed at a later stage. It can also be erased for reprogramming but only by a special system such as exposure to UV light.

*floppy disk*    A flexible plastic magnetic disk. Cheaper than hard disks and often used as backing store on microcomputers.

*hardware*    Physical equipment as opposed to computer programs or methods of use: e.g. CPU, printer, disk drives and VDUs.

*high level language*    A means of communicating with and programming computers which is designed for ease of programming. Programs written in a high level language have to be converted into machine code (machine language) before the CPU can understand them. This conversion can be done in one go before the program is run (compiled) or it may be done step by step as the program is running (translated) e.g. BASIC, FORTRAN, COBOL, PASCAL.

*IC*    Integrated Circuit or chip.

*kbyte*    1024 bytes: sometimes abbreviated to just k.

*language*    A set of rules used for writing a computer program e.g. Machine language, BASIC, ALGOL, FORTRAN.

*LED*    Light Emitting Diode display. Used for displaying output from calculators and computers.

*LSI*    Large scale integration. This term is usually applied to integrated circuits containing 100 to 5000 logic gates or 1000 to 16 000 memory bits.

*machine code or machine language*    A language for writing programs that the computer can understand directly. Writing programs in machine language is much more difficult and time consuming than writing in a high level language.

*magnetic bubble memory*     A relatively new method of backing memory storage which contains tiny moveable magnetized regions in a thin film of magnetic garnet crystal fabricated similar to an IC. Such devices provide very dense serial access storage of information.

*Mbyte*     1024 kbytes.

*microprocessor*     An IC that contains all or nearly all the circuits required for a simple CPU.

*modem*     (MOdulator—DEModulator): A device that modulates and demodulates signals transmitted over communication facilities.

*noise*     Any signal that is not supposed to be there. Noise is often picked up by induction from neighbouring conductors or when small sparks occur as a switch is opened or closed.

*parallel operation*     The transmission of a piece of information in which different parts of the information are transmitted simultaneously on separate lines to speed up transmission: e.g. the separate "bits" of a 16 "bit" word may be transmitted simultaneously along 16 wires of a data bus.

*program*     A set of instructions written for a computer according to the rules of a particular computer language.

*PROM*     Programmable read only memory: a memory into which information can be written after the device is manufactured, but thereafter cannot be altered.

*RAM*     Random access memory: a memory from which all information can be obtained with approximately the same time delay by choosing any address at random and without searching through a vast amount of irrelevant information.

*ROM*     Read only memory: a memory in which information has been programmed at the factory and cannot be amended later.

*serial operation*     The transmission of information where individual bits of that information are transmitted one at a time along a single conductor. Serial operation is slower than parallel operation.

*software*     Programs, procedures and associated documentation.

*UART*     Universal Asynchronous Receiver Transmitter: a device that converts a serial input into a parallel output, and may simultaneously convert a parallel input to serial output.

*VDU*     Visual Display Unit: a television screen type output.

*volatile storage*     A storage device in which information is lost when the power is removed.

*word*     A number of bits of information that are considered as an entity by a computer. Different computers can have different word lengths. Most microcomputers have a word length of 8 bits.

*write*     To record information into a storage device.

# Index

Synthetic fibres, 2, 3, 6–8, 10–13, 18–20, 22, 27, 28, 30, 52, 53, 66, 82, 86, 93, 104, 116, 130, 137, 164, 168, 173–175, 184–186, 190, 205, 210, 211, 222, 291

**T**

Take up roller, 301, 302
Tail of yarn, 212–214, 220
Tamarix data monitoring system, 421, 424, 431, 432
Tapes drives, 78, 158–161, 169
Tappet motions, 298, 331
Telematic weaving machine, 316
Telescopic rapier, 315
Teletype paper tape terminal, 429
Temples, 301, 306
Tenacity of fibre, 18–20, 191–192 of yarn, 19, 208, 213, 222
Tension, 80, 94, 120, 122, 159, 161, 164, 170, 198, 208, 301, 322
loom, 252
yarn, 78–82, 94, 167, 170, 198, 252, 272, 274, 279, 286–288, 296, 299, 301, 306, 312, 321, 322, 378, 383, 387, 395, 396, 398, 402–404
Terry fabric, 378–379
weaving machine, 305, 307, 378
Terylene, 32, 175
Tex, 74, 90, 93, 124, 129, 131, 164, 168, 179, 191, 207, 208, 214, 230, 239–241, 253–255, 314
Texture, 15, 27, 31, 105, 209, 267, 268, 272, 273, 283, 284, 287, 288
Texturing, 20, 31, 120, 271–274, 282, 284
air jet, 273, 274, 283
draw, 288
speed, 282, 284, 288
Thermal fabric, 389
Thermobending, 362
Thermoplastic fibres and yarn, 274, 284, 301
Thibeau comb, 52–54
Tissue, unwoven, 24
Titan II machine, 362–363, 369
Titanium oxide, 268, 287

TNO spinning, 92
Top, 2–7, 27, 30, 37–39, 41, 55, 63, 66, 67, 69, 71, 120, 157, 193, 422
Topham box, 11, 12, 19–21, 83
Tow, 27, 30, 66, 106, 115–125, 193, 284
Toyoda weaving machine, 322
TP 300 weaving machine, 296, 317
TPC 1330 weaving machine, 325
Transducero, 420, 422
Transport, 40, 46, 102
Travellers, 16, 80–82, 170
Tricot process, 392–393
Trumpets, 108, 113, 114
Tuck stitch, 402
Tuftu Hydra shift, 354–355
graphics, 354–364
Tufting, 289, 350, 359, 370, 376
machine, 289, 347, 348, 350–360
speed, 357
Turbo stapilizer, 30
convertor, 120
TWILO process, 223
Twin spun, 87, 88
Twist, 6, 7, 14, 15, 31, 32, 75, 77, 83, 84, 85, 89, 93, 94, 105, 106, 114, 115, 129, 131, 165–169, 185, 198–202, 207–209, 212, 214, 219–223, 241, 249
false, 7, 15–18, 20, 31, 90, 166, 167, 169, 200, 273–284, 287, 422
folding, 94
self, 91, 92, 199
twistless, 75, 92, 93, 199, 222
two for one, 31, 94–95, 209
Two-fold yarn, 15, 86–90, 129

**U**

Unidryer, 46–47
Unifil, 307, 335
Universal sensor, 258
Unwoven tissue, 24
UNISAS biological scouring, 44–46
UNSW, Hot acid flocculation scouring, 44
Upholstry fabrics, 93, 356, 378, 389
Uster standards, 208, 236–238